Aristotle's Philosophy of Biology

In addition to being one of the world's most influential philosophers, Aristotle can be credited with the creation of both the science of biology and the philosophy of biology. He was the first thinker to treat the investigation of the living world as a distinct inquiry with its own special concepts and principles. This book focuses on a seminal event in the history of biology – Aristotle's delineation of a special branch of theoretical knowledge devoted to the systematic investigation of animals. Aristotle approached the creation of zoology with the tools of subtle and systematic philosophies of nature and of science that were then carefully tailored to the investigation of animals.

The papers collected in this volume, written by a preeminent figure in the field of Aristotle's philosophy and biology, examine Aristotle's approach to biological inquiry and explanation; his concepts of matter, form, and kind; and his teleology. James Lennox's introduction to the volume provides the reader with the interpretive vision that unifies the essays. Gathering important essays written over a span of twenty years, this volume will be of special value to historians of science and philosophers of science.

James G. Lennox is Professor of History and Philosophy of Science and Director of the Center for Philosophy of Science at the University of Pittsburgh.

T0296435

CAMBRIDGE STUDIES IN PHILOSOPHY AND BIOLOGY

General Editor

Michael Ruse   *University of Guelph*

Advisory Board

Michael Donoghue   *Harvard University*
Jean Gayon   *University of Paris*
Jonathan Hodge   *University of Leeds*
Jane Maienschein   *Arizona State University*
Jesus Mosterin   *University of Barcelona*
Elliott Sober   *University of Wisconsin*

Published Titles

Aristotle's Philosophy of Biology

Studies in the Origins of Life Science

JAMES G. LENNOX
University of Pittsburgh

CAMBRIDGE
UNIVERSITY PRESS

CAMBRIDGE UNIVERSITY PRESS
Cambridge, New York, Melbourne, Madrid, Cape Town,
Singapore, São Paulo, Delhi, Tokyo, Mexico City

Cambridge University Press
The Edinburgh Building, Cambridge CB2 8RU, UK

Published in the United States of America by Cambridge University Press, New York

www.cambridge.org
Information on this title: www.cambridge.org/9780521659765

© James G. Lennox 2001

First published 2001

A catalogue record for this publication is available from the British Library

Library of Congress Cataloguing in Publication data
Lennox, James G.
Aristotle's philosophy of biology : studies in the origins of life science / James G. Lennox.
p. cm. — (Cambridge studies in philosophy and biology)
Includes bibliographical references and index.
ISBN 0-521-65027-5 — ISBN 0-521-65976-0 (pbk.)
1. Biology–Philosophy. 2. Biology–History. 3. Aristotle. I. Title. II. Series.
QH331.L528 2000
570'.1–dc21   00-026070

ISBN 978-0-521-65027-4 Hardback
ISBN 978-0-521-65976-5 Paperback

To Patricia and Cressida

Read Aristotle to see whether any [of] my views very ancient?

Charles Darwin, Species Notebook C 267

Contents

Contents

Preface

The essays collected in this volume were written during a twenty-year period when a small group of European and American scholars were engaged in an exciting re-evaluation of Aristotle's scientific study of animals – his zoology, as it might be called. It has been my great good fortune to share in this work.

My interest in this subject goes back to my final year as an undergraduate at York University in Toronto. My honors thesis was to be a philosophical study of Aristotle's *De Anima*. My faculty supervisor, Allan Cobb, pointed me to the *Generation of Animals* for interpretive clues, and I gradually recognized that Aristotle's zoological treatises had a complex relationship with his metaphysics and natural philosophy that bore investigation. A mutual friend suggested that I share that honors thesis with Allan Gotthelf, who was himself just completing a dissertation at Columbia University on Aristotle's conception of final causality. From that day to this, his work has been a model and a source of inspiration, and I am proud to call him a friend.

As I was nearing the end of a dissertation, written under the supervision of Father Joseph Owens at the University of Toronto, on the relationship between Aristotle's *Metaphysics* and his biological treatises, Allan suggested that I attend the 1976 Princeton Colloquium on Aristotle's biology. Among the speakers was David Balme, whose recent translation of some of the biology for Oxford's Clarendon Aristotle Series had been my constant companion while working on my dissertation. I marshalled the courage to ask him to be the outside reader on my dissertation. He agreed, and his perceptive comments on my first effort at serious scholarship were full of hints about work needing to be done. From then until his untimely death in 1989, I shared my work with him and tried hard to learn from him how to let Aristotle speak for himself.

Thanks in large measure to the entrepreneurial skills of Allan Gotthelf, the years from then to the present have seen continuous international cooperation in this scholarly adventure. An NEH-sponsored summer seminar organized by Allan at Williams College in Williamstown, Massachusetts, in 1983 brought a group of interpreters of Aristotle's philosophy together with philosophers of biology to explore Aristotle's zoological works, with exciting results. A selection of the papers from that meeting formed the core of the volume Gotthelf and I edited, entitled *Philosophical Issues in Aristotle's Biology*. It was at this conference that I first met two important contributors to this scholarly enterprise, Pierre Pellegrin and Wolfgang Kullmann.

Many conferences on related topics were to follow – in 1985 at King's College, Cambridge; in 1987 on Isle d'Oleron in France; an NEH Summer Seminar in 1988 at the University of New Hampshire; in 1990 at Trenton State College (now The College of New Jersey); and in 1995 in Bad Homburg, Germany. The proceedings of two of these events were published, while other papers first presented at these events found their way into Festschriften for David Balme and Wolfgang Kullmann.

In 1988, with the help of an NEH translation grant, I began work on a translation and commentary of the *Parts of Animals*. A number of the more recent papers reprinted here grew out of insights gained as I worked on that project. Often, those insights were due to the careful comments of John Ackrill, Myles Burnyeat, Mary Louise Gill, Allan Gotthelf, Lindsay Judson, Aryeh Kosman, and my research assistant Katherine Nolan. I did not always accept their suggestions, but I always learned from them.

Nearly all of the essays in this collection are contributions to this cooperative enterprise, and they have been immeasurably improved by the interactions with everyone involved. I would like to single out for special mention those with whom I have taken part in various reading groups and seminars focused on Aristotle's biology over the years: Robert Bolton, David Charles, David Depew, Mary Louise Gill, Allan Gotthelf, Aryeh Kosman, and Tim Maudlin.

Many of the ideas in these papers were first tested in graduate seminars at the University of Pittsburgh. The students who made helpful suggestions are too many to mention and must be recognized collectively. However, for his enthusiastic and meticulous efforts in helping to prepare the manuscript for submission, it is a pleasure to here acknowledge my research assistant, Tiberiu Popa.

It is a special pleasure to thank three people who directly contributed to the publication of this collection. First and foremost is my friend Michael Ruse, whom I have known since graduate school days. He has over and again taken bold steps to familiarize his audience with the growing, and now quite extensive, body of high-quality work on Aristotle's philosophy of biology, such as commissioning an essay-length review of *Philosophical Issues in Aristotle's Biology* for *Biology and Philosophy*. It was Michael who suggested that I propose this volume to Cambridge University Press for their Cambridge Studies in Philosophy and Biology, of which he is general editor. I am deeply grateful to him, and to the Syndics of Cambridge University Press, for allowing me the opportunity to gather these papers together in a volume targeted to an audience of historians and philosophers of biology.

They agreed to do so partly because of the reports they received from two readers who have shed their anonymity and whom I therefore can thank publicly. The first is David Depew, a member of the history and philosophy of biology community who has worked tirelessly to make Aristotle's philosophy of biology a mainstream topic in that community. His enthusiasm for the project was most heartening. The second is Geoffrey Lloyd. It is perhaps an understatement to say that we have had our differences. But he has been invariably supportive of my work, and his detailed comments on the initial proposal for this collection led (as only he will know) to many positive changes.

My deepest debts of gratitude are to the two women to whom I dedicate this volume. I began writing a dissertation on Aristotle's philosophy and biology when Cressida was one year old, and wrote the most recent essay in this collection the year she graduated from college. She good-naturedly put up with more than a little grief from her dad as he tried to balance the challenges of parenthood and an academic career. (Of course, there were occasional advantages for her, such as six months in Cambridge followed by a week on Isle d'Oleron!) When Pat and I first met I had yet to discover philosophy, let alone Aristotle or his biology. Without her steadfast support and encouragement I could not have accomplished half of what I have, and without her to share life with, it would not have meant half as much.

Pittsburgh, 1999

Acknowledgments

"Between Data and Demonstration: The *Analytics* and the *Historia Animalium*" is reprinted by kind permission of the Institute for Research in Classical Philosophy and Science.

"Aristotelian Problems" and "Are Aristotelian Species Eternal?" are reprinted by kind permission of Mathesis Publications, Inc.

"Putting Philosophy of Science to the Test: The Case of Aristotle's Biology" is reprinted by kind permission of the Philosophy of Science Association.

"The Disappearance of Aristotle's Biology: A Hellenistic Mystery" is reprinted by kind permission of Academic Printing and Publishing, Edmonton.

"Teleology, Chance, and Aristotle's Theory of Spontaneous Generation" is reprinted by kind permission of the *Journal of the History of Philosophy*.

"Aristotle on Chance" is reprinted by kind permission of *Archiv für Geschichte der Philosophie*.

"Nature Does Nothing in Vain . . ." and "Material and Formal Natures in Aristotle's *De Partibus Animalium*" are reprinted by kind permission of Franz Steiner Verlag, Stuttgart.

"Theophrastus on the Limits of Teleology" is reprinted by kind permission of Transaction Publishers.

"Plato's Unnatural Teleology" is reprinted by kind permission of Catholic University of America Press.

Abbreviations

AELIAN
*De nat. an. (De natura
  animalium)*

ALEXANDER OF APHRODISIAS
in APr. (In Analytica Priora)
in Met. (In Metaphysica)
in Top. (In Topica)

PSEUDO-ARISTOTLE
Mech. (Mechanica)
Probl. (Problemata)

ARISTOTLE
APo. (Posterior Analytics)
APr. (Prior Analytics)
Cael. (De Caelo)
Cat. (Categories)
de An. (De Anima)
EE (Eudemian Ethics)
EN (Nicomachean Ethics)
FR (Fragments)
GA (Generation of Animals)
GC (Generation and Corruption)
HA (Historia Animalium)
IA (De Incessu Animalium)
Int. (De Interpretatione)
Juv. (On Youth and Old Age)

*Long. (On Length and Shortness
  of Life)*
MA (De Motu Animalium)
Mem. (On Memory)
Metaph. (Metaphysics)
Mete. (Meteorology)
PA (Parts of Animals)
Ph. (Physics)
PN (Parva Naturalia)
Po. (Poetics)
Protrep. (Protrepticus)
Resp. (On Respiration)
Rh. (Rhetoric)
SE (Sophistici Elenchi)
*Sens. (On Sense and Sensible
  Objects)*
Somn. (On Sleep)
Top. (Topics)

H. DIELS, W. KRANZ
*DK (Die Fragmente der
  Vorsokratiker, 7th ed.)*

DIOGENES LAERTIUS
DL (Lives of the Philosophers)

HERODOTUS
Hist. (Historiae)

Abbreviations

HIPPOCRATIC CORPUS
De vet. med. (De vetere medicina)

MICHAEL OF EPHESUS
in MA (In De Motu Animalium
  and *De Incessu Animalium)*

PHILOPONUS
in APo. (In Analytica Posteriora)
in de An. (In De Anima)
in Ph. (In Physica)

PLUTARCH
Quaest. Conv. (Quaestiones
  Convivales)

PROCLUS
in Eucl. (In Euclidem)

SIMPLICIUS
in de An. (In De Anima)
in Ph. (In Physica)

THEMISTIUS
in APo. (In Analytica Posteriora)
in de An. (In De Anima)

Introduction

You must let me thank you for the pleasure which the Introduction to the Aristotle book has given me. I have rarely read anything which has interested me more; though I have not read as yet more than a quarter of the book proper. From quotations which I had seen I had a high notion of Aristotle's merits, but I had not the most remote notion what a wonderful man he was.

Charles Darwin to Dr. William Ogle, Feb. 22, 1882[1]

Even in the study of animals disagreeable to perception, the nature that crafted them provides, for those who are able to know their causes and are by nature philosophers, extraordinary pleasures. Surely it would be unreasonable, even absurd, for us to enjoy studying the likenesses of animals because we are at the same time studying the art, such as painting or sculpture, that made them, while *not* prizing even more the study of animals constituted by nature, at least when we can behold their causes.

Aristotle, *Parts of Animals* I.5 645a4–15

The papers collected in this volume focus on the seminal event in the history of biology – Aristotle's delineation of a special branch of theoretical knowledge devoted to the systematic investigation of animals. Aristotle approached the creation of zoology with the tools of subtle and systematic philosophies of nature and of science that were then carefully tailored to the investigation of animals. He thus also deserves to be considered the first, and one of the greatest, *philosophers* of biology. Or so I have been arguing for the past twenty years. The papers collected in this volume were written between 1978 and 1998. This introduction provides the reader with the interpretive vision that unifies them. There are also brief introductions to each section of the volume that discuss the papers in that section, scholarly reactions to them, and the current status of research on their topics.

Introduction

MOTIVATIONAL PRELUDE

Why should a historian and philosopher of biology study Aristotle's philosophy of biology? There are two basic reasons. First, Aristotle essentially created both the science of biology and the philosophy of biology. Second, his influence on the history of biology through the 19th century is pervasive.

A strong case can be made that Aristotle created the idea of a general scientific investigation of living things. For the historian of science, careful study of a scientific investigation so distant in time and outlook from our own is an exciting prospect in its own right. In addition, however, Aristotle's animal studies profoundly influenced the origins of modern biology as they were recovered and translated in medieval and Renaissance Europe, and the complicated story of that influence is just beginning to be told.[2] (How far this influence stretches can be seen by looking at Richard Owen's *Hunterian Lectures in Comparative Anatomy* of 1837, recently edited by Phillip Sloan.)[3] Understanding Aristotle's influence on the history of biology and its philosophy must begin, of course, with a detailed understanding of the original.

The distance between Aristotle's philosophical outlook and that which underlies modern biology also stimulates philosophical reflection. The essays collected here make the case that Aristotle's philosophical understanding of nature in terms of substantial being, matter, form, nature, essence, definition, division, explanation, teleology, chance, and necessity shapes and informs his way of doing biology. What does biology look like when it is based on such understanding? Are there affinities between his approach and our own? There are, no doubt, deep and far-reaching differences. Understanding these differences can, I believe, be as philosophically valuable to us as understanding the affinities.

One theme to which I have occasionally alluded in my writing will quickly illustrate the point. Modern biology develops in the inhospitable Christian-Cartesian philosophical context of inert material body and active, immaterial soul, and with teleological explanation rooted in natural theology. This had a profound impact on its foundations. Aristotle's biology, by contrast, flourished in a very different philosophical environment. Animals are unities of matter and form – souls are simply forms (read 'functional capacities') of animate bodies. Aristotle's teleology is based on recognition that animate bodies are structured as they are, and develop as they do, in order to perform the functions that make up an animal's life. Thus he gives explanatory primacy to function, while insisting that a complete biological explanation must take detailed account of the material basis of function.

Aristotle's biology provides an example of a fully developed biology with philosophical foundations innocent of Christian/Cartesian presuppositions.

ARISTOTLE'S PHILOSOPHY OF BIOLOGY: AN OVERVIEW

Aristotle's concept of nature is developed against the background of two centuries of predecessors who wrote down their reflections 'on nature'. Many of those predecessors saw it as their task to discover the simplest and most enduring entities in the universe and to provide compositional explanations of everything else. Aristotle's philosophy of nature leads him to reject this 'bottom-up' approach to understanding. Every natural object has a material nature, certainly; but they owe their distinctive modes of being and acting to their *formal* natures. It is thus important that the student of nature investigate both an object's matter and its form. In his words, ". . . if art imitates nature, and it belongs to the same branch of knowledge to know the form and to know the matter up to a point (thus the doctor has knowledge of health, and also of bile and phlegm, the things in which health resides; and the builder knows the form of a house, and also the matter – that it is bricks and beams; and it is the same with other arts), then it belongs to the study of nature to know both sorts of nature" (*Physics* II.2 194b21–26).

Aristotle's arguments for his anti-reductionism often begin, as here, with examples from the world of craftsmanship: statues and axes may both be composed of bronze, but to understand what *distinguishes* them – what makes a statue a statue and an ax an ax – you need to know what they are for, which provides insight into their structure, into their production, and finally, into their materials. What sanctions these occasional appeals to the crafts is the view, expressed above, that *art imitates nature,* not that nature is a craft product. Aristotle saw this latter view articulated in Plato's *Timaeus,* and he rejected it utterly. For Aristotle, nature is the starting point for the natural scientist.

Natural things have their own intrinsic origins of change; craft objects, on the other hand, are produced, and used, by us. That difference extends to their teleological origins. The ends served by craft products – statues or axes – are *our* ends; the ends served by the parts of an animal are *the animal's* ends. A predator springs into pursuit in order to fulfill *its* nutritional needs; an ax is designed and used to satisfy human needs.

There are extremely simple natural things – bits of earth or water or air – and Aristotle devotes considerable attention to their natures, in *Generation and Corruption,* in *Meteorology* IV, and in *De Caelo* III-IV. These 'elements' also move naturally: earth, and things predominantly earthen, fall toward the center of the cosmos unless prevented from doing so, while air has a natural tendency to rise. Neither requires an external mover to make this happen. Aristotle is clearly more fascinated by the complex, multi-layered natures of animals and plants, however, where these elements constitute tissues and organs that are organized to serve myriad different activities. Both in its development and in its day-to-day life, there is an amazing coordination in an animal's construction and movements that could not be reduced to the simple natural capacities of the four elements. All of this amazing, coordinated complexity gives the impression of serving the continued life of the animal. The simple capacities of the elements – to heat and cool, to soften or harden, to evaporate or condense – are the tools used by an animal's formal nature to achieve its goals. In fact, in one very important case, reproduction – important enough for Aristotle to devote the five books of *On the Generation of Animals* to it – the goal served appears to be extending the life of the animal beyond itself to a formal replica of itself.

THE ORIGINS OF BIOLOGY

Aristotle was impressed enough by the uniqueness of living natures to treat their study as a distinct branch of natural philosophy. The book that has come down to us as the first book of the *Parts of Animals* is actually a series of essays which lays the philosophical foundations for a science of living nature. This fact, together with the fact that Aristotle devoted four books *(Prior Analytics* I-II and *Posterior Analytics* I-II) to a philosophical investigation of scientific knowledge, raises an exciting possibility: that his zoology reflects the recommendations of his *philosophy* of zoology; and that his philosophy of zoology reflects his *general* philosophy of science. In one way or another, every chapter in this volume – even those that discuss Plato and Theophrastus – aims to contribute to the case for these connections.

A number of problems stand in the way of making this case. Foremost among these is that the philosophy of science in the *Posterior Analytics* is devoid of a number of concepts, and conceptual distinctions, that are central to Aristotle's science of animals. It never distinguishes between matter and form, nor does it once mention the special concept of necessity that Aristotle develops especially for the study of living things, hypothetical or conditional necessity. Further, while he talks about biology having a "different manner of demonstration" in *Parts of Animals* I, there is no systematic attempt to relate zoological demonstration to demonstration as described in the *Analytics*.

Similarly, the logic of division is discussed at some length in Book II of the *Posterior Analytics,* and Plato's method of dichotomous division is subjected to a scathing critique – and replaced by a complex multi-differentia division suitable for biology – in *Parts of Animals* I.2–3. Yet Aristotle makes no effort to connect these two discussions, and it is not obvious that the goals of division enunciated in the *Analytics* can be achieved if one assumes that division is to be practiced simultaneously on a multitude of unrelated differentiae.

Similar problems arise when one begins to compare Aristotle's actual biological explanations with the theoretical foundations presented in *Parts of Animals* I. *Parts of Animals* I stresses the importance of distinguishing material and formal natures in animal investigations, and of distinguishing what is conditionally necessary from what is necessary without qualification. Yet in his actual biological study of the parts of animals in Books II–IV, in which he uses the term 'nature' *(physis)* 229 times and 'necessity' *(anankē)* and its cognates 136 times, he never acknowledges either distinction. Nor is it obvious that his actual zoological explanations conform to *any* sort of demonstrative ideal, nor that his organization of zoological information conforms to any of his theoretical pronouncements on division.

Each of the chapters in this volume contributes to the task of overcoming these various problems, and thus to presenting a unified vision of Aristotle's philosophy of biology as carrying out, in a specific domain of natural science, the philosophical program of the *Posterior Analytics.* This involves considerable enrichment of that program, but not an abandonment of it.

It is my hope, having gathered these essays together and published them in a series intended primarily, in the editor's words, for "a broad cross-section of biologists and philosophers," that many others, like Charles Darwin, will come to appreciate "what a wonderful man [Aristotle] was" – and in particular, what a wonderful philosopher of biology he was.

NOTES

1. Our understanding of the correspondence between Darwin and Ogle on the occasion of Ogle's sending Darwin a copy of his translation of Aristotle's *Parts of Animals* has been enriched recently by Allan Gotthelf in Gotthelf 1999.
2. See Cunningham 1985, in Wear, French, and Lonie 1985, 195–222.
3. Sloan 1992.

I

Inquiry and Explanation

> In fact this is what it is for something to be a first principle (*archê*): to be
> the same cause of many things, with nothing higher being a cause of it.
>
> Aristotle, *Generation of Animals* V.7 788a14–16

A fundamental question for philosophers of biology is how the concepts,
theories, explanations and methods of the biological sciences are related to
those of the other sciences. At one extreme is the view that a perfectly
general philosophy of science is adequate for all science, at least all natural
science; at the other is the view that there is no general philosophy of science,
only philosophies of this or that science. Most philosophers of biology, I
hazard to guess, hold a view intermediate between these extremes. The
chapters in this section argue that Aristotle does as well.

More precisely, they make the case that Aristotle's zoological treatises,
and the investigations upon which they rest, reflect ideas about scientific
inquiry and explanation expressed in the *Prior* and *Posterior Analytics*. But
the application of those ideas to zoological inquiry and explanation is accomplished through the introduction of a variety of concepts and distinctions that
are absent in the *Analytics*.

When I began investigating this subject, the prevailing view in Anglo-
American scholarship was just the opposite: the *Posterior Analytics* provides
us with an austere, formal, deductive model of explanatory proof, while the
biological treatises are a set of informal and tentative explorations. These
essays, along with the work of David Balme, Robert Bolton, David Charles,
Wolfgang Detel, Allan Gotthelf, Wolfgang Kullmann, and Pierre Pellegrin
have significantly altered the landscape. The debate is now over the extent
and the nature of the relationship. The most steadfast – if friendly – critic of
this work has been Geoffrey Lloyd (see the first four chapters of *Aristotelian
Explorations* [1996] and his "Aristotle's Zoology and His Metaphysics, the

1

Status Quaestionis: A Critical Review of Some Recent Theories" [in Devereux and Pellegrin 1990]). I will first outline the thematic continuity among these papers, and then briefly characterize the current state of the art.

"Divide and Explain: The *Posterior Analytics* in Practice," argues that principles enunciated in the *Prior* and *Posterior Analytics* regarding the proper form of a scientific proposition, and about the nature of true as opposed to sophistic demonstration, lie behind the pervasiveness of certain practices in the biological treatises. This chapter discusses five practices that are puzzling until one recognizes their philosophical roots in the *Analytics:*

   (i) the lack of concern for a taxonomically complete classification;
  (ii) the structure and variety of biological explanations;
 (iii) the sustained attention to finding differentiae that are coextensive (or 'commensurately universal');
  (iv) the practice of the method of division;
   (v) the nature of the eight (or nine) great kinds *(megista genê)* in the biology and their role in the organization of the different treatises.

The conclusion that Aristotle's zoological science was driven by the epistemological norms of the *Analytics* has special implications for how to understand the *History of Animals.* "Divide and Explain" touches on these implications. The second chapter in this section, "Between Data and Demonstration," argues for them in detail. David Balme's work convinced me of the theoretical integrity of this great treatise. In this chapter I seek to nail down the case that the overarching purpose of *HA* was to organize information about animal likenesses and differences in precisely the form required for the modes of explanation we find in the explanatory biological treatises – *Parts of Animals, Generation of Animals, Progression of Animals, On Respiration,* and so on.

I also begin to explore a minefield of problems I carefully avoided in "Divide and Explain." *Posterior Analytics* II opens by distinguishing two pairs of inquiry which, if successful, issue in four kinds of knowledge. Successful inquiries into whether some kind of thing exists lead naturally to inquiries into that kind's nature; successful inquiries into questions regarding the truth of some fact (such as whether it is true that all broad-leafed trees drop their leaves) lead to inquiries into why that fact is as it is. But almost immediately after presenting these distinctions, Aristotle begins to collapse them. In particular, he begins to question the distinction between existential and predicative inquiries. Eventually, he argues that a successful scientific explanation of why some feature belongs to every member of a kind will also

2

be a scientific account of what that feature is. To know *why* broad-leafed trees lose their leaves seasonally is to have a scientific definition of leaf loss, to cite one of his examples.

Both of these chapters discuss briefly how Aristotle may have seen the method of division (discussed in *Prior Analytics* I.27–30 and *Posterior Analytics* II.4–6 and 13) as related to the activity of identifying scientific 'problems'. They assume – but do not prove – that *scientific problems* are propositions prepared for demonstrative explanation. Aristotle's concept of a 'scientific problem', and its relationship to the concept of a *dialectical* problem, is the subject of "Aristotelian Problems." These three chapters together provide a comprehensive view of Aristotle's theory of scientific inquiry – that is, inquiry aimed at producing demonstrative explanation. The view throughout is that this is the principal subject of the second book of the *Posterior Analytics* (the first book being focused primarily – though far from exclusively – on providing a theory of demonstrative proof). Together, they make a case for the influence of this theory on the biological works.

The penultimate and ultimate chapters in this section have distinct but closely related aims. "Putting Philosophy of Science to the Test: The Case of Aristotle's Biology" argues that the first book of the *Parts of Animals* outlines a zoological research program. In this context I stress the need for a conceptual and methodological machinery which, while *consistent* with the *Posterior Analytics,* is specifically and distinctively *biological.* "The Disappearance of Aristotle's Biology: A Hellenistic Mystery" argues that this research program was, with the exception of his colleague and friend Theophrastus, either ignored or misunderstood by others in the classical world.

These papers were not produced in isolation. They constitute part of the general reappraisal of Aristotle's scientific work in the light of his philosophical work on the theory of science – the *Prior* and *Posterior Analytics* and the *Topics,* in particular. On the subjects central to the chapters in this section, the following works are especially valuable: Balme 1987b, 1991, 1992; Bolton 1987; Charles 1990, 1997; Detel, 1997; Gotthelf 1985b, 1987a, 1997; Kullmann 1974, 1990, 1997; Pellegrin 1982, 1986. The two-tier concept of demonstration developed in "Divide and Explain" and modified in "Between Data and Demonstration" has received broad acceptance in recent literature on the *Posterior Analytics* (Ferejohn 1990, 160; Kosman 1990, 359; Marcos 1996, 59, 105, 270, 272; McKirahan 1992, 296). Similarly, even our harshest critics have accepted the idea, developed by Allan Gotthelf and me, that the discovery and formulation of commensurately universal propositions is a central feature of Aristotle's biological works that derives from the *Analytics* ideal of scientific knowledge.

And speaking of critics, there is no substitute for a careful reading of G. E. R. Lloyd 1990 and 1996. He has followed the arguments in this literature with care and responded with caution and skepticism. Lloyd has been especially critical of attempts to bring the practices displayed in the actual biological works in line with the austere demands on scientific knowledge issuing from the *Posterior Analytics*. I will take a moment here to move the dialectic on this question along another step or two.

The problem of fitting theory to practice can be viewed as having layers. As Lloyd himself stresses, even within the *Posterior Analytics*, Aristotle often departs from the axiomatic ideal of science described in its first five chapters. There are demonstrations that don't have causes as middle terms, demonstrations that don't have primary causes as middle terms, demonstrations in which the subject is a particular, demonstrations of predications that hold only for the most part, demonstrations that mix mathematical and natural premises – one can go on and on.

Then, moving beyond the *Analytics*, Lloyd notes that this heterogeneity extends to discussions of 'demonstration' in the *Rhetoric, Topics, Nicomachean Ethics*, and the *Parts of Animals*. The most important texts for those of us who wish to see the demonstrative ideal at work in the biology are those in *Parts of Animals* I.1 and *Physics* II.9. It is in these passages that Aristotle insists on another 'mode' of demonstration to accompany another 'mode' of necessity, hypothetical necessity, operative in teleological contexts. Yet *Posterior Analytics* II, chapter 11, also discusses necessity and teleology and whether natural things can be demonstrated by reference to both. And though that discussion is ultimately inadequate, and inadequately tied to the theory of demonstration, it nevertheless seems to be the context for revisiting these questions in *PA* I.

In my work, on the other hand, I have been stressing that the *Posterior Analytics* is about both demonstration and inquiry. The second book in particular – but not just the second book – is a study of scientific inquiry and progress. The discussion, including its examples, is of science in progress, moving gradually toward the principles, which will serve as explanatory starting points. As I note in "Aristotelian Problems," the discussion in Book II of the search for a basic, causal explanation for leaf loss in all broad-leafed plants reveals the same motives as the discussion in Book I of the basic explanation for why all triangles have interior angles equal to two right angles.

Nevertheless, the *Analytics* is a very general account of scientific knowledge and inquiry. There is every reason to think it will provide only the outlines for a philosophy of biology. In many ways, it does: the two pairs of inquiry that structure the argument of *APo.* II structure both the overall zoological program and the details of each discussion. For example, many of Aristotle's accounts of biological parts close with a formulaic summary: "That P is present in all the Ss was shown in the *History of Animals;* here we have said what P is, and why it is present in all Ss." *(*For variants on this theme, see *PA* 646a8–12, 650b8–12, 651b18–19, 652b20–23, 667b12–14, 673a32-b4; *IA* 1.704b7–10; *Somn.* 3.458a26–32.) These are the summations of someone with a particular view of how factual and explanatory claims are related to one another, and how definitions and explanations are interrelated. In fact, it is the view argued for in *Posterior Analytics* II.

If one now turns to the biological works, there are similar problems, and Lloyd's work has made me more sensitive to them. There are a number of important methodological discussions in those works – most obviously, in *Parts of Animals,* Book I. That it is related to the *Posterior Analytics* seems clear – there appear to be direct references to the strictures of the *Analytics* at 639a12–15, 639b8–11, 640a1-9, 642a5–6. Moreover, David Balme has made a convincing case that the discussion of division in *PA* I.2–3 is a development of the discussion of the same topic in *APo.* II. In every case, however, Aristotle introduces concepts and distinctions that are absent from the *Analytics.* The question that Lloyd has persistently and tellingly raised is this: How much of the model of science found in the *Analytics* remains once you introduce such ideas as that organisms are active, goal-oriented unities of matter and form; that for this reason there are special forms of necessity and of demonstration operative; and that the only way in which division can be of any use in biology is if it is practiced simultaneously on multiple differentiae?

In particular, does syllogistic remain? I agree with Lloyd that even if Aristotle did have a theory of demonstration prior to inventing syllogistic, the theory presented in the *Analytics* is a thoroughly syllogistic theory. Yet there is no attempt to present biological explanations in syllogistic form. Why not? One reply, which Lloyd seems dissatisfied with, is that the *Analytics* demands only that a good scientific explanation be *capable* of syllogistic recasting, not that it actually be written as a syllogism. Not only am I satisfied with this, I can't imagine a reasonable philosopher of science asking anything more. Even when providing mathematical examples in the *Posterior Analytics,* Aristotle doesn't claim that the actual geometric proofs, which he clearly understands, are in syllogistic. He insists only on the following points.

5

1. Typically in geometric proofs, a certain property (A) is shown to belong universally and necessarily to a geometric figure of a certain sort (C).
2. Typically, that property belongs to the sort of geometric figure it does because of some more basic property of that figure (B), perhaps a property essential to being that kind of figure.
3. Thus, however complicated the actual geometric construction displaying this connection, the ingredients for a syllogistic representation of the proof are in place: AaB, BaC, AaC.

I continue to believe that many of the explanations of the *De Partibus Animalium* meet this standard well.

Even on strict *Analytics* principles, however, important content will be lost in the process of revealing syllogistic form. The middle term must, in however tortured a manner, *refer* to the cause which explains the presence of some part in some group of animals (and be convertible into an account of that part). But there are a variety of kinds of cause, as Aristotle reminds us. Syllogistic by itself cannot reveal whether the middle term is the matter common to the group, or some goal or function subserved for the group by the part in question. Aristotle's *informal* explanations make the nature of the causation clear; reducing them to syllogistic would leave the mode of causation implicit. (This can be seen clearly in the reductions in *APo.* II.11.) This is a problem already at the heart of, and internal to, the theory of demonstration in the *Analytics*. Were Aristotle attempting to capture the entire task of explanation with logical syntax, as early logical positivism was, this would be a serious problem for his enterprise. But it is clear he was not attempting to do that.

I close with a rhetorical question: Is it plausible that a philosopher as systematic as Aristotle could formulate the first rigorous theory of scientific inquiry and demonstration, pepper the treatise in which he does so with biological examples, and then *not* aim to structure his science of animals in accordance with that theory?

1

Divide and Explain:
The *Posterior Analytics* in Practice

I

A longstanding problem concerning Aristotle's philosophy of science is the extent to which there is a serious conflict between the account of scientific explanation and investigation in the *Posterior Analytics* and the explanations and investigations reported in treatises such as the *Historia Animalium, Parts of Animals,* and *Generation of Animals.*[1] I shall not here mount a frontal attack on this question, preferring instead a rearguard action. This will consist of (i) articulating a familiar epistemological distinction between unqualified and sophistic or incidental understanding, which plays a fundamental role in Aristotle's philosophy of science; (ii) relating that distinction to the methodological recommendations of *Posterior Analytics* II.14; and (iii) presenting three sorts of evidence from the *PA* and *HA* indicating that the zoological works owe a great deal both to the above epistemological distinction and to its methodological implications. The evidence for (iii) will consist of the organization of information found in the *HA*, the relevance of that organization to the explanations of *PA*, and the theoretical and practical concern with the way in which lack of an appropriate zoological nomenclature can hamper the achievement of understanding. The evidence points to a much more direct relationship between the *APo.* and the biology than recent commentators have suggested.[2]

II

The distinction between incidental and unqualified understanding first appears in the preliminary account of understanding (*epistēmē*) which opens *APo.* I.2.

7

We think we understand a thing without qualification and not in the sophistic fashion, that is, incidentally, whenever we think we are aware both that the explanation because of which the fact is is its explanation, and that it is not possible for this to be otherwise.

This distinction is clarified in *APo.* I.5, which is primarily devoted to indicating circumstances under which we will fail to demonstrate "universally and primitively" that a predicate holds of some subject. There are three sources of such failures discussed and exemplified (74a6–32), and the chapter closes with a brief methodological suggestion for determining "when you know without qualification."[3]

I shall confine myself to the second case in Aristotle's list. Here, while the same feature is predicated distinctly of a variety of different sorts of entity, and that feature actually belongs to each of them in virtue of their being of some wider kind, this fact about them goes unrecognized, owing to the more general kind lacking a name (74a8–9, 74a17–25). In addition, 74a26–33 provides an example where there is a name designating the common kind in virtue of which the feature belongs, but where it belongs to each sub-kind because of its being an instance of the general kind. That is, there would seem to be two distinguishable sources of error of this sort: lack of a general term may be the source of a failure to grasp the wider kind; or the wider kind may be grasped, without the realization that the predicate in question belongs per se to the wider kind and only incidentally to the sub-kinds.

Later in the paper I will take up the issue of the epistemological and methodological problems which arise from the lack of general terms when they are needed, and will discuss 74a17–25 in that context. Prior to doing so, however, I want to discuss the way in which the distinction between unqualified and incidental understanding is developed from 74a26–33.[4]

For this reason even if one demonstrated, for each triangle, either by one or by a different demonstration, that each has two right angles, separately for the equilateral, scalene, and isosceles, you would not yet know that the triangle has two right angles, except sophistically; nor do you know it of the triangle as a whole, even if there is no other triangle besides these. For you do not know [that it has two right angles] as triangle, nor do you know it of every triangle, except in a numerical sense; but you do not know it of every one according to kind, even if there is none which is not known.[5]

The general notion of a subject to which a predicate belongs universally (*katholou*), per se (*kath' hauto*) and as such (*hēi auto*), developed in the previous chapter, is assumed here (cf. 73b25–74a3). There are many expressions which may truly identify the bearer of some property such as having interior angles which are equal to two right angles (*2R*). For example, it may be a bronze object, an isosceles figure, a triangle, and a plane figure. Which of these expressions identifies that about it in virtue of which it possesses *2R?* Answering this question is crucial in advancing toward unqualified understanding. For, if *2R* is in fact an immediate consequence of being a triangle, we will only have a sophistic sort of grasp of why every isosceles figure has it if we do not realize this. There are, as Aristotle notes, extensional marks that can guide one to the correct identification (74b1, 73b39); things which aren't isosceles also have *2R,* so it can't be because it's an isosceles figure that it has this feature.[6]

But as Aristotle makes clear in the above passage, this is not ultimately an issue that can be understood in extensional terms. It is, as the language of 'as such' and 'in virtue of' makes clear, the intensional grasp, the identification of the subject *as* the appropriate sort of thing, that is crucial to achieving unqualified understanding. Knowing of every sort of triangle that each has *2R,* while missing the fact that it is *as* triangle that each has it, is to have only an incidental grasp of the predication in question. You grasp it incidentally, because you know all the things which have *2R* but you don't know that they have *2R* because they are triangles. As *APo.* I.1 71a17–29 implies, you would be better off knowing that *2R* belongs to whatever it does *as* triangle while lacking experience of certain sorts of triangles – at least then, once these sorts are familiar, you will instantly grasp the fact that they have *2R.* (This implication is drawn out explicitly at 86a26–8.)

The realization that different forms of a kind have certain features in virtue of those features belonging primitively to that kind points toward a sort of demonstration, which I will simply refer to as 'type A'. An A-type explanation of the above predication would be,

> A. *2R* belongs to every triangle.
> Being a triangle belongs to every isosceles figure.
> _____
> *2R* belongs to every isosceles figure.[7]

The syllogistic formulation paints over the explanatory force here, but it can be brought out as follows. *2R* flows from the very nature of triangularity *as such,* and this is the general nature of isosceles figures. To reveal the isosceles figure as a sort of triangle, and *2R* as a primitive attribute of triangles, implies that the isosceles figure must, in the nature of things, have *2R.*

I wish to contrast such explanations with another sort, which will be labeled 'type B'. Let me highlight the differences between them:

A. 1. The predication to be explained is the predication of a feature which belongs to its subject necessarily, but to other subjects as well, what Aristotle describes in *APo.* as the least restricted sort of universal predicate: as *2R* belongs to isosceles triangles.
   2. The predication is explained by showing that the subject is an instance of the kind to which the predicate belongs primitively and as such: as *2R* belongs to triangle.
   3. Thus, were one to syllogize the explanation, the middle term would identify the proximate kind of the subject, with respect to the predicate in question.
B. 1. The predication to be explained is the predication of a feature which belongs to its subject primitively and as such: as *2R* is predicated of triangle.
   2. This primitive predication is explained by identifying some aspect of the subject's specific nature as responsible for it.
   3. Thus the middle term identifies, *not* a wider kind of which the subject is a sub-kind, but an aspect of that subject's *specific* nature, i.e., something proper to it which makes it that sort of thing.

Aristotle is ambivalent toward A-type explanations. There are times when he seems to reserve the concept of demonstration for explanations of per se predications.[8] But in these contexts he is typically contrasting per se predications with incidental predications in the weakest sense, i.e. those which hold only contingently. He gradually develops a terminology for the predications which figure as the explananda of A-type explanations: he speaks of certain predicates which belong in virtue of something common,[9] or which stretch beyond their subject, but not beyond their subject's kind.[10]

APo. I.9 is a prime example of this ambivalence toward A-type explanations. A strong contrast is drawn between unqualified and incidental demonstrations, the former being B-type alone. That is, only those explanations in which the conclusion states a primitive predication (of the subject *qua* that subject), and is explained by reference to starting points of that subject, are unqualified demonstrations (cf. 76a4–9).

On the other hand, even here 'incidental' demonstrations include those in which harmonic properties are explained by reference to their purely mathematical features (76a9–10; cf. 75b15–21), and such demonstrations share at least some of the features of A-type explanations.

This ambivalence may arise from an ambiguity in Aristotle's language. Take the remark, "That bird has a beak in virtue of being a bird." There are two ambiguities here. First, it is unclear whether the subject is being identified as a *sort* of bird or simply as a *bird*. Second, it is equally unclear whether 'in virtue of being a bird' implies knowledge of those features of a bird's life which necessitate its having a beak, or merely knowledge that beaks, for whatever reason, are a feature unique to the birds.

When the subject is viewed as a sort of bird, and 'in virtue of' implies nothing more than knowledge of the kind to which the feature in question belongs per se, A-type explanations are a possibility; when the subject is considered as a bird, and 'in virtue of' implies knowledge of what it is about being a bird that necessitates a beak, the above remark may point toward a B-type explanation.

The distinction can be put in historical terms as well. We may occasionally learn that a feature belongs to every *A*, then that it belongs to every *B*, and *C* . . . , and not yet notice that *A*, *B*, and *C* share a common nature (*K*) in virtue of which they all have that feature. In fact, I learned that chimpanzees, great apes, humans and orangutans were each tailless before I realized these were all members of a group, the hominoids, of which this is a peculiar feature. I learned this, in turn, before I learned that the hominoids share a common ancestor in the late Miocene era.

Now it is sometimes supposed that the intermediate stage of cognitive development in the above sort of sequence only gives the illusion of the capacity for explanation. At the zoo, my daughter asks why chimpanzees don't have tails, and I say it is due to their being hominoids, one of the oddities of which is lacking tails (always with the implied context of a wider group for which tails are the norm, of course). The person next to us, overhearing this conversation, may silently accuse me of sophistry – but Aristotle, I think, would not. For I have accomplished two very important tasks with this explanation. First, I have identified hominoids at the appropriate level of generality with respect to gaining a further understanding of the lack of tails – my daughter now knows the kind which has the feature *as such*. Second, even without pursuing this further question, she has gained a picture of the unity of the primates which the partial knowledge that chimpanzees and humans don't have tails did not provide. A-type explanations, then, are a crucial stage in the acquisition of understanding about a domain.

These two sorts of explanation are distinguished quite clearly in *APo*. II.16–17. This is not surprising, for these chapters explore just those issues of counter-predication and of the role of definitions in explanation where the distinction between A- and B-type explanations becomes important. I simply wish to draw on this discussion to mark the distinction on which I have been insisting.

The primary goal of *APo.* II.16–17 is to establish that, within a given subject kind, a scientific demonstration involves a middle term which is extensionally equivalent with, causally prior to, and part of the definition of, the predicate to be demonstrated (98a35–98b4, 98b14–24, 99a1–4, 99a21–2). The following passage summarizes the main features of this discussion.

> . . . the middle term is an account of the first extreme: that is why all under-standing comes about through definition. E.g. shedding leaves follows together with the vine [kind] and extends beyond it; and with the fig [kind] and extends beyond it – but it does not extend beyond all, but is equal in extent to them. Thus if you were to take the primitive middle term, it is an account of shedding leaves. For there will be a middle term in the other direction (that all are such and such); and then a middle for this (that the sap solidifies or something else of this sort). What is shedding leaves? The solidifying of the sap at the connec-tion of the seed. (99a21–9)[11]

This complex and difficult passage is an attempt to distinguish the two sorts of explanation I have been discussing. Shedding leaves is here the feature to be explained. It is noted that it belongs to all of various kinds, extends beyond each kind, but is equal in extension to their conjunction.

Thus there are two initial problems: why does shedding leaves belong to every vine?, and why does shedding leaves belong to every fig tree? Phi-loponus suggests that the enigmatic reference to "a middle term in the other direction" is a reference to an earlier example, in which the middle term is a 'such and such' over various sorts of plants which shed leaves, namely 'broad-leaved plants' (cf. 98b5–16).[12] If this suggestion is correct, the expla-nations which emerge through this middle term are A-type explanations of why fig trees or vines shed their leaves.

Now, however, we are introduced to the idea of a more basic middle *for this,* that is, for why broad-leaved plants shed their leaves (as such). This will, we are told, be a definition of shedding leaves. The parallel explanation we might build on his hints would look like this:

B. Shedding leaves belongs to everything which undergoes solidification of sap at the seed connection.
Solidification of sap at the seed connection belongs to every broad-leaved plant.

———————

Shedding leaves belongs to every broad-leaved plant.[13]

In this explanation, the explanandum involves a predicate which belongs to its subject 'in itself.' But it doesn't belong primitively; that is, there is a cause of its belonging. Shedding leaves is a process resulting from a more basic process of solidification.[14] In such explanations, that close relationship between answers to 'what-is-it?' and 'on-account-of-what?' questions argued for in *APo.* II.1–10 is clearly revealed. For this causally basic process is both what shedding is, and the cause of certain plants shedding their leaves.[15]

To reiterate then: in A-type explanations, a universal predication of a subject is accounted for by noting the primary kind of that subject relative to the predication to be explained, i.e. by reference to the kind which has the predicate per se. In B-type explanations, such per se predications are accounted for by understanding what it is about being that kind which immediately necessitates it having that predicate. This sort of explanation answers the primitive question relative to the predicate under scrutiny – why does it belong to the things it belongs to *as such?*

III

The distinctions between incidental and unqualified understanding, and between A-type and B-type explanations, have a number of methodological implications. If there are ways of organizing our true propositions about a subject so that A-type explanations will be more readily apparent, which in turn will reveal those per se predications which will need to be explained by reference to peculiar, differentiating features of a subject, then one ought to so organize one's true propositions. This, I suggest, is the subject of *APo.* II.14. One of the features of that chapter which gives prima facie plausibility to this suggestion is that, as in *APo.* I.5, the subject of how kinds are related to their forms leads directly to the issue of forms which have much in common but no common name. Again, the issue of nomenclature will be put off until a later section of the paper.

APo. II.14 suggests how to use 'dissections and divisions' in grasping problems (98a1).[16] 'Problem' is used here in a technical sense defined at *Topics* I.101b17: "Problems are the things about which there are deductions." Scientific problems are in the form of requests for explanations – why *p?* The discussion in *APo.* II.14 recommends, developing arguments of *APr.* I.27, that the investigator is

to select thus, positing the kind common to all; e.g., if the subjects of study are animals, [select] what follows all animals, and having grasped these, again what follows all the first remaining things, e.g. if this is bird, what follows every bird, and thus always [ask what belongs] to the nearest kind.[17]

A full evaluation of this chapter would involve placing it in the wider context of the relevance of division to definition and explanation, and of the general rules for constructing scientific syllogisms in *APr.* I.27–30.[18] For the moment, I merely wish to draw attention to its relevance to the issue of formulating A-type explanations and hitting on per se predications.

In considering the above quotation, it is crucial to remember that it assumes a set of divisions of a subject domain ready at hand, and divisions, the example suggests, organized along Aristotelian lines. But as Aristotle makes clear in *APr.* I.31, the investigation of a subject S cannot move to an explanatory stage via division alone. We may recognize that S belongs to a kind K, and know that K is differentiated into sub-kinds F_1 and F_2; but that only allows us to infer that S is either F_1 or F_2. We need first to grasp empirically, as a problem, which of the differentiae belongs to our subject (F_2aS), ask why every S has F_2, and search for a middle not wider in extent than F_2. Divisions are useful, however, as a potentially exhaustive *source* of predicates from which to select appropriate predications (cf. *APr.* I.31 46a31–b19).

APo. II.14 is thus suggesting how best to make use of such information. The suggestion is that we begin with the most common kind relative to the object being investigated and select or pick out what follows necessarily and what it follows necessarily before proceeding. Again, assuming our divisions have been properly made, we can proceed to the *first* remaining group, i.e. to the immediately next level. As an example we are given an extensive kind which is common coinage, bird, and we are told to collect all the features common to *every* bird. We are to apply this methodology iteratively throughout our division, presumably until we get to the indivisible kinds.

Why are we doing this?

For it is clear that we shall now be in a position to state the reason why what follows the items under the common kind belongs to them – e.g. why it belongs to man or horse. Let A be animal, B what follows every animal, and C, D, E sorts of animals. Well it is clear why B belongs to D, for it does so because of A. (98a8–12)

To use a concrete example: suppose we are examining the peregrine falcon, and are asking why it possesses a hooked beak. We recognize it is a bird, and, consulting our divisions, we see that having a beak follows on being a bird, that beaks are differentiated in many ways, that having hooked beaks follows on having crooked talons, strong wings, soaring flight, being carnivorous (all under *other* divisions). We are now in a position to formulate an explanation:

> Having a hooked beak belongs to every crook-taloned bird.
> Being a crook-taloned bird belongs to every peregrine falcon.
> _____
> Having a hooked beak belongs to every peregrine falcon.[19]

Further, should one find features which belong to all and only one sort of bird such as hooked beaks belonging *only* to birds with crooked talons, one will know that, while these may be demonstrated to belong to less extensive kinds, such predications cannot themselves be explained through the subject's membership in a wider kind. Nor can we know, simply by consulting this information, whether a middle term is just part of what it is to be this kind of thing, or a per se incidental feature to be B-explained, but we know that determining *that* is what is at issue.[20]

What this downward division/correlation methodology accomplishes, then, is to provide a collection of propositions which will play various roles in explanation, to reveal in certain cases the A-type explanation for kinds having certain features, and to direct, or redirect, inquiry to the appropriate level for B-type explanations of certain more primitive predications.

But this, it shall now be contended, is rather like what we find in the *Historia Animalium,* and thus this chapter may provide clues to the aims of that work, and to its relationship with the other zoological works. It is now time to look to the biological works for indications that the epistemological and methodological recommendations of *APo.* to which I have drawn attention play an important role in the organization of investigation and explanation in that domain.

IV

About to launch into a discussion of the bloodless animals in *HA* IV, Aristotle looks back self-consciously on the preceding discussion.

> We have said previously, about the blooded animals, which parts they have in common and which are peculiar to each kind; and about the non-uniform and the uniform parts, which they have externally and which internally. Now we must discuss the bloodless animals. (523a31–b2)

15

This is a careful summary. The second half of the sentence reflects precisely the order of the composition of *HA* I.7–III, which within extensive kinds has surveyed first the external non-uniform and internal non-uniform, and then the external uniform and internal uniform, parts. Likewise, the first part of the sentence points to a pervasive feature of the earlier discussion. For within each extensive kind, there is a tendency to move from the consideration of what this kind has in common with previously considered kinds, to the isolation of the distinctive features of that extensive kind, and then to the noting of what is distinctive to various sub-kinds.

Considered within the context of contemporary biological classification, the purposes of the *HA* are difficult to fathom. Many attempts to construct an organized taxonomy out of its materials have failed, and for the first half of this century it came to be viewed as a loosely organized natural history.[21] David Balme began serious re-evaluation of *HA*'s scientific function by stressing the way in which it draws attention to closely related groups of differentiae:

> . . . if the *HA* is read, not as an encyclopaedia nor as a collection of simple natural history, but as being itself to some extent a theoretical treatise – namely a study of differentiae – then much of its apparent incoherence disappears.[22]

Both Balme and, recently, Pierre Pellegrin have stressed the role of the *HA* in providing the materials for scientific definition. Pellegrin, in addition, has furthered the work on the meaning of Aristotle's classificatory concepts, kind, form and difference begun by Balme, showing how the apparent chaos in the use of these terms in the biology is largely dissipated when they are viewed within the context of the logic and metaphysics of division.[23] With these insights as background, Professor Allan Gotthelf and I decided to read continuously through the *HA* with two eyes (one each) on the details of organization and structure of that work. Part of our concern was to test some preliminary speculations I had ventured at an NEH Research Conference directed by Gotthelf, about the relevance of the *APo.*'s theory of explanation to Aristotle's biological practice.[24] The provisional results of this joint venture are reported here and in Professor Gotthelf's "*Historiae* I: *Plantarum* et *Animalium*" (cf. Gotthelf 1988), and a first attempt shall be made to tie them to those wider concerns about explanation and scientific methodology.

In outline, the argument of this section of the paper is that a reasonably constant focus of *HA* is the indication of the widest class to which a feature belongs universally, and drawing attention to cases where a feature predicated universally of an extensive kind belongs to other extensive kinds as well. There is a persistent concern to distinguish these features from those which are proper to the kind in question. And where some common feature of a kind is differentiated in unique ways in sub-divisions of that kind, this is noted, and a list of other features peculiar to the sub-kind is usually provided. The organization, that is, is sufficiently akin to the ideas in *APr.* I.27–30 and *APo.* II.14 to make one suspicious that that sketch is in the background.

And not too far in the background, perhaps. Consider *HA* I.6 491a7–14:

> These things have been said in this way now as an outline, to provide a taste of the things, however many they may be, about which we must study, in order that we may first grasp the differences and the incidental features in every case. (We will speak in more detail in what follows.) After this we must attempt to discover the causes of these [differences and incidental features]. For to pursue the study in this way is natural, once there is an investigation about each kind; since it becomes apparent from these investigations both about which things and from which things the demonstration ought to be carried out.

This is a pivotal text. The previous chapters have given a theoretical account of the nature of similarity and difference of animal parts, distinguished four types of zoological differentiae (parts, lifestyle, activities, and character traits), and given a carefully selective analysis of how various widely different animals can be said to differ according to features of these four sorts. (It is also shock therapy for those who would use such differentiae as wild/tame, winged/wingless, land-dweller/water-dweller, indiscriminately.) The first half of chapter 6 then lists what he takes to be established as extensive kinds, and argues for additions and limits to that list. It closes with the above account of this investigation's place in the zoological enterprise.

What does this account tell us? It apparently alludes to the first five chapters as providing a taste of the sort of investigation which will be carried on in a more precise fashion afterwards. On the other hand, it could be referring to the immediately preceding discussion of the extensive kinds, for there is explicit reference to a discussion of the number of things to be studied, i.e. to the question how many kinds there are.

The first task, we are told, is to grasp the differences and the incidental features that belong in each case. This may allude to two aspects of the 'divide and select' methodology of *APo.* II.14, that is, the division of a kind into its immediate sub-kinds and then the 'grasping' of what belongs to each.

But the phrase is admittedly enigmatic, and is best left relatively unin-terpreted until the actual method has been examined.

Whatever precisely this first stage comes to, it is called a *historia* and is said naturally to precede the attempt to discover the causes of these, which I take it refers to these differences and incidental features of kinds. The *histo-ria* will make apparent the 'about whiches' (the *explananda*) and the 'from whiches' (the *explanans*) of our scientific explanations. This last sentence is self-consciously worded in the language of the *Posterior Analytics*.[25] Notice that it does not say merely that the investigation of difference and incidentals is a preliminary to causal investigation. Rather, it says that differentiation of *explanans* and *explananda* becomes clear as a consequence of this investiga-tion. This, I suggest, makes it more likely that the organization of *HA* will reflect the needs of Aristotelian explanations.[26]

Is this passage *post hoc* methodological window-dressing, or does it actually describe *HA*'s procedures? And is the methodology of *APo.* II.14 behind *HA*'s organization, if it has one?

Our preliminary results suggest that it is. The evidence from the *HA* alone is suggestive, but not conclusive. But when it is combined with an examina-tion of explanations in *PA*, that evidence gains in power. In the following sections I shall focus exclusively on *HA* II–IV and *PA*.[27]

HA II opens[28] with a seemingly innocuous remark which nonetheless takes on significance against the background of the methodology of *APo.* II.14:

> Some of the parts of the other [i.e. non-human] animals are common to all of them, just as was said before, while others are common to certain kinds. (497b6–9)

Select what's common to the widest kind first, and then to the next widest, may be the suggestion. He then reviews the earlier discussion of sorts of likeness and difference, and proceeds according to the summary quoted earlier (p. 15) from the opening of Book IV. Let us begin with a wide-angle shot of the terrain and then "zoom" in on certain sections to study details.

Summary: HA II–IV
II.1–7: external non-uniform parts of viviparous quadrupeds
   8–9: apes, baboons, monkeys – tending both toward biped and toward quadruped
    10: external non-uniform parts of oviparous quadrupeds

11: chameleons

12: external non-uniform parts of birds

13: external non-uniform parts of fish

14: external non-uniform parts of serpents

15–17: the internal non-uniform parts considered one part at a time in all the blooded kinds.

III.1: genitalia (which are not neatly divisible into internal or external, and which differ considerably between male and female)

2–4: blood vessels

5–22: sinew, fibre, bone, cartilage, nail, hoof, claw, horn, hair, skin, feathers, membrane, omentum, bladder, flesh, fat, suet, blood, marrow, milk, semen

IV.1: comparison of four bloodless kinds
cephalopods, external and internal parts

2: crustacea, external and internal parts

3: testacea, external and internal parts

4: hermit crabs which tend both toward crustacea and toward testacea

5: sea urchins and odd testacea

6: the odd ones, sea squirts and sea anemones

7: insects, external and internal parts
rare unclassified creatures

8: sense organs across the animal kingdom

9: noises across the animal kingdom

10: sleep and dreams across the animal kingdom

11: male and female across the animal kingdom, which neatly points toward V–VII on reproductive behavior

Even at this level of resolution there are some interesting variations to be noted. While the discussion of the external parts of the blooded animals takes the form of an account of each part in one kind, and then a new account of each part in the next kind, each of the internal and the uniform parts is discussed as it appears *throughout* the blooded kinds, and then discussion moves to the next *part*.

Yet another shift in method occurs in *HA* IV. Aristotle returns to discussing the parts of each bloodless kind, kind by kind, again. But now *both* the external and internal parts are discussed for one group before proceeding to the next.

These shifts, and more subtle ones that occur more regionally, are, I suspect, thought out, not arbitrary.[29] But, as David Balme has so forcefully brought home, if one were looking for a systematic treatment of kinds of animals, kind by kind, organized by one uniform method of classification, one could only be disappointed.

Nonetheless, the above summary does suggest that the 'extensive kinds' do have an importance as a locus of a number of similarities and differences among parts. Yet if understanding of the sort envisaged in *APo.* is Aristotle's ultimate goal, these can't be the primary focus. What we will want to know is the widest group to which a particular *feature* belongs, and how that feature differs or is differentiated in sub-kinds, and what the widest group is to which each of these differentiated features belongs. Focusing on such questions will not suggest an encyclopedia of animals with a full and complete non-comparative description of each kind. Nor will it recommend remaining with one principle of organization as the features being examined change.

Let us now narrow our focus in on one passage from each of the methodologically distinct discussions mentioned. *HA* II.12 discusses the external organs of birds. It begins with a review of parts which belong to birds but extend more widely:

> the birds as well have some parts similar to the animals already discussed [viviparous and oviparous quadrupeds]; for all birds have a head, neck, underbelly, and an analogue of the chest. (503b29–32)

In addition, they are bipedal, like humans, but the legs bend more as those of the quadrupeds do (503b33–4). Various features are marked as proper attributes of birds – feathered wings (503b35), a long haunch bone (504a1), beak (504a21), quilled feathers (504a31–2). Certain features, while common to the birds, are found in distinctive ways in various sub-kinds. Crook-taloned birds have the largest and strongest chests (504a4–5); all have numerous clawed toes, but swimmers have them with webbing (504a8), and high flyers all have *four* toes. Of this latter group, most have three forward and one behind ("instead of a heel"), but a few, such as the wryneck, have two front and two rear, a proper attribute of theirs (504a10–13). All long-legged or web-footed birds have short rumps and extend their legs when in flight (504a33–6); good flyers have no spurs, while the heavy birds do (504b7–10). Some have feathery crests, but the domestic cock's is peculiar in being flesh-like (504b11–12).

What prevents this material from being an extremely spotty, random assortment of observations is its consistent focus on the following questions:

1. What features of birds 'extend beyond' to a wider class?

2. What are the proper attributes of birds, features belonging to all and only the birds?
3. What differences does one find among the birds – can we find a group to which such differences belong uniquely?[30]

The first concern is indicated by the fact that every separate consideration of an extensive kind begins by taking up the issue of which parts it has in common with other such groups – the opening of the discussion of the external parts of the oviparous quadrupeds, for example, is a review of features which are "just as the vivipara among quadrupeds" (502b29–32). The focus on the extensiveness of a feature is often remarkable.

> All the quadrupeds have bony, fleshless and sinewy limbs; in fact, this is generally true of all the other animals which have feet, excepting man. (499a31–2)

This feature not only extends to all the quadrupeds (this remark occurs during a discussion of viviparous quadrupeds), but to birds as well – to all blooded, footed animals except man.

Let us turn briefly to a consideration of a discussion of the internal organs. Remember that Aristotle has here abandoned the methodology of remaining within a kind for each part and is reviewing a number of related organs throughout the blooded animals.

> Now those which are quadrupedal and viviparous all have an esophagus and windpipe, and it is positioned in the same way as in humans; and whichever of the quadrupeds are oviparous are alike in this respect; in fact the situation is similar among the birds; but there are differences in the forms of these parts. Generally speaking, all which take in air by inhaling and exhaling have a lung, windpipe and esophagus . . . Not all the [blooded] animals have lungs; for example, fish do not, nor if there are any other animals which have gills.[31]

When one compares this with the discussion of these organs in *PA*, its acausal, a-explanatory nature is truly remarkable. But notice – he is seeking the highest level of generality possible for these features; schematically the passage reads "in *K1*, in *K2*, also in *K3*, in fact, in all the *K*s that breathe: not in *all* the blooded *K*s, however, for fish don't have lungs." Now at this point he could easily have said. "For gills perform the same function as lungs, namely cooling blooded animals, but fish, living in water, cool themselves with water rather than with air." This he clearly avoids. Notice, however, that we are told what else, besides lungs, breathers have in common, and are provided with a mark of the widest unnamed class with these features. We have moved from knowing that isosceles, equilateral and scalene each have *2R*, to knowing that triangles do, so to speak. Thus, while *HA* stops short of actual explanations, it has organized the facts in such a way as to direct us toward A-type explanations for sub-kinds having a lung (by identifying the kind to which it belongs primitively); and B-type explanations for the possession of a lung (by noting its correlation with the inhaling and exhaling of air). In fact, Aristotle devotes an entire work to these features, or perhaps part of a treatise.[32]

Certain innards are dealt with rather summarily.

> All the viviparous quadrupeds have kidneys and a bladder; none of the non-quadrupedal ovipara have them, neither the birds nor fish, while the sea turtle alone among the oviparous quadrupeds has these in proportion to its other parts. (506b25–8)

Occasionally, however, while a part may be common to all the blooded kinds, it will be differentiated in a markedly different manner at some level: 507a30–508a8 is an extensive discussion of the stomach and gut of the ruminants (as we should say), and of features correlated with these differences.

The first uniform part taken up (for it is most common in the blooded animals and has the appearance of being a first principle) is the system of blood vessels. This discussion is atypical, both because it contains long quotations and extensive discussion of previous views, and because his own account is explicitly generalized so as to be applicable to all blooded kinds.

> The facts about the starting-points and the great blood vessels are thus in all the blooded animals, though the other numerous blood vessels are not alike in all; for they neither have the parts in the same way, nor do they all have the same ones . . . (515a16–19)

Thus the account provided can be applied to *any* of the blooded kinds, *qua* blooded. Any more detailed description would have to take into account the different organization of vessels reflecting the different organization of the other parts, kind by kind.

Finally, one passage will be examined from Aristotle's account of the bloodless animals (*HA* IV.1–7). The discussion begins with an overview of the four main kinds, especially focused on which have the hard parts inside and which outside, and the nature of those parts. Let us focus on the *cephalopods.*

> First is the kind of the so-called softies or cephalopods – this consists of those which are bloodless and yet have their fleshy part outside, while if they have any hard part it is inside, just as with the blooded animals – for example, the kind consisting of cuttlefish. (523b1–5)

The specific discussion of the cephalopods first notes four external parts which they all have in common ("select the common kind first, and what follows this"), and proceeds,

> Now all have eight feet, and all these have a double row of suckers, except one kind of octopus. But the cuttlefish and the small and large calamary have, as a proper attribute, two long tentacles, at the end of which is a rough part having two rows of suckers, by means of which they acquire nourishment and take it into the mouth . . . (523b28–32)

Again, there is clear movement from those features which belong to the cephalopods as such, to those which belong to a group which again has no single name, but a number of features in common (cf. 524b22–4). When Aristotle moves on to a discussion of the octopuses (a kind with many kinds – 525a13) he again notes first common features and then those peculiarities of the sub-kinds, e.g. the *heledōnē* which is "alone in having a single row of suckers [on each tentacle]" (525a17).

Professor Balme has pointed out the oddity, from the standpoint of natural history, of Aristotle's treatment of the blind mole, mentioned often, but as if it had only two features, being viviparous and having rudimentary eyes below the skin.[33] But from the standpoint of Aristotelian explanation, it makes perfect sense. As with the *heledōnē* above, only features *not* common to the wider group are worth noting – all others are to be explained at the wider level of generality.

Balme has also drawn our attention to discussions which read as reports about various kinds which have not been parceled out to the appropriate section of *HA.*[34] One of these is the discussion of the apes and baboons (502a17–b27). In some sense, Balme is clearly right; both character traits and internal parts are referred to (although only by the way), and there is no discussion of their behavior or character elsewhere in *HA*. On the other hand, this material is certainly well placed, for it appears after creatures which are

23

either clearly viviparous bipeds or clearly viviparous quadrupeds, and opens by announcing

> Some animals are allied in nature both to man and the quadrupeds, for example, apes, monkeys and baboons. (502a17–18)

Accordingly, throughout the discussion, their features are allied either with the bipeds or with the quadrupeds. The focus thus remains on identifying the widest class in virtue of which a feature belongs. Let me direct the discussion toward actual explanation by comparing a statement about the ape's hindquarters with an A-type explanation from *PA*.

> . . . and as a quadruped [the ape] has no buttocks, while as a biped, it has no tail. (502b21–2)

> But the ape, owing to its tending to both sides with respect to its shape, that is, to its being neither sort and yet both – owing to this, it has neither a tail nor buttocks; as a biped no tail, as a quadruped no buttocks. (*PA* IV.10 689b31–4)

Now of course, as it stands, this is not a fully satisfactory explanation of either feature – it is, as I have characterized them, an A-type explanation. In order to understand why quadrupeds lack buttocks, a B-type explanation is required, which might point to aspects of the life of quadrupeds which would make buttocks useless or harmful. (In fact such an explanation precedes the above-quoted passage.) To explain why *apes* lack them, one needs only to be reminded that they are, *in this respect,* built like quadrupeds.

More will be said about such explanations in a moment. Here I want to close the discussion of *HA* by stressing the relationship between these two texts. The *HA* gives us the relevant descriptions of the features of the animal in question, assigns those features under the appropriate wider kind, and leaves it up to *PA* to do just what it says it will:

> From what parts and how many parts each of the animals are composed has been shown in greater detail in the histories about them. Now we must investigate the causes through which each is the way it is, separating these off by themselves from what was said in the histories. (646a8–12)

The hypothesis that *HA* has, as one of its primary aims, the purpose of finding and characterizing the groups to which various features belong primitively, even when those groups have no common name, makes its organization reasonable, and squares well with the recent work of both David Balme and

Pierre Pellegrin showing its non-taxonomic nature. In addition, the use in the *PA* of the information that one finds in *HA* strongly suggests that these generalizations were used as interim explanations of lower-level kinds having certain features, and as the focus of the more basic explanations of that work.

V

The *Parts of Animals* is a work of sustained scientific explanation. Its explanations are typically worded as answers to 'problems', i.e. questions of the form, 'Why do these animals have this part?' They can be fairly cleanly divided into A-type and B-type explanations, the latter subsuming both explanations by reference to a creature's material make-up and by reference to its lifestyle or activities.

The A-type explanations are exemplified well in the following example:

> But even the dolphin has, not fish spine, but bone; for it is viviparous. (664a16–17)

There are two facts specified about dolphins: they *are* viviparous, and they *have* bones. These important facts about dolphins are remarked on in the *HA* (that they have bones at 516b11; that they are viviparous at 489b2 and 566b3–26). Having bones rather than fish spine is a necessary, but not a primitively universal, feature of dolphins. Bone *tout court,* however, *is* a primitively universal feature of the vivipara; indeed, it is while pointing this out in the *HA* (516b12) that Aristotle there remarks, "But the dolphin also has bone, *not* fish spine." The above explanation, then, directs us to the lives of vivipara for an understanding of their having bony skeletons. But it also tells us that, whatever that explanation might be, *dolphins* have them because they are a sort of viviparous animal.

The interaction between these two types of explanations can be seen clearly if we remain with cetaceans for a moment. *PA* III.6, on lungs, opens with the following broad explanatory generalization:

> Any *genos* of animals has a lung on account of its being a land-dweller (668b33–4)

As we saw earlier, four of the five extensive kinds of blooded animals have lungs. It thus can't be because they are *blooded* that they have this structure, for then *all* blooded ones would. Aristotle here sanctions an A-type explanation for any of them in terms of their being 'land-dwellers'.

The term translated land-dweller can also mean walker, but either way the above generalization would seem to be in serious trouble when confronted with cetacea. How does Aristotle handle this problem?

> The animals that breathe, on the other hand, produce cooling by air, so that all the breathers have a lung. Further, all the land-dwellers breathe, and so do some of the water-dwellers, as well, e.g., whales, dolphins and all the spouting sea monsters. For many animals have a nature which tends to both sides and while among the land-dwellers and those receptive of air, they spend most of their time in the water owing to the blend of materials in their body. And some of those which live in the water participate to such an extent in the land-dwelling nature that the decisive factor in life consists in their breathing. (669a6–13)

There is a thorny nest of problems here, having to do with how Aristotelian division deals with 'dualizers', problems extensively discussed in *HA* VIII.1–2, and alluded to in Balme 1987b (85–6). Most of them I will not discuss, but will focus on the form of explanation being offered here.[35] The chapter opened, as we saw, with the bald claim that any *genos* of animal has a lung due to its being a land-dweller. In the passage just quoted, however, it is said that (among blooded animals) both land-dwellers *and some water-dwellers* breathe. We seem to be in deep trouble here.

The solution which emerges in *HA* VIII.2 lies in distinguishing different criteria for being a water-dweller: for our purposes, a creature may be a water-dweller with respect to nutritive requirements, or with respect to the way it cools itself. Thus Aristotle imagines two lines of division, one under 'cooling organ' and one under 'nutritional nature', with a water-dweller/ land-dweller division within each line. In this way, the same animal may be both a land-dweller (insofar as it cools itself by means of inhaling and exhaling air) and a water-dweller (insofar as its bodily nature requires sea food). The porpoises, then, partake in the nature of land-dwellers to the extent that they cool themselves as land-dwellers do, though from the standpoint of their material makeup and nutritive requirements they are water-dwellers. But as they are each according to different criteria, and under different lines of division, we will not have the same animal appearing in more than one line of the same division.

The issue here concerns the nature of lunged creatures *as such*. Aristotle, that is, moves beyond noting the kind to which anything which has lungs must belong to an exploration of the nature of this wider class. These animals all cool themselves by means of inhaling cool air and exhaling warm air – that is they partake in the nature of land-dwellers (*qua* cooling).[36] Here is a clear B-type explanation of why it is that all these creatures have lungs, for lungs are necessary if a creature is to breathe – that is what lungs are for.[37] On the other hand, to apply this understanding to the cetacea, one may simply point out that, with respect to cooling, they participate in the land-dwelling nature.

To indicate that this distinction between 'forms' of explanation is neutral with respect to what sort of causation is involved, it is useful briefly to examine examples of the very common double-barreled material/teleological explanations found throughout the *PA*. At the same time, an awareness of what is and is not said about the same features in the *HA* will help focus attention on the nature of the explanations in *PA*. In each of these cases I will look at explanations of features which are proper attributes of their kind.

The *HA* has the following to say about the bushiness of the hair on man's head:

> All those animals which are quadrupeds and viviparous are hairy, so to speak, unlike man, who has a small amount of minute hairs except on his head, which is the hairiest of all the animals. (*HA* 498b17–18)

The *PA* tells us why:

> The human head is the hairiest of all the animals *from necessity because* of the moistness of the brain and *because* of the sutures in the skull (for wherever there is a lot of moisture and heat, there growth will necessarily be greatest), but also *for the sake of safety,* in that the shelter it provides can protect against excess of cold and warmth. (*PA* 658b2–6, emphasis added)

As a human attribute, one cannot provide an A-type explanation for the fact that it belongs to humans by showing that it belongs to a wider kind. (One could use knowledge of its being a human attribute to account for Persians having it, however.) It is by delving further into man's nature that one will come to understand why it's there. And that is just what Aristotle does, rooting its presence in facts about human physiology (facts provided elsewhere in *HA*: the size and moistness of man's brain at 494a28, the sutures at 491b2–5, 516a12–20 and 517b13–21), and the special need to protect the human brain from climatic extremes so that it may perform its proper function.

Both explanations, however, point to basic facts about human nature in accounting for a per se attribute of human beings. In this respect, they are both complementary B-type explanations of the same fact, one invoking material necessity, the other invoking the function for the sake of which it is there.[38]

The same relationship between *HA* and *PA* is clear in their respective discussions of the shedding of deer horns.

> Deer alone among those having horns have them solid throughout . . . (*HA* 500a6–7)

> Deer alone shed their horns in season, beginning from the age of two, and grow new ones again . . . (*HA* 500a10–11)

> In deer alone are the horns solid throughout, and deer alone shed their horns, on the one hand for the sake of the advantage gained in having their load lightened, on the other hand from necessity, due to their weight. (*PA* 663b12)

The solidity of the horns of the deer is a per se feature, and in fact is not further explained, but rather plays a role in explaining the shedding, another proper attribute of deer. For, as in the previous example, this feature happens necessarily, given the weight of solid horns, *and* because the deer benefit from this happening. Again, there is reference both to the peculiar material nature of deer, and to the requirements of their lives, in accounting for the shedding of horns. The shedding of horns is explained to belong to deer in virtue of more basic features of their nature; in the context of this discussion, it has been B-explained.

My last example is an explanation of why the strange wryneck has one of its peculiar features (see p. 20, above). It is, of course, a soaring bird, and were I to wonder why it has four toes, it would be appropriate to simply point that out.[39] If, however, I wish to understand the peculiar arrangement of *its* toes – two fore and two aft – knowing something about *its* nature will be necessary, such as the following.

> In other birds, then, the position of the toes is thus [three fore, one aft] but the wryneck has only two in the front and two behind; a cause of this is that its body is less inclined in the forward direction than the body of other birds. (*PA* 695a22–6)

Thus, we may imagine, this particular feature is needed to insure proper 'balance'. Similar explanations of features found only among the swimming, wading or soaring crook-taloned birds are also found throughout this chapter: in each case, some peculiar feature of their lifestyle is shown to require, or at least be improved by, a structural difference in parts common to birds.[40]

These texts are representative of a pattern that one finds throughout the *PA*. Often, a part is said to belong to some sub-kind by noting that this part is a common feature of some more general kind. The message of such explanations, which I have referred to as A-type, is that a deep theoretical explanation for this part should be sought at a more general level, but that an explanation for its belonging to this particular kind is that *it* is a form of the kind to which this part belongs primitively and as such. On the other hand, if a feature is identified as peculiar to a kind, the explanation does not take the form of subsuming the kind under a wider kind, but of exploring that kind itself, to see if something basic to its life or material constitution can account for it as a consequence. This seems to be a treatise thoroughly in the spirit of the philosophy of science of the *Posterior Analytics*.

VI

PA I, Aristotle's 'philosophy of zoology', begins by considering whether we should study the most common kinds first or the most specific. It offers what at first sight appear to be rather trivial reasons for choosing the former strategy. But, as David Balme has hinted, *APo.* II.14 is in all likelihood in the background here too.[41] I should like to develop these hints in two directions. First, I want to note in detail the parallel between this strategic recommendation and the ideas I earlier discussed in the context of *APo.* about unqualified understanding. Then I shall return, as promised more than once, to the problem of groups with common features and no common name.

Having distinguished between the actual practitioner of a science and the person with a methodological concern with the principles "to which one will refer in appraising the method of demonstration" (639a13), Aristotle goes on to consider questions of this latter sort in *PA* I. The first is:

> should one take each being singly and clarify its nature independently, making individual studies of, say, man or lion or ox and so on, or should one first posit the attributes common to all in respect of something common (639a16–19, tr. Balme)

The question here is so reminiscent of *APo.* II.14–18 that it is difficult not to see its methodological suggestions in the background. That background would recommend the latter approach, on grounds that if one knew, for example, that lion and ox were each covered with body hair, but hadn't yet recognized that each of these kinds had a common nature, both being viviparous quadrupeds, one would lack a true understanding of why these particular kinds possess this feature. In fact, he goes on to say that, pursuing the first option, a researcher would consistently describe in partial terms what belongs universally (639a23). The alternative is summarized as "studying that which is common to the kind first, and then later the proper attributes" (639b5).[42] This is indeed the method he goes on to recommend in chapter 4, wherever there is a common kind available (644b1–7; 645b1–12). Further, as I have argued, a plausible construal of the *HA* is that it is organizing animal differentiae according to this method.

The same background in the *Analytics,* which I will now consider, explains both a theoretical and a practical concern over naming animals so that one refers to their nature at the correct level of generality with respect to the problem to be solved.

As I said earlier, one pervasive source of a failure to achieve anything more than incidental understanding according to *APo.* I.5 is the lack of a name which characterizes a subject at the appropriate level of generality.

> And it might seem that proportions alternate for things as numbers and as lines and as solids and as times, as when it was proved separately, though it was possible to prove it of all cases by one demonstration. But because all these – numbers, lengths, times, solids – were not some one named group, and differed in form from each other, they were grasped separately. But now they are proved universally. (74a17–23)[43]

Lacking the appropriate name for the universal common to these various forms of continua, an investigator may fail to recognize that this property belongs to them all in virtue of something common. Nor will he direct his search toward the reason why all continuous magnitudes have alternating proportions.

This will have methodological ramifications, which again are the concern of *APo.* II.14.

> We are currently speaking according to the traditional common names, but it is necessary to investigate not only in these cases, but also by selecting anything else which is seen to belong in common, and then to investigate what this follows and what follows this; for example, the possession of a third stomach follows on having horns as does being without both rows of teeth; and again possessing horns follows after something. For it is clear why the thing mentioned will belong to them, for it will belong on account of the possession of horns. (98a13–19)[44]

This suggests a method of 'scanning' a set of universal predications to observe correlated sets of differences – to my mind, it sounds much like a description of those passages in the *HA* which read, in effect, "All *S* has *P*, and indeed all *T* and all *U* — in fact generally whatever is a *K* has *P*," a couple of which were quoted in section IV above. And the above passage notes the A-type explanation that displays this understanding by means of an example which appears in *PA*.[45] To see deer, antelope and oxen as all horn-bearers is to recognize in them a common nature in virtue of which a number of features common to them can be understood.

This particular concern surfaces with some regularity in the biology. *PA* I.4 opens with a puzzle about whether the water-dwellers and the flyers should be brought under one common name. Aristotle answers that those kinds which differ only by the more and less can be linked under one kind (644a18). He then answers the question posed in chapter 1.

> The right course is to speak about some affections in common by kind, wherever the kinds have been satisfactorily marked off by popular usage and possess both a single nature in common and forms not far separated in them — bird and fish and any other that is unnamed but like the kind embraces the forms that are in it; but wherever they are not like this, to speak of particulars, for example about man or any other such. (644b1–7)

The *HA* often raises the issue of whether a variety of kinds ought to have a common name,[46] and again, given the importance of this problem to the facilitation of explanation and therefore unqualified understanding of the animal world, this is to be expected.

Allan Gotthelf's recent work on substance and essence in the biological works has focused attention on a passage in *PA* III.6 which indicates the variety and complexity of the issues which the lack of common names can raise.[47] In part, the question of the extent to which the various forms need to be similar before they can be viewed as varieties of a kind is at stake. Does Aristotle make a rigid distinction between true kinds and forms which can be grouped together only for narrow explanatory purposes? Can 'horn-bearers' or 'crook-taloned birds', for example, be considered a true kind, or is it merely a convenient grouping in a narrowly prescribed context? How about 'lung-possessors', a 'kind' we have already discussed?

> Now speaking generally, the lung is for the sake of breathing, but there is also a bloodless sort for the sake of a certain kind of animal; but that which is common in these cases is without a name, unlike 'bird' which names a certain kind. So, just as the being for a bird is from something, possessing a lung also belongs in the being of these. (669b8–12)

The first sentence here suggests Aristotle has in mind a general kind, those with lungs for the sake of breathing, which has two forms, a bloodless-lunged sort and a blooded-lunged sort.[48] Implicitly, it is the common kind – what is common to these sub-kinds – which lacks a name. This is contrasted with birds where there *is* a name which applies to the common kind. There are features which constitute the being for a bird, in virtue of which the name applies. The suggestion seems to be that the possession of a lung likewise may (perhaps partially) constitute the being for breathers. It of course is not suggested that 'lung-possessor' is the name, any more than we name birds 'feather-possessor'. What does seem at least to be under consideration here, however, is the possibility that the lunged group, which has a good deal in common though diverse in so many ways (e.g., some fly, some live continuously in the water; some have feathers, some hair), ought to be considered a kind with its own name.[49]

From the standpoint of explanation, the value in recognizing the commonality among these extensive kinds, which Aristotle sees as crucially bound up with having an expression which refers to this fact, cannot be denied. The last chapters of *De Juventute,* often referred to separately as *De Respiratione,* treat as fundamental the distinction between breathers (air-coolers) and gilled water-coolers. Aristotle may here be facing the issue of conditions under which explanation requires (or is at least facilitated by) more general referring expressions. The account of the development of a general theory of proportional magnitudes in *APo.* I.5 shows that Aristotle was sensitive to this issue *as* a general issue in the philosophy of science. These passages in the biology show him to be facing the practical question of when a more general concept ought to be introduced.

VII

A number of lines of evidence have now been presented suggesting deep affinities between the theory of explanation and understanding in the *APo.* and the zoological treatises. The strongest evidence against this suggestion is, I believe, to be based on the claim that the *Posterior Analytics* recommends a picture of science, at least a science at the stage of providing explanations, at odds with the structure of explanation found in the biology. I have been unwilling to adopt the expedient that *APo.* is primarily a treatise in pedagogy. At the very least it is an exploration of those themes central to the *Theaetetus:* what is scientific understanding, how do we distinguish the person possessing it from the person with opinions, what sorts of facts can we hope to have understanding of, what place do sense perception, causal explanation, definition and division have in our account of understanding? As Myles Burnyeat has pointed out, *APo.* is innocent of our distinction between philosophy of science and epistemology – it explores, as one, issues contemporary philosophy has tended to treat separately.[50] Nor is it easy to distinguish philosophy of science from methodology in this treatise; *APo.* I.5 at once explores the questions of what unqualified understanding is, the role of the formation of new concepts as it relates to explanation, the close relationship between predication, reference and explanation, closing with methodological recommendations for searching for the appropriate referring expression for a subject relative to identifying it as the bearer of a certain predicate.

The argument of this paper leads me to think that we need to reassess the evidence that has been used to claim that Aristotle would have pictured a science of zoology as an axiomatic system *à la* Euclid's *Elements,* had he based it on the *APo.* Jonathan Barnes has recently suggested that the claim, defended earlier in this century by Solmsen, that the *APo.* was initially innocent of the syllogistic, which was then grafted onto it after its discovery, has a good deal to be said for it.[51] On the other hand, it has also been recently argued that the function of the syllogistic was not primarily viewed by Aristotle as that of structuring explanation, so much as that of testing the logical properties of explanations presented in a natural language.[52] Surely there are considerations which, even after the 'discovery of the syllogism', would lead us to reject the claim that Aristotle would have recommended putting scientific explanations into syllogistic form.[53]

All these issues, as well as questions about the nature of hypotheses and definitions as starting-points of explanation and the relationship between definition, division and explanation, are worth exploring within the context of the zoological treatises. It is not safe to assume that those treatises have little to do with the exploration of *epistēmē* carried out in the *Analytics.*[54]

NOTES

1. For two recent, but quite different, attempts to deal with this problem, along with valuable surveys of the history of the problem in recent times, cf. Barnes 1975b, 1981.
2. Cf. Kullmann 1974; Barnes 1981, 58; and Bolton 1987.
3. Cf. Barnes 1975a, 122; Ross 1949, 324.

4. The importance of the distinction between incidental and unqualified *epistēmē* to the argument of the *Analytics* is brought out forcefully by Myles Burnyeat in his 1981. I am following the original suggestion of Kosman 1973, 374, insisted upon recently by Burnyeat, that *epistēmē* is better served translated as 'understanding' than as 'knowledge'.

5. This last sentence is interesting for reasons I explore in chs. 6 and 7 below, namely its use of the distinction between numerical and formal sameness in characterizing the relationship between three *kinds* of a more general kind. This is the only passage I know of in which a kind, as opposed to a countable individual, is said to be one in number relative to the unity constituted by it and related kinds.

6. These 'marks' are interestingly similar to the constraints on what can be an adequate explanation in the *Phaedo;* cf. 97a7–b3, 101a6–b2, and the discussions in Gallop 1975, 186–7, and Nehamas 1979, 93ff.

7. For now I must sidestep the issue of how well this example fits the account of predication in chapter 4. A useful recent survey of the issues and literature is Tiles 1983, 1–16. I will refer to predicates which belong to all and only some kind but which can be explained by reference to what the kind is as 'primitive' or 'belonging primitively to the kind'. The complex and fluid language Aristotle uses to describe various sorts of predicates and modes of predication in the *Topics, Analytics* and *Metaphysics* is nicely discussed in Ferejohn 1981.

8. E.g., 75a18–22, 75b1–2, 76a5–16.

9. I.23 84b7–8.

10. I.22 84a25; I.24 85b7–15; II.13 96a24–31; II.17 99a18–21, 24.

11. This passage applies a set of principles argued for in *APr.* I.27 concerning selecting predicates for premises of syllogisms (esp. 43b11–38). As here, the context is how to select terms relative to a specific problem (*APr.* I.26 42b29, 43a18; 27 43b34, 28 44a37), and the first principle enunciated is, "It is necessary to select *not* the things which follow some part, but whatever follows the whole thing, e.g. not what follows a sort of human, but what follows *every* human" (43b11–13). The passage goes on to counsel against selecting predicates which follow from a kind *as* following from a form of it – in such cases, one should select proper attributes of the form, for the predicates of the kind apply to its forms implicitly (43b22–9). What is entirely missing in the *APr.* discussion is the identification of the middle term as identifying the cause of the predication to be explained. Cf. chapters 2 and 3 below.

12. *CIAG* XIII.3 429.32–430.7. John Ackrill suggests that in such cases 'the only explanation why *S*s are *P*s is the explanation why *G*s in general are *P*s' (1981b, 380). But *APo.* I.5 and *APo.* II.14 state that, while we will ultimately want the explanation why *G*s are *P*s, noting that *G*s in general are *P*s and that *S*s are sorts of *G*s is a respectable explanation of why *S*s are *P*s. This issue is discussed at length in the next two chapters of this volume.

13. It is worth noting, because of its affinity to the actual practice of the biology, that this explanation is not stated syllogistically in the text, but rather as the answer to a *problem* about why leaves are shed. Stated in this way, the appropriate answer is just to provide the reason why (e.g. "This is so, because certain processes take place in these plants"), not an argument with "Shedding leaves belongs to every broad-leaved plant" as a conclusion.

14. The research strategy this theory of explanation suggests is to ask, when one has the same predicate belonging to distinct subjects, whether these distinct subjects share in a common nature in virtue of which the predicate belongs. An important question, approached repeatedly in *APo.* II.16, 17, is whether this *must* be the case, and thus whether this strategy is *always* appropriate (cf. 98b25–39, 99a1–16, 99b4–7); see chapter 3 below.

15. For discussion of these issues cf. Ackrill 1981b and Bolton 1987, 139–46.

16. Jonathan Barnes (1975a, 240) rightly argues that the likely reference of 'dissections' here is a lost work referred to regularly in the biological treatises. Cf. *PN* 456b2, 474b9, 478b1; *HA* 497a32, 509b22–4, 510a29–35, 511a13, 525a8–9, 530a31, 565a12–13, 566a14–15; *PA* 650a31, 666a9, 668b29–30, 674b17, 680a2, 684b4, 689a19; *GA* 719a10, 740a23–4, 746a14–15, 764a35, 779a8–9. *DL* v. 25 lists a *Dissections* in eight books and one book of *Selections from the Dissections.* The notion of a problem here is paralleled at *GA* I. 724a7 and II. 746b16; cf. *APo.* 88a12–16 and *APr.* 43a18; cf. chs. 2 and 3 below.

17. *APo.* II.14 98a2–8; cf. Balme 1972, 72, Kullmann 1974, 196–202.

18. Allan Gotthelf and Myles Burnyeat independently pointed out the relevance of these chapters to *APo.* II.14. It indicates that the method of II.14 may be simply an application of a more general theory about the use of systematically organized information in framing explanations. Cf. note 11 above.

19. This explains the constant stress that division is valuable as a means of insuring that the intermediate kinds are not left out in articulating the nature of a kind, e.g., 96b35–97a6. This is simply a methodological consequence of the point made at *APo.* I.1 71a18–71b9, that, if one knows the predicates which hold at the universal level, acquaintance with a new instance of the universal will directly provide understanding of it. The centrality of division in the overall program of the *APo.* is stressed in Ferejohn 1982/3.

20. In either case, even if it is known that the predicate belongs to all and only the members of a certain kind, it does not follow that one will know whether it is a primitive incidental or proper attribute of the subject, or an aspect of what the subject is.

21. The most painstaking study is that of Meyer 1855; his argument has been carefully critiqued by Balme 1987b, 81–5.

22. Balme, 1987b, 89.

23. Pellegrin 1982; 1987. Balme 1987a has in general endorsed Pellegrin's conclusions on this matter.

24. "Aristotle's *Posterior Analytics* and the Biological Works."

25. Besides the reference to *apodeixis,* the description of the components of demonstration as that *about which* and that *from which* is reminiscent of the description of demonstrative understanding at *APo.* I.10 76b12–23, esp. 76b22–3. Cf. I.12 77a40–b3. The use of *historia* here as the appropriate name for the pre-demonstrative investigation of a subject is likewise reminiscent of *APr.* I.30 46a18–28; esp. 46a24–8. Cf. chapter 2 below.

26. I don't intend this to suggest that Aristotle conceived of science in a Baconian fashion. Division is a way of organizing information for the sake of explanation/definition, not a method of discovering information. Furthermore, it is often

discussed in *APo.* as if it were as much a method of *testing* antecedently organized information for completeness as it is a method for organizing previously unorganized propositions; cf. 96b27ff.

27. *HA* II–IV.7 deals exclusively with parts-differences.

28. The beginning of a new 'book' does not signal a lack of continuity with I.17, as the syntactic link of the first sentence of Book II with the last sentence of Book I shows.

29. It seems plausible that these shifts are dictated by differences in the manner in which different sorts of parts are distributed and differentiated among the extensive kinds. For example, given that the introduction to the bloodless kinds stresses the fact that hard and soft parts are external/internal in some, internal/external in others, it makes sense both to stress this mode of difference and to discuss what is internal and what external in each kind before proceeding to the next.

30. One might compare the discussions of elephants (497b22–498a12), camels (499a13–30), seals (498a32–b4), or river crocodiles (503a8–15). Discussion of passages of similar import can be found in Gotthelf 1987a.

31. *HA* II.15 505b32–506a11. While he insists that no animal will have both lung and gills, he is quite open to the possibility of kinds other than fish turning up which have gills. In fact he treats the newt as such; cf. *HA* VIII. 589b27–9, *PA* IV.695b25, *Juv.* 476a6.

32. *De Juventute* 7–27 (*De Respiratione* 1–21). I agree with Ross (1955a, 50–60) that there is no good reason to view this as a separate treatise.

33. Cf. Balme, 1987a, 9.

34. Balme, 1987b, 88.

35. The line developed somewhat dogmatically in my text has been questioned by Allan Gotthelf in a private communication. For the purposes of this discussion, however, I don't think the differences are crucial.

36. The rendering given of 669a13 above, and the interpretation of it here, are suggested by a similar turn of phrase at *Juv.* 480b19–20.

37. Notice that any attempt to syllogize this explanation would eliminate the causal content of the explanation. *APo.* II.11 suggests that the middle term in an explanation may *refer* to a goal. But in order to know *how* a middle term explains, one must understand *the way in which* its referent is responsible for what is being explained. The effect of formulating all predications in a neutral 'belongs to' language is to obscure distinctions between various types of predications (e.g., between 'essential', primitive, and merely 'universal' predications), and distinctions between various causal relationships holding among the predicates in question.

38. The background to these joint explanations in terms of material necessity and 'the good' is, of course, *Phaedo* 96–100 and the *Timaeus*. This specific explanation is a development of that at *Timaeus* 76a–d (as Cornford 1937, 301 notes). On the differences between Plato and Aristotle on the way these two sorts of explanations ought to be integrated, cf. Balme 1987c, 276–8. On the role of such explanations within Aristotle's teleological framework, cf. Cooper 1987. I am inclined to agree with Cooper that, where such explanations are contrasted, as here, with teleological explanation, we need to distinguish the *subsumption* of the necessity associated with a thing's material/elemental make-up under the necessity relative to a goal, from a *reduction* which would eliminate explanations by simple 'element poten-

tials' (Gotthelf 1987b, 211–13) from biology altogether. For here, excessive hair growth is seen as an instance of what necessarily happens to certain materials under conditions of excessive moisture and heat; and in the examples to follow, deer horns are said to fall in part simply because of the weight of their dense, earthy nature. That is, the force of *these* explanations ('because it is heavy stuff, it moves down'; 'because there's excessive heat and moisture here, things grow') is not vitiated by showing that it is also true that this must happen if an organism of a certain sort is to flourish. And in fact, it is not clear that these cases are hypothetically *necessary:* that is, it is not clear that Aristotle believed that if there are to be deer, there must be horn-shedding, or that if there are to be human beings, they must have excessive head hair. (Cf. Cooper 1987, 256)

39. Cf. 695a15–22.
40. Cf. chapter 7 below, 174–177. *PA* I.5 645b15–20 suggests that certain actions tend to be fundamental in explaining both various parts and other, less fundamental actions. Nonetheless, the evidence gathered by Gotthelf 1985b (and summarized in 1987a, 190–2) suggests one can't take the further step of supposing that only these fundamental actions will be given in an account of an organism's nature. On the basis of Bolton's suggestions (142–6 below), one might attempt to characterize different sorts of accounts which approach, to different degrees, a scientific/explanatory definition. Balme has suggested that what might be included in an abstract account of an animal's essence will be quite different from an account of the animal's form at any level of generality (Balme 1987d, 294–8).
41. Balme 1972, 72; cf. 1987b, 86.
42. Cf. Balme 1972, *ad* 639a15–b7.
43. A scholium to Euclid, *Elements* V attributes the discovery of a general theory of proportion to Eudoxus (cf. Heiberg 1919, V 280.8), which would account for Aristotle's 'now': for the discovery would have been made in recent times and within the Academic circle. The scholium is suggested by Heath (1956, 112–13, 1949, 43) to be due to Proclus. Aristotle doesn't give a general name to the subject to which the alternando property belongs per se. For Euclid it was 'magnitude', but there are good reasons for supposing Aristotle would have resisted this (cf. *Metaph.* Δ 13 1020a7–15). As Heath (1949, 44) suggests, Aristotle would likely have preferred quantity.
44. Compare Socrates' injunction to Theaetetus, *Theaetetus* 148d4–7: " . . . try to imitate your answer about the powers. Just as you collected them, many as they are, in one class try, in the same way, to find one account by which to speak of the many kinds of knowledge" (tr. McDowell). Cf. *APr.* I.27 on selecting what follows x and what x follows.
45. Cf. *HA* II. 507a30–b13, where Aristotle adopts 'the horn-bearers' as a semi-technical expression; cf. *PA* 674a31. For the relevant background explanations, cf. *PA* 662b35–664a7; 674a30–674b5. The four stomachs are named at *HA* II. 507b1.
46. E.g., at 490a13–14, 505b30, 531b20–5, 623a3. This issue, not surprisingly, comes up in other disciplines as well. Cf. *Po.* 1447b9–12, *Mete.* IV.9 387b1–5.
47. Cf. Gotthelf 1985b. My understanding of this passage has been greatly helped by discussion with Gotthelf.

48. This distinction is developed in the immediately preceding passage, 669a24–669b8, and at *Juv.* 7 (*Resp.* 1) 470b13–27.
49. Whether Aristotle can be seriously recommending this group as a kind is discussed by Balme (1987b, 84–5; 1972, 120–1) and by Gotthelf (1985b, 31–3).
50. Burnyeat, 1981, 97.
51. Barnes 1981.
52. Lear 1980.
53. See note 40 above and Gotthelf 1987b, esp. 194–6.
54. This chapter is a revision of a paper delivered at a most exciting NEH Research Conference, Philosophical Issues in Aristotle's Biology, July 1983, directed by Allan Gotthelf. I would like to thank all those who offered constructive criticism on that occasion. In addition, I should like to thank L. A. Kosman, Allan Gotthelf and Michael Ferejohn for their careful comments on an earlier draft of this revision. The material in section IV of this paper owes much to the discussions of texts in *HA* I–IV which Gotthelf and I, and occasionally Kosman, engaged in during the fall of 1983, at the Center for Hellenic Studies. Finally I should like to thank Bernard Knox, the staff at the Center and the Trustees of Harvard University for support of the project of which this is part while I was a Junior Fellow at the Center in 1983–84; and the National Science Foundation for support during the summers of 1983 and 1984. The errors and imprecisions which remain are, of course, my responsibility. Some of the ideas of this chapter are developed further in chapters 2 and 3.

2

Between Data and Demonstration:
The *Analytics* and the *Historia Animalium*

There is a subset of Aristotle's treatises which we usually refer to as his biology or zoology. Aristotle himself occasionally mentions the investigation of animals and plants, although seldom in a way that marks it off decisively from the study of coming to be and passing away in general (*Mete.* 339a5–8, 390b19–22; *PA* 644b22–645a10). When we turn to these works individually and as a group, a number of simple but important questions arise regarding the manner in which the study of animals is partitioned and about the ways in which the different works are related to each other. Of course, we may have ready responses to these questions, but such responses are in part influenced by recent developments in the biological sciences that have little to do with Aristotle or his conception of nature or of science (cf. Balme 1987a, 9–11). So, to answer these questions in a way that will increase our understanding of Aristotle's science, we need to understand better his aims and methods.

Take the case of the *Historia animalium.* This treatise stands apart from those aimed at offering various explanations for the parts, development, motions, and so on, of animals. Much of the information it contains is duplicated in these other treatises,[1] and there are numerous casual references to it in them. Further, unlike many of Aristotle's works, the title of the *Historia animalium* is used to refer to it within the corpus itself.[2]

Why, then, are the 'researches concerning animals' distinguished from these other studies? Obviously, there is danger that in answering this question we will obscure Aristotle's own purposes. Accordingly, the aim of this paper is to answer the 'what-is-it' question about the documents Aristotle refers to as 'the animal histories' or sometimes simply as 'natural histories'. I shall argue that attending to Aristotle's wider logical and epistemological vision can help us to understand better his approach to the systematic study of living things.[3]

I MODERN VIEWS OF THE *HISTORIA ANIMALIUM*

From a post 17th-century perspective, there are two standard ways to conceive the function of the *Historia animalium* within a systematic study of the animal kingdom. The first takes it as the classificatory ground work of Aristotelian zoology. Yet, so viewed, the treatise is hopelessly inadequate (cf. Balme 1987b, 80–85; Pellegrin 1986, 1–12). The taxonomic vocabulary is restricted to the two terms 'kind' (*genos*) and 'form' (*eidos*), and these terms refer to groups of animals at all levels of the taxonomic hierarchy. There appears to be no concern for finding or consistently using certain features as classificatory markers in order to provide either a classification which is exhaustive or a hierarchy of *taxa* from widest to narrowest. The second treats the *Historia animalium* as a collection of 'natural histories', that is, as a series of more or less complete descriptive studies of each of the kinds of animals discussed. But from this standpoint the work is even more disappointing (cf. Balme 1987a, 9; 1987b, 85–88). As David Balme (1987b, 88) has put it, "To any reader looking for information about given genera or species, the *HA* seems an incoherent jumble."

The *Historia animalium* is clearly not organized according to a rigorous taxonomic scheme, nor as a reference work on the various kinds of animals discussed. To expect such an organization is to rely in part on the etymological tie between the name given to modern works of this character and Aristotle's (Pratt 1982). But since the *Historia animalium* is so disappointing when looked at in these two ways, we must either dismiss it (with an excuse, perhaps) or ask in charity whether we have misunderstood it. It is one of the lasting achievements of David Balme's scholarship to have provided us with the framework for a serious re-evaluation of the *Historia animalium*.

> Yet Aristotle does state his purpose: 'first, to grasp the *differentiae* and attributes that belong to all animals; then to discover their causes' (*HA* I. 491a9). The *HA* is a collection and preliminary analysis of the differences between animals. The animals are called in as witnesses to *differentiae,* not in order to be described as animals. (Balme 1987b, 88)

The thesis developed in this paper is fully in the spirit of this reassessment: like Balme (1987b, 80), I shall insist that the *Historia animalium* is a work directed "toward a methodical apodeixis of living nature."

II PRE-DEMONSTRATIVE SCIENCE

In this section I wish to accomplish two interrelated tasks: to present evidence that Aristotle distinguished a pre-demonstrative yet theoretical scientific inquiry in his philosophy of science; and to show that he was inclined to refer to this pre-demonstrative inquiry as *historia*.[4]

The *Posterior Analytics* is well advertised by scholars nowadays as the first attempt in the history of philosophy to provide a rigorous theory of explanatory proof. Its first six chapters do characterize scientific understanding of a fact in terms of deductive proof from premises which are true, unmediated, and primary, and which state facts more familiar than, prior to, and causative of the fact stated in the conclusion (71b19–23). But this advertising has been so successful that the work is often discussed now as if its second book – which announces itself as an extended account of different sorts of inquiry and their interrelations – did not exist.[5] The result is a general interpretation of the *Posterior Analytics* which makes it seem oddly out of touch with Aristotle's substantive scientific and philosophical works.[6] The subject of this section, then, is *APo.* II and, specifically, its concept of a stage of inquiry aimed at establishing that a predication holds, an inquiry preliminary to investigation of the reason why it holds.

"The things we seek are equal in number to the things which we understand" (89b23–4). So opens Book 2 of the *Posterior Analytics*. Aristotle claims (89b24–35) to be able to reduce the objects of scientific investigation to four: the fact, the reason why, whether something exists, and what it is. These inquiries are paired, in the following way:

(1) Is it the case that *S* is *P?* → Why is it the case that *S* is *P?*
(2) Are there *S*s (or *P*s)?   → What are *S*s (or *P*s)?

where the arrows indicate that the first question in each pair must be answered before the second, as Aristotle's remarks in *APo.* II suggest. Like so many of Aristotle's introductory sentences, this apparently straightforward, sensible division of investigations opens up a Pandora's box of difficulties which the rest of the book aims to resolve. In *APo.* II, perhaps the most important concern the ways in which the two pairs of inquiries, as well as their respective results, mesh with one another. I will, however, overlook most of these problems, since I am here primarily interested in whether this broad picture of types of inquiry has any parallels in Aristotle's various works which record the results of his investigation of animals.

De incessu animalium presents itself as a work concerned with why each of the parts involved in animal locomotion is as it is, and lists a large number of specific causal questions it aims to answer. Toward the end of this list of why-questions, which constitutes most of the first chapter, Aristotle states:

41

For *that* these things are in fact thus is clear from our inquiry into nature, but *why* they are thus we must now examine. (*IA*, 704b9–10: cf. *PA* 646a8–12, *HA* 491a7–14)[7]

It seems probable that the wording of this remark intentionally reflects the distinction between the two stages of the first pair of inquiries (1) listed above. If so, it indicates that the 'natural inquiry' mentioned is supposed to establish that certain predicative relationships hold true and, thus, that it is a necessary preliminary to the inquiry aimed at establishing the causal basis for these predications.

A similar distinction is defended as a matter of principle in *PA* I, which is sometimes referred to as 'Aristotle's philosophy of zoology' (Balme 1972, 69; Le Blond 1945, 51–72). This book begins by distinguishing two sorts of 'proficiency' relevant to a given study: a first order proficiency in understanding the subject-matter, and a second order proficiency in judging whether the study is well presented. The rest of the book is then organized around a series of questions bearing on the second type of proficiency, since

. . . the inquiry about nature, too, must possess certain principles of the kind to which one will refer in appraising the method of demonstration, apart from the question of how the truth has it, whether thus or otherwise. (*PA* I.1 639a12–15)[8]

The second question[9] in the series is,

Should the natural philosopher, like the mathematicians when they demonstrate astronomy, first survey the appearances in regard to the animals and their parts in each case, and only then go on to state the because-of-what (i.e. the causes), or should he proceed in some other way? (*PA* I.1 639b7–10)

The question is answered in the affirmative at 640a14–15: "[natural philosophers are] first to take the appearances in respect of each kind, and only then go on to speak of their causes." Now the stage of natural inquiry in which one surveys the appearances regarding a kind before stating their causal explanations is not here described as *historia*. Still, the connection between the use of this term in referring to the *Historia animalium* and the sort of survey of the appearances that is discussed in *PA* I.1 can be made more secure in two steps.

First, Aristotle introduces the distinction as familiar from the domain of astronomy. *APo.* I.13 records that astronomy is one of the sciences which have mathematical and physical aspects, where the latter are called the phenomena (78b39).[10] The general point Aristotle makes about such sciences is that to establish the facts one attends to the appearances, whereas one considers the appropriate mathematical principles in order to demonstrate the reasons why the facts are as they are (79a2–6). That is, he sees this difference in aspects as an instance of the more general distinction between the two sorts of inquiry given as pair (1) above. Second, *PA* II.1 opens by noting that "in the histories" it was made clear from which parts each of the animals is constituted, while the present work will investigate the causes through which each of the animals is so constituted (646a8–12).[11] This would appear to be the same distinction as that found in *PA* I.1, with the *Historia animalium* serving as the treatise which reports on the first investigation.

I need not rely entirely on indirect evidence of this sort, however. For there are two passages, closely allied in language, one in the *Prior Analytics* and one in the *Historia animalium* itself, which explicitly describe the predemonstrative stage of inquiry as *historia*. The first passage insists that, just as demonstrations in astronomy were discovered only after the principles were supplied by astronomical observation, any craft or science has its principles supplied by experience (*APr.* 46a17–22).

> So that if the predicates about each thing have been grasped, we will be well prepared to exhibit their demonstrations. For if none of the predicates which truly belong to the subjects have been left aside by our inquiry (*historia*), we will be able, with respect to everything for which there is a demonstration, to discover the demonstration and carry it out; but of that which in the nature of things has no demonstration, we will be able to make this apparent. (*APr.* 46a22–27)

The function of the *historia* is to enable one to 'grasp' the predicates which hold of each item in the general subject being investigated. This is apparently intended to explicate the way in which experience with the phenomena of a subject sets the stage for demonstration.

Treating this passage in isolation does not, however, give one a sense of how detailed Aristotle's recommendations in fact are. For this one must turn to the beginning of chapter 27, regarding the proper method to be used in 'picking out' or 'selecting' premises appropriate to the deduction of a given predication (*APr.* I.27–29). Very briefly, the method involves taking as given the subject and predicate of the predication at issue, and developing a list of everything that the predicate belongs to universally as well as a list of all the things that belong universally to the subject.

For those wishing to establish something of some whole, they must look to the subjects of what is established, that is, the subjects of which it happens to be predicated, and to whatever follows that of which it is to be predicated. For if any of these are the same, the one must belong to the other. (*APr.* 43b39–43)

Suppose the predication we wish to prove deductively is that A belongs to every C (i.e. *AaC*). Aristotle recommends that we generate lists of propositions of the form,

Predicate: A	Subject: C
AaD	**FaC**
AaE	*GaC*
AaF	*HaC*

in the hope of finding, as here, a middle term which can 'unite' the terms (cf. *APr.* 41a11–13). For ". . . no syllogism can establish the attribution of one thing to another unless some middle is taken, which is somehow related to each by predication" (41a2–4). Chapter 27 is careful to state that this rather algorithmic procedure is only relevant to demonstration insofar as the lists identify other true predications, and it provides a set of rules for identifying predications at the appropriate levels of generality and specificity as well as distinguishing what is in the essence, what is predicated as a property, and what is predicated as an accident (cf. esp. 43b1–32). That is to say, this is a recipe for organizing information in such a way as to identify middles: it is not a description of how the credentials of the information are established.

Since nothing is said here about how one is to establish the truth of a predication, or about how one is to determine which among a set of universal predicates are predicated in the essence and which are not, this is clearly not a method which will simply allow us to read off demonstrations. When Aristotle concludes his account of selecting premises and division by remarking that "it is apparent from what things and in what way demonstrations come about and to what sorts of things we should look concerning each problem" (46b38–39), we must take him to mean, I think, that the foregoing method is a necessary condition for the production of demonstrations. Suppose, for example, that in addition to the statements *AaF* and *FaC* (which yield *AaC*), one's selection also provides *FaA;* in other words, that according to our divisions, F and A are commensurately universal terms. Thus, we have two syllogisms in Barbara:

$$
\begin{array}{cc}
AaF & FaA \\
FaC & AaC \\
\hline
AaC & FaC
\end{array}
$$

There is nothing here that will help us determine which of two commensurate predicates of a subject is explanatory of which, or even whether there is any explanatory relationship between them at all. This is, of course, the problem which Aristotle raises in *APo.* I.13 when he distinguishes demonstration *that* and demonstration *why;* and it is the problem that he returns to and discusses in detail in *APo.* 98a35–b24. There, we are told that the predicate which is in the account of the other is the explanatory middle, though nothing is said about how one acquires this knowledge.

The description of *historia* as an inquiry establishing which predicates truly belong to which things is consistent with Aristotle's occasional claims that this algorithmic procedure for selecting premises is relevant to demonstration – which one would expect, since demonstration is a species of deduction. The organization of true propositions in this way is presented as facilitating the development of a demonstrative science. And one can see why: it aids in using information imbedded in divisions to identify commensurate universal predications and gives us a "short list" of candidates for demonstrative middles.

The *Historia animalium* characterizes its task in terms very similar to those used in *APr.* 46a22–27.

> These things have now been said thus in outline to give a taste of the number of things that one must study and how far – we will speak in detail later – so that first we may grasp the *differentiae* present and the attributes in every case. After this, we must attempt to discover the causes of these. For it is natural that the study be carried out in this way, when there is an inquiry (*historia*) concerning each thing. For about which things and from which things the demonstration should be becomes apparent from these. (*HA* 491a7–14)[12]

The technical language of the theory of demonstration in this passage is hard to deny; and equally clear is the distinction between an investigation aimed at establishing the *differentiae* as well as the incidental features of each kind of animal and a search for causes based on this. It is as a result of the first investigation that the elements of demonstrations become apparent. As in the *Prior Analytics* this investigation is called a *historia*.

It is time to take stock of our progress thus far. A number of passages from the biology and the *Analytics* agree in detail that there is a distinction to be made between an investigation aimed at establishing that *p* is the case and one aimed at establishing why *p* is the case. The former is, apparently, a pre-demonstrative inquiry, that is, an inquiry devoted to organizing empirical information in such a way that the identification of middle terms is facilitated. This suggests that the causal inquiry may not actually be a search for new, more basic entities so much as an inquiry into the causal relationships which hold among the predicates established during the initial inquiry.

Aristotle, as we have seen, is inclined to restrict the range of the term *historia* to the first stage of natural inquiry, that is, to a particular sort of pre-demonstrative investigation. The second stage of inquiry is directed toward scientific demonstration.[13] This suggests that the way to understand the distinction between the *Historia animalium* and such works as the *Parts of Animals* or *Generation of Animals* is in terms of Aristotle's own distinction between two stages of inquiry into predications, one involved in grasping that the predication is the case, another involved in establishing the reason why. The reasons for dividing up the investigation in this way are to be found in his theories of explanation and inquiry in the *Posterior Analytics*.

III THE *ANALYTICS* ON 'PROBLEMS'

These passages, however much they may cohere, are all theoretical in nature, even those in the biology. In order to show that we *must* understand the distinction between the *Historia animalium* and the other biological works in terms of the distinction between factual and causal inquiries, we need to look carefully at the *Historia animalium* to test the claim that it does in fact respect these theoretical ideals. But to perform such a test, we need to know what to look for. Now, to determine whether the *Historia animalium* is an inquiry of the sort described in *APo.* II as a 'fact-establishing' investigation, we require some idea of what the report of the results of such an investigation would look like. The key here, as I argued in chapter 1, is to work back from Aristotle's concept of demonstrative understanding. For that concept places constraints on how empirical information is to be organized if it is to be converted by demonstration into science (see pp. 8–15 above).

What, then, are these constraints? First of all, we must recognize that not just any universally true predication can be the subject of demonstration. Two sorts of predications are distinguished at the end of *APo.* I.4.

> If, then, any chance triangle is proved primitively to have two right angles or whatever else, it belongs universally to this and primitively, and the demonstration of this [universal primitive predication] holds universally in itself; but it holds of the others in some other fashion, not in itself; nor does it hold universally of the isosceles but extends further than it. (*APo.* 73b39–74a3)[14]

The use of 'universally' is restrictive and based on a stipulation made in this chapter, viz. a predication is universal if the predicate belongs to the subject "in every case, in itself and as such" (73b26–7). That is, the subject and predicate of the proposition to be proved must be coextensive, and the predicate must belong to the subject in virtue of its being that subject, not incidentally. Notice, for example, that when Aristotle says that the property of having interior angles equal to two right angles does not hold of isosceles universally, he does not mean merely universally, because all isosceles triangles do in fact have this property. His point is that the property is true of other sorts of triangles as well ("extends beyond isosceles"), and so does not belong to the isosceles *qua* isosceles. Rather it belongs to the isosceles triangle *qua* triangle. Thus, it is the proof showing why this property belongs to triangles as such that is basic.

Aristotle does, however, allow for demonstration of the weaker predication, though he insists that the demonstration holds in some weaker fashion. In later passages, it becomes clear that Aristotle has in mind a special class of non-coextensive universal predications, namely, those cases where a predicate belongs coextensively to a kind and, consequently, belongs to all the differentiated forms of that kind. To describe such predications he will occasionally remark that a predicate "extends beyond (this form), but not beyond its kind" (see *APo.* 85b7–15, 96a24–31, 99a18–21 and 24). In such cases Aristotle admits partial demonstrations (see *APo.* I.24), meaning, I take it, demonstrations covering a part of the kind.

Thus we come to the first constraint on *historia* imposed by Aristotle's theory of demonstration: it must aim for predications in which the predicate is coextensive with its subject. Furthermore, the subject kind must be differentiated into its immediate sub-kinds or forms if there are to be 'partial demonstrations' asserting that the sub-kind has the feature in question because it is of the kind that has the feature primitively.

APo. I.5 discusses extensively the types of ignorance that can prevent one from having unqualified rather than sophistical understanding, and each type turns on failure to recognize the primitive level at which a predication holds (cf. 74a25–32). But in *APo.* II.13–18, there is a marked concern with the manner of acquiring predications at the primitive level, and it is to this problem that we shall now turn.

APo. II.14 opens with a cryptic statement of method.

Relative to grasping problems, one should select from both the dissections and the divisions, and do so by positing the kind that is common to all of them. For example, if the objects of study are animals, select what belongs to every animal; and, having grasped these, once again select what follows all the first of the remaining kinds. For example, if this is Bird, select what follows every bird; and in this way always select what follows the proximate kind. For it is clear that we will immediately be able to say why the things which follow belong to those kinds under the common one, for example, why they belong to Human Being or Horse. Let A stand for Animal; B for the things which follow every animal; and C, D, and E for certain animals. It is quite clear why B belongs to D, for it is because of A; similarly with C and E. And the same account always applies in the case of subordinate kinds. (*APo.* 98a1–12; my trans.)

The recurrence of the process of singling out certain features that 'follow' indicates that *APr.* I.27–30 is the formal background for this chapter (cf. Barnes 1975a, 239–240; 1993, 250–251; chapter 1 above, pp. 13–14). Further, that one must select from divisions suggests (as does *APr.* I.31) that division is at best a preliminary stage of the method described here.[15] Clearly, it is from divisions already made that one selects, at each level of generality, what belongs universally; and equally clear is the fact that this division presupposes a division of the kind, animal. So it looks as if *APo.* II.14 describes a procedure for using information imbedded in divisions to produce propositions of the sort required for a demonstrative science.

This procedure directs attention to predications at the level of commensurate universality. If, among the things which follow, that is, which belong to every S, one finds a feature also belonging to T, the natural question to ask is whether S and T are both forms of some kind K which has that property primitively. (Or, if one is already aware that S is a K, one would note that the property in question belongs not just to S but to K in general.) So if, for example, one finds that having a heart belongs to birds in virtue of the fact that hearts belong to all blooded animals, the next step is to ask why hearts belong to all blooded animals. In fact, one would not really understand why birds have hearts until this more basic question is answered: saying that birds have hearts because they are blooded animals means that they have hearts for the same reason all blooded animals do.[16] Notice that the grasp of the problem one ends up with in the above passage has the same form as that achieved by the move from incidental to unqualified understanding in *APo.* I.4, 5. Aristotle's method, then, is intended to identify the widest kind to which a predicate selected from a division belongs. Once this has been done, that predicate will immediately show itself as belonging to immediate forms of that kind: the subject in question will have this feature just because it is (a form of) the kind to which the feature belongs universally.

The continuation of *APo.* II.14 also echoes concerns of I.5. In the latter it was noted that while there once were distinct proofs that proportionals alternate in the case of numbers, lines, solids, and times, now it is proved universally in a single demonstration. The original failure to see the universal demonstration was because "all these things . . . do not constitute a single named item and differ in form from one another."[17] The lack of a name to signify the universal contributed to the mathematicians' failing to see that "it did not belong to things as lines or as numbers, but as this which they suppose to belong universally" (*APo.* 74a23–25).

APo. II.14 says more concerning what it is to 'grasp problems'.

> Now at present we argue in terms of the common names that have been handed down; but we must not only inquire in these cases, but also if anything else has been seen to belong in common, we must extract that and then inquire what it follows and what follows it . . . (98a13–16)

Searching for "what follows that which belongs in common and what follows it" is the method recommended in *APr.* I.28 for finding a middle term that is relative to a problem, and the example which follows, an example familiar from *PA* 674a23–b18,[18] clarifies the strategy:

> having a third stomach and not having incisors ⟨follow⟩ having horns; again, ⟨we should inquire⟩ what having horns follows. For it is clear why what we have mentioned will belong to them, for it will belong because they have horns. (*APo.* 98a16–19)

Here two *differentiae* are noted (one negative, incidentally) which follow 'the possession of horns'. Thus, given

(1) having a third stomach belongs to every horned animal
(2) lacking upper incisors belongs to every horned animal,

we are asked to inquire, to what does 'the possession of horns' belong universally?, that is, to find a value for P such that

(3) horns belong to every P.

For, given this, we may then infer that

(4) having a third stomach belongs to every P.

As Aristotle says, having a third stomach belongs due to the possession of horns, the possession of horns being the middle through which the link between the possession of a third stomach or the absence of a second row of teeth and the third item, *P,* is established.

The choice of this example in *APo.* II.14 is interesting. Aristotle needs a case that clearly goes beyond the common nomenclature, for that is the point he is making. Thus, his example requires the use of specialized descriptive phrases[19] to refer to items predicated of one another. Apparently, Aristotle wishes to emphasize that this predication is established by realizing that the possession of these other two features follows from the possession of horns, and that they will thus belong to anything which possesses horns because that thing is horned.[20] The final lines of *APo.* II.14 extend the method even further, and likewise use an example familiar from the biological works: "Again, another way is by excerpting in virtue of analogy; for you cannot get one identical thing which both pounce and spine and bone should be called; but there will be things that follow them too, as though there were some single nature of this sort (*APo.* 98a20–23)."

Let us try to reconstruct the steps in the process here sketched. The three sorts of 'skeletal' parts referred to are related by analogy. But there may be predicates within the divisions being used which belong to all of them (as if they had a single nature). Thus, 'excerpting in virtue of analogy' means searching 'the dissections and divisions' for *differentiae* common to subjects related by analogy. At *PA* 653b33–36, Aristotle says that "among those animals having bones, the nature of the bones, being hard, has been devised for the sake of the preservation of the soft parts; and in those not having bones the analogue ⟨has been devised for this⟩, for example among some of the fish, fish-spine, among others, cartilage." Accordingly, *APo.* 98a20–23 may propose that certain passive capacities, e.g., hardness or brittleness, belong to each of these analogous parts in virtue of a common function that each plays in the life of its respective kind, or in virtue of a common material nature (since all these parts are earthen). Again, it might also mean that all three analogous hard parts are associated with soft flesh and viscera, an association which would naturally suggest the idea that the analogous parts play an identical functional role in their respective kind's life (so Barnes 1975a, 240; 1993, 251).

Here, then, is a characterization of methods for achieving predications of the sort that prepare the investigator for acquiring understanding through causal explanation. The methodology is clearly an application of the more formal methods of *APr.* I.27–31 to the specific goals of the scientific investigator, which should be no surprise given that the *Analytics* is introduced as an investigation of demonstration and demonstrative understanding (*APr.* 24a10–11, 25b26–31).

APo. II.15–17 explore the complexities involved in finding a causal account relative to a pre-established problem, where a problem is here essentially a why-question asked of accepted facts: That *P* belongs to all the *S*s is clear; why then does it belong? A botanical example will allow us to tie the concerns of *APo.* II.14 to the second stages of each pair of the investigations with which we began, the whether-it-is/what-it-is pair and the that-it-is/why-it-is pair.

In *APo.* II.8–10, Aristotle finally comes to grips with the question of the relation between definition and demonstration within a science. In the process, he has much to say about the way in which why-inquiries and what-inquiries are related to one another. The interpretation of these chapters is controversial (see Bolton 1976, 1985, 1987; Sorabji 1980; Ackrill 1981b). But certain features of the debate may be lifted from the fray for present purposes. At least in cases relevantly similar to the examples in these chapters, the middle term in a demonstration of why some predicate belongs to some subject will also serve to account for what the predicate is (*APo.* 93a3–5, 93b3–14, 94a1–10, 95a16–21, 99a3–4, 99a25–9). An investigation of why those occasional noises in the clouds occur – an investigation based on our awareness that they do – is completed when we have grasped the most fundamental causal explanation of those noises. Aristotle claims that not only is 'quenching of fire' a candidate for the middle which accounts for the occurrence of thunder in the clouds, but that it is also a possible answer to the question, What is thunder? Thus, the familiar account of thunder as a certain characteristic noise in the clouds is underwritten by a more basic account – more basic in that it serves to explain the perceptually familiar features by which we became acquainted with thunder initially (see Bolton 1976).

Can one take this view of the way the results of these two inquiries converge in the case of the sorts of facts biologists wish to explain? Aristotle seems to think so.

> The middle term is an account of the first extreme, and thus all the sciences come about through definition. For example, loss of leaves follows the vine while exceeding it, and follows the fig while exceeding it; but it does not exceed all, but is equal in extent with them. Now if one takes the primary middle term, it is an account of shedding leaves. For there will be a first middle in the other direction, that all are such; then a middle of this, that sap coagulates or some other such thing. But what is shedding of leaves? It is the coagulation of the sap at the connection of the seed pod. (*APo.* 99a21–29; my trans.)

At the outset we see once more the language of 'following' (i.e. belonging to all), and the idea of a predicate which both follows and exceeds two kinds of plant while being coextensive with (one must suppose) all the kinds that shed their leaves. Aristotle is not making the less-than-startling point that all the kinds which shed their leaves shed their leaves; rather, he is indicating the necessity of identifying the group of plants with *differentiae* coextensive with this one if we are going to account for it scientifically. As a matter of fact, in the previous chapter at 98b4–21, Aristotle identified just such a group by noting a feature common to them all and coextensive with the shedding of leaves, namely, being broad-leafed. The fact that these *differentiae* (being broad-leafed, shedding leaves) are coextensive is noted (98a35–b3) within the context of the question, Does the causal basis of something have to be coextensive with that of which it is the cause? The question is then raised (as it has been ever since of attempts to describe explanation in purely extensional terms) as to whether either of the coextensive terms can be used to prove the other – there being no question that one can construct a valid and sound syllogism in Barbara with either term in the middle position.[21]

Throughout this discussion Aristotle assumes that being broad-leafed is the cause of any plant's losing its leaves, and the sample explanation we are given 'demonstrates' that vines lose their leaves because they are broad-leafed (*APo.* 98b5–16). In the language used elsewhere, this is a 'partial' or 'A-type' demonstration, given that the 'problem' being explained predicates loss of leaves to one of the sub-kinds which loses its leaves. *APo.* II.16 closes by correcting the impression that this is a primitive scientific explanation.

> Or if problems are always universal, must the explanation be some whole and what it is explanatory of universal? E.g. shedding leaves is determined to some whole, even if it has forms, and ⟨it belongs⟩ to these universally (either plants or plants of such and such a sort); hence in these cases the middle term and what it is explanatory of must be equal and convert. E.g. why do trees shed their leaves? Well, if it is because of solidification of their moisture, then if a tree sheds its leaves solidification must belong to it, and if solidification belongs – not to anything whatever but to a tree – ⟨it must⟩ shed its leaves. (*APo.* 98b32–38)

Problems need to be universalized, in the sense, as Ross notes, of *APo.* I.4. But when this is done, the kind that took the middle position in the partial demonstration is now that to which shedding leaves belongs, the subject of the predication to be explained.

We are now in a position to make sense of *APo.* II.17. Aristotle tells us that if one takes the primary middle term, it is an account of shedding leaves: "for first there will be a middle in the other direction, that all are such; then a middle of this, that sap coagulates or some other such thing" (99a25–28). Two middle terms are mentioned here, only one of which is identified as an account of shedding leaves. The other is called a "middle in the other direction." Suppose that this is the property of being broad-leafed. Being broad-leafed serves as a middle in the direction of the various forms of plants which shed their leaves – "Shedding leaves belongs to all the vines because they are broad-leafed." But there will now be a primary middle for this, where 'this' indicates the proposition which predicates shedding leaves of being broad-leafed.[22] Thus, only when one has elevated problems (or why-questions) to the level of the commensurate or primitive universal does the middle term also become an account of what the predicated property is. Here is one further way in which the methods outlined by Aristotle are important in setting the stage for a demonstrative understanding of a subject. These methods move us to the stage where further exploration can aim for the primitive definitions that may serve as the starting-points of our explanations.

IV   THE *HISTORIA ANIMALIUM* AS PRE-DEMONSTRATIVE SCIENCE

In section 2, I reviewed the evidence that the *Historia animalium* was offered by its author neither as a report of a systematic taxonomy of the animal kingdom nor as a series of natural histories of them, but as a rendering of the true propositions currently known about animals for the purpose of causal demonstration. The theory of problems and the methodology for working with them that are presented in the later chapters of *APo.* II and discussed in section 3, indicate that Aristotle would have something quite specific in mind, when he came to organize information into propositional form for scientific purposes. He would, for example, make use of information imbedded in divisions. This would mean that the terms he was working with would refer to *differentiae* which were ordered so as to reveal how a general feature could be specified or determined (e.g., wing→feathered wing v. membranous wing v. dermatous wing). Thus, Aristotle would seek to identify the coextensive relationships among *differentiae* from different divisions, that is, to identify groups all of which and only which had certain *differentiae*.[23] This process would have to begin by identifying universal predications of a given subject, and what the subject was itself universally predicated of. But doing this would lead to the recognition of coextensive or 'primitive' universal predications; for example, that wing follows bird but not vice versa, and that feathered-wing follows bird and bird follows feathered-wing. Aristotle would not be concerned to stick with the popularly designated kinds.[24] If he identified a feature belonging invariably to all the animals with some other feature, he would want to see what these features followed and what followed them – even to the extent of seeking features predicated of all of a group of analogically related features. In doing this, he would prepare the ground for the sort of causal understanding that he regarded as the goal of science.

We have seen that the *Historia animalium* aims to "grasp the *differentiae* and the attributes which belong to all the animals," since, after this is done, one can try to discover their causes. In this treatise, Aristotle maintains that this is the appropriate way to proceed on the grounds that, if the inquiry has been carried out properly, one should be able to distinguish the things from which demonstration proceeds from the things about which we want demonstrations. I have argued that Aristotle's *APo.* II and a number of related texts provide us with the theoretical background for viewing *historia* as a pre-demonstrative preparation for causal explanation – just the sort of study that the *Historia animalium* introduces itself as.[25]

We are now in a position to ask whether the *Historia animalium* is a work which aims to organize information found in divisions in a way that is preliminary to demonstration as Aristotle understood it. Research carried out in collaboration with Professor Gotthelf and with just this question in mind indicates that in fact it does, though this should not be taken to imply that this is all it does or that it reflects the workings of a mind mechanically following a set of formal rules.[26] But before exploring one passage in detail, I should like to draw attention to the support for this contention that is provided by the work's overall organization.

As the important transition near the end of *HA* I.6 indicates, the first six chapters of Book 1 are in some sense introductory. At least five theoretical preliminaries are addressed. (1) We are introduced to the distinction between parts that are uniform (flesh, bone), simple and non-uniform (eye, finger), and complex and non-uniform (head, limb). (2) Aristotle then distinguishes sameness in form, sameness in kind, and sameness by analogy, as they apply to animals, to the parts of animals, and to the degree of sameness and difference exhibited by animals and their parts. Special attention is paid to the ways in which things differ when they are the same in kind but not in form (see chapters 6 and 7 below; Pellegrin 1987, 331–336). (3) Having introduced these distinctions in the context of parts alone, Aristotle next says that animals are differentiated according to their lives, activities, dispositions, and parts.[27] (4) The use of these ideas in the study of animals is then clarified by a series of examples organized under the categories mentioned in (3), differences of the first three kinds being discussed down to 488b28 and differences among parts from there to the beginning of chapter 6. (5) Finally, Aristotle establishes a number of extensive kinds, that is, kinds embracing a significant variety of forms sufficiently alike to be treated together.

Aristotle introduces (3) and (4) by writing, "the differences among the animals are with respect to their lives, their activities, their dispositions and their parts, about which let us first speak in outline, while later we will speak attending to each kind" (*HA* 487a11–13). As we saw, he later refers back to this outline as a preliminary sketch that is intended merely to give us a flavor of what is to come. One difference between this outline and the remainder of *Historia animalium* is that the later account will be about each kind. We will see the force of this contrast shortly.

Now the preliminary material that Aristotle mentions in *HA* 487a11–13 clearly draws on divisions. Let me simply quote two brief passages which can stand for dozens of a similar character in these chapters.

> . . . some of these animals are water dwellers, others are land dwellers; water dwellers are of two sorts: some live and feed in water, take in and expel water, and are unable to live if deprived of it (e.g., many of the fishes); others take nourishment and pass time in the water, yet do not take in or expel water and give birth out of water. Many of these are also footed, such as the otter, beaver, and crocodile; others are winged, such as the diver and the grebe; and still others are without feet, such as the watersnake. (487a15–23)

> Among fliers, some are feather-winged (for example, the eagle and hawk), some membranous-winged (e.g., the bee and the cockchafer), and some are dermatous-winged (for example, the flying fox and the bat). (490a5–8)

A number of features of these passages are relevant to our earlier discussion. First, even at the most abstract level, we begin with the assumption of four broad categories of *differentiae*. No kind of animal can be adequately characterized without a study of the life it leads in its environment, the activities it performs (locomotive, generative, perceptive, nutritive), its dispositional differences (Is it gregarious or a loner, timid or brazen, predator or prey?), and its parts. Further divisions are indicated under each category: thus, under manner of life, water-dweller/land-dweller; under water-dweller, those which do/do not take water in or generate in water;[28] under winged, feathery, dermatous, membranous. Specific kinds of animal are presented to illustrate the differences mentioned: these kinds are not themselves the subjects of the division. This does not mean that the universe of division is *differentiae*, however. It is often said explicitly, and it is nearly always implicit in the method used, that we are first to identify *animals* by a common feature (e.g., all the ones with wings rather than all the wings) and then to divide according to the way in which that common feature is differentiated (see note 23 above). Notice what strange bedfellows this method produces: at the common level, birds, bats, and bees are united as winged. Moreover, depending on which general *differentia* is chosen, animals will be grouped and re-grouped, to use David Balme's phrase. Such a methodology will indeed perplex the reader bent on taxonomy.

One must be careful about what one views as a division in these texts. In the first of the pair of passages just translated, Aristotle notes that some of the 'partial' water-dwellers are footed, whereas others are winged and others footless. Is this a further subdivision of this group? If so, since he appears to insert a division by locomotive organs into a division according to mode of life, Aristotle would seem to violate a basic rule of division, namely, that one should never divide by something accidental to the axis of division.

In fact, however, sketching out divisions is only a part of what one finds in these pages. Aristotle is also concerned to correlate animals grouped and divided according to one sort of *differentia* with those grouped according to others. He indicates that animals sharing one mode of life are diverse when viewed from the standpoint of locomotion. In a similar vein, he adds that among land-animals all those with lungs inhale and exhale air (487a29–31); that all insects live and find their food on land (487a31–32); that no creature which inhales and lives in water finds its food on land, though some that inhale and live on land find their food in the water (487b1–2); that all animals have a mouth, stomach, the sense of touch (and an unnamed organ for same), and a life-sustaining liquid (with container) (488b29–31, 489a17–19, 489a20–24); that all animals with stomachs have bladders, though not every animal with a bladder has a stomach (489a3–6) – and so on. The correlations are occasionally disjunctive (animals with feathered or membranous wings have either two feet or none [490a10–12]), and occasionally conjunctive (the feathered and the dermatous winged flyers are all blooded [490a9–12]). Such material, then, combines a sketch of how one should lay out divisions under various broad categories of animal differences with sample identifications of positive and negative correlations between groups in different divisions.

Aristotle himself constantly points out that he is here merely giving us a sense of the method and the sorts of differences that occur under the four categories he is discussing, and that the more systematic study to follow will need to center these methods on each kind. Thus, the last step (5) in this introductory stretch of text, the articulation of nine 'extensive kinds' (which are themselves broadly grouped according to whether they have blood or its analogue) is very important.

None of this preliminary material is mere window dressing. The overall structure of *Historia animalium* owes much to the principles articulated and discussed in these chapters. At the broadest level, the entire work is organized around the four categories of *differentiae:* a study of parts (I.7–IV.7), activities and lives (V–VIII), and dispositions (IX). Within the study of the parts, the investigation of the blooded animals (I.7–III.22) is distinguished from that of bloodless animals (IV.1–7). And the investigation of the parts of the blooded animals is divided into an account of the external non-uniform parts (I.7–II.14), the internal non-uniform parts (II.15–17), the genitalia, which are not always clearly internal or external (III.1), and, finally, the uniform parts (III.2–22). In the bloodless animals, parts external in one group are often internal in another; and this may be at least part of the reason why Aristotle investigates the internal and external parts together in each kind before moving on to the next.

We have seen that part of the more systematic nature of this work will involve its studying animal *differentiae* 'concerning each kind'. What role do the extensive kinds play in the way in which the information in the *Historia animalium* is presented? The study of the external non-uniform parts of the blooded animals moves from man to viviparous quadrupeds, through those which are biped in one respect and quadruped in another (apes and baboons) to the oviparous quadrupeds, birds, fish, and serpents.[29] Yet parts which extend (at some level of description) across more than one extensive kind are said to do so when they are first introduced; consequently, the later in the discussion a kind comes the less tends to be written about it. The study of the bloodless kinds is organized similarly. In both cases, numerous groups are noted which either do not fit into these extensive kinds at all or fit into one of them in one respect but not in another. On the other hand, the investigation of the internal non-uniform parts and the uniform parts of the blooded animals is organized, not kind by kind, but part by part. This may reflect Aristotle's belief that viscera do not differ as radically as external features from kind to kind, and so may be considered across the entire blooded clan.

The method of these passages can be seen clearly in the following discussion of the lungs and related organs.

As many animals as are quadruped and viviparous, all have an esophagus and windpipe,[30] placed in the same way as in humans; the placement is similar among the quadrupeds which are oviparous, and among the birds; but these kinds differ in the forms of these parts. Generally, all those which taking in air inhale and exhale, have a lung, a windpipe, and an esophagus, and the position of the windpipe and esophagus is similar, but these organs are not the same in all, since the lung is neither alike in all nor similar in position. (*HA* 505b32–506a5)

Aristotle goes on to note that not all blooded animals have lungs, and identifies those that do not (e.g., fish and any other animal with gills) (506a11–12).[31] Differences in these three organs are regularly referred to during the discussion of the other groups mentioned here (507a11–12, a24–27; 508a17–21, 32–33; 508b30–509a16).

This passage first establishes correlations between three distinct extensive kinds and three organic parts, and goes on to record a more general correlation between the animals that breathe and these three organs.[32] This is a move to the identification of the animals that have these organs as such, that is, to the 'primitive universal'. Such a move is achieved by uniting the kinds with these features by means of another common *differentia,* their breathing. Aristotle then discusses the differentiation of these organs under two headings, position and 'similarity'. Throughout the entire group the windpipe and esophagus are alike in position, though both differ both in their 'affective' qualities and their quantitative dimensions from kind to kind. The lung, on the other hand, differs in all these respects from kind to kind.

What is not said is of equal interest – for instance, the functional relationships among these organs, the reasons why all breathers have all three, the reasons for the differences in their character and position, and the reason why animals with gills have none of them. But the level of generality at which the search for such explanations should proceed is made progressively clearer, as the coextensive *differentiae* at that level (including the activities of inhaling and exhaling) are noted.

Aristotle's procedure in discussing the lungs has a number of features which recur to a greater or lesser degree throughout his description of the viscera of the blooded animals:

(1) the specification of an organ's pervasiveness among blooded kinds,
(2) the identification of coextensive organic structures,
(3) the attempt to identify the entire class with the feature in a unified rather than a conjunctive manner,
(4) in combination with (3) an emphasis on the diversity in the forms of the parts in the variety of kinds which share them,
(5) the distinction between the qualitative and the positional differences among the groups with regard to these organs,[33]

(6) the correlation of those differences with identified kinds of animals, and
(7) the identification of features coextensive with the differences primarily under consideration.

These features call to mind the ideas discussed in section 3 above, regarding the organization of information in a manner suitable for demonstration. That this was Aristotle's intent is evident if one compares this discussion with those of the same structures in the *Parts of Animals* II–IV. Does Aristotle there offer explanations at the level of primitive universality that he has here identified, and by reference to the activities and parts at the same level? It seems clear that he does. For he begins by noting that only animals which have windpipes and esophaguses have a neck, since the neck is simply a device for their protection (*PA* 664a12–17; cf. *HA* 495a18–20). He also remarks that the windpipe exists for the sake of breathing, since it is through this that the air passes on its way to and from the lung (*PA* 664a16–20: compare the unexplained universal that all animals that have a lung have a windpipe at *HA* 495a20–22; and the equally unexplained claim that all animals which breathe have all three organs at 506a1–5). Moreover, he asserts that the esophagus is not required for nutritional reasons (witness that fish get along without one). It is rather a by-product of the fact that those animals with lungs must have a windpipe of some length. The presence of a windpipe in turn produces a certain distance between mouth and stomach; consequently, there must be an organ to connect them. This explains why all and only breathing animals have an esophagus, an organ that seems to have little to do with breathing. (Compare *PA* 664a20–31 with the purely descriptive discussion at *HA* 495a22–30). Next in this chapter Aristotle reviews the relative placement of the windpipe and esophagus (cf. *HA* I.16, II.15). He points out that having the windpipe in front of the esophagus seems less than optimal, since food must pass over the windpipe when such animals eat. Such organisms must have a means of closing the windpipe when eating: in vivipara, this is accomplished by the epiglottis; in ovipara, by a windpipe that can open and close at the top. By contrast with this rich explanatory discussion, the *Historia animalium* is content simply to describe these organs and their locations: it never says that they are necessary or what they are for.

The discussion of lungs in *Parts of Animals* III.6 presents an array of interesting difficulties. I shall focus only on its conclusion.

> Generally, then, the lung is for the sake of breathing, while it is also bloodless for the sake of a certain kind of animal. But what is common to animals with lungs is without a name, that is, unlike 'bird' which names things in a certain

kind. Wherefore, just as the being for a bird is constituted from something, the possession of a lung likewise belongs in the being of these. (*PA* 669b8–12)[34]

Aristotle begins here by alluding to a teleological explanation for the possession of a lung, and for its possession in a different form in a sub-kind.[35] Not only do lungs belong to all the animals which breathe; they belong for the sake of breathing. The discussion leading up to this passage, in fact, shows concretely how closely tied together are the explanation of why certain animals have a certain organ and the account of what that organ is. Not only does understanding why animals breathe explain why they must have a lung; it also provides us with an account of what the lung is.[36]

The remainder of these concluding lines is puzzling at first. But, reminding ourselves of the following points may help to remove some of the problems. First, recall that the lung is neither limited to one of the extensive kinds identified by Aristotle, nor does it extend to all the blooded kinds. And yet, as we have seen, there is a complex network of anatomy and physiology related to breathing and the possession of lungs. Apparently, the 'universal' common to all these animals has not been named (unlike 'bird', which picks out those feather-winged, beaked, two-legged creatures). But that, we must remember, should not stop us from seeking scientific understanding: "we must not only inquire in cases where there is a common name, but also if anything else has been seen to belong in common, we must extract that and then inquire what it follows and what follows it" (*APo.* 98a14–16). The basic account for lung, windpipe, esophagus, and neck must show why all the animals which have them do in fact have them. The studies of these interconnected organs and the animals that possess them in *Historia animalium* and *Parts of Animals* and the relationships between these studies, display the methodological concerns of the *APo.* II. There appears to be a common conception of the activity and aims of scientific inquiry underlying Aristotle's science and his theory of science.

V *HA* IV.1–7: A CASE STUDY

In order to display the structure and aims of the *Historia animalium* concretely, I will conclude with a study of Aristotle's account of the parts of the bloodless organisms. Aristotle identifies four 'extensive kinds' among bloodless animals: the soft-bodied, the soft-shelled, the hard-shelled, and the insected. These groups correspond roughly to our cephalopods, crustacea, testacea, and insects, and in due course I shall use these terms. But it is important to emphasize a number of points about these Greek names. First, when they appear alone they are always in the plural.[37] Moreover, all are derived from vividly descriptive adjectives – literally, these kinds are the softies, the soft earthenwares, the earthenware-skinned, and the divisible; and Aristotle's initial differentiation of them remains close to this basic sense. Aristotle distinguishes these kinds on the basis of whether their hard parts are inside or outside (or all the way through!), and the nature of that hardness. The cephalopods, if they have a hard part, have it inside; the crustacea have a hard but crushable exterior, a soft interior; testacea have a hard fragmentable exterior and soft interior; insects are uniformly hard throughout. This descriptive terminology is introduced in *HA* 490b10–16, with an occasional remark that suggests there are no common names for these groups as such. But throughout the *Historia animalium* these terms consistently identify kinds with a variety of forms, forms possessing the general features of the kind differentiated 'in degree'.

The parts of the bloodless kinds are discussed in *HA* IV.1–7, beginning with cephalopods. The discussion opens in a manner typical of the entire work with an account of the external parts.

> The following are the external parts of the so-called softies: first, the so-called feet; second, the head which is continuous with these; third, the sac, which contains the internal parts and which some erroneously call the head; and again the fin which encircles the sac. But it so happens that the head is between the feet and the belly in all the cephalopods. Now all have eight feet, and all have a double row of suckers, except in one kind of the octopuses. But, distinctively, the cuttlefish and the small and large calamary have two long tentacles, with rough tips and two rows of suckers. . . . (*HA* 523b21–31)

Those external *differentiae* that can be predicated in general of the cephalopods are given first. Aristotle then remarks on a peculiar feature common to the cuttlefish and calamary, and goes on (524a3–19) to discuss features peculiar to octopuses as a group. Next, he describes differentiations among cuttlefish and calamary, and then among octopuses (524a20–32). After describing the head, the eyes, and the mouth (with its two teeth and tongue-analogue), he moves down the esophagus and discusses the internal parts: features common to the cephalopods (524b1–22), a hard part peculiar to the cuttlefish and calamary (524b22–23) which is nonetheless differentiated, the sepia of the cuttlefish being harder, bonier, and flatter than the calamary's firmer, thinner, more cartilaginous 'pen' (524b22–28). The octopuses as a group have no hard external part. Aristotle then discusses sexually related differences at various levels and, finally, certain features which distinguish a number of kinds of octopus (524a14–28).

Likewise, Aristotle begins his account of crustacea with "Now common to all these is, first . . ." (*HA* 526b21), and adds "But now the distinctive *differentiae* must be studied with respect to each kind" (526b34–527a1). The insects are introduced as a kind with many forms (531b21), and two groups are identified which have numerous forms closely akin to one another but which are not bound together by a common name (531b22–232): bees, hornets, wasps, and the like (cf. 623b23), and those insects with encased wings, which Aristotle refers to as coleoptera (cf. 552b30, 601a3). Then, as in his study of the cephalopods, Aristotle turns to features common to all insects: to the articulation of the body into head, thorax, and abdomen (531b26–28); that they all live when divided (531b30–532a5); and that all have eyes (532a5). Yet only some have stings (532a14–17) and wings (532a19–22); and among these latter some have encased wings, while others do not; but all their wings are membranous, lacking the stock and divisions of feathers. In sum, as he says, ". . . the parts of all the animals, both external and internal, both those peculiar to and those common to each kind, belong in this manner" (532a27–29).

In this passage we see a mind striving to identify the widest group of animals to which a feature belongs universally. But 'feature' is ambiguous. How widely one predicates a feature often depends on how generally or specifically it is described. No one for whom division was a scientific tool will forget this – all the calamaries and cuttlefish have a hard structure; so if one wishes to understand why they do (and the octopus does not), this is the predication that is crucial. But if one wishes to understand why the cuttlefish has a sepium rather than a pen, that hard feature must be described and identified more specifically. Aristotle's method throughout the texts we have surveyed is tailor-made to achieve these explanatory goals, that is, to provide propositional descriptions of the animal world that meet his strictures on proper explanation.

There are, however, important differences between the chapters devoted to the parts of the bloodless animals and the earlier discussion of the parts of the blooded ones. The differences related to the order in which the internal and external parts are surveyed has already been mentioned. In addition, there are very few attempts to identify features which extend beyond a given extensive kind. And the universal predications are primarily of the form which correlate a *differentia* of some sort with a named group, rather than with another *differentia*. Thus, these chapters are virtually devoid of the 'as-many-as-are-X,-all-(most, some, or none)-have Y' form of proposition that is found with such regularity in the earlier books. Attending to these sorts of differences among the various discussions that make up the whole of *Histo-*

ria animalium is not likely to invalidate the claims I have been making, but it will in all likelihood lead to a yet richer and more complex picture of this great work.

An independent test of the soundness of this view of the *Historia animalium* is to consider its ability to treat naturally those features which are anomalies on other accounts. As David Balme (1987a, 9) has stressed, one such anomaly for anyone who reads the *Historia animalium* either as systematics or as natural history, is that various animals are mentioned regularly, but only in order to point out some oddity. Balme's favorite example is the blind mole-rat: we are told on a number of occasions of its peculiar, rudimentary, subcutaneous eyes, though, for all else we are told it could also have wings, scales, gills, and ten feet (cf. 491b27–34). Similarly, there is a variety of octopus, the *heledōnē*, which Aristotle mentions only once and only to tell us that it has a single (rather than a double) row of suckers on its narrow tentacles. Such selectivity in Aristotle's treatment is to be expected if the majority of the features of such animals are in fact more appropriately discussed as features of the wider kind of which they are one example. As we have seen, the features common to all octopuses are predicated of octopus, while those shared by all cephalopods are discussed at this more general level. Only that peculiarity of the *heledōnē,* the single row of suckers on its narrow tentacles, is termed a proper attribute (525a16–19). From the perspective of the theory of explanation in the *Posterior Analytics,* this feature will be explained, if at all, in terms of other features peculiar to the *heledōnē.* And as a matter of fact, the discussion of the cephalopods in *PA* IV does just that.

> The other cephalopods have two rows of suckers, but one kind of octopus has only one. The cause of this is the length and slimness of their nature; for their being narrow necessitates a single row. Now they have these things arranged thus not because it is best but because it is necessary due to the proper account of their being. (*PA* 685b12–16)

Both the property identified as the cause and the property identified as the effect in the *De partibus animalium* are said in the *Historia animalium* to belong to the *heledōnē* alone; but there is not the slightest hint of a causal relationship between these features in the latter. The *heledōnē* has a long and slim nature and, thus, long and narrow tentacles. Accordingly, it must have a single row of suckers. It is not that having one row of suckers is better than having two. If, however, a case could be made out that either one or two rows were equally possible and that one of these possibilities were better, Aristotle might be inclined to say of the *heledōnē* that its possessing one row is better than its having two rows, on the ground that "we see that nature does nothing pointless, but always the best for each being among the possibilities." But, in this case, he takes the possession of one row as necessitated by the antecedently given nature of animal.[38]

VI CONCLUSION

The methodological unity of the chapters in the *Historia animalium* which are devoted to the parts of the bloodless animals is clearly not one imposed by the aim of a hierarchical classification. There is, for example, no effort to introduce a vocabulary for *taxon*-categories of different extension. *Genos* is the all-purpose word for animal-kinds at any level of generality. The cephalopods as a whole are a 'kind' (523b3), but so are the large calamaries (524a29), and there are many 'kinds' of octopus (525a13). The crustacea are a 'kind', but so are the crabs and carids; and there are many 'kinds' of each of these (525a33–b1). The testacea are a 'kind', but so are the snails, oysters (528b11–12), and the sea-urchins (528a2). Finally, animals which do not fall into any of the extensive kinds, such as sea-anemones, are 'kinds' as well (531a31). The inclusion of this last 'kind' indicates another fact about the *Historia animalium* which points to its lack of interest in systematics – the untroubled recognition of kinds which do not fit into the wider kinds Aristotle has identified.

Finally, at the most general level, this is a work organized as a study of the *differentiae* which belong to animals; it is not presented as a classification of animals. This explains the fact that information regarding any particular kind of animal is sprinkled, in a seemingly haphazard manner, throughout the nine books of this treatise. But the appearance of happenstance is in the eyes of the modern biological beholder. It is in general a well-ordered treatise – given its (declared) demonstrative aims, aims that are revealed in a variety of ways. For there is a persistent concern to identify groups which share a number of features as a group and yet have not been identified as a single unit, i.e. that have not been united by a common name. Identifying such groups, I have suggested, is an activity fostered by the scientific ideals of the *Posterior Analytics*. Moreover, there is a recurrent effort to find the widest group to which a given organ or tissue belongs universally; to note how these organs or tissues are differentiated qualitatively, quantitatively, and positionally, in different subgroups; and to identify the widest group which possesses the various differentiated forms of the general type of organ or tissue being discussed. Again, this is what *APo.* II.14–18 would lead us to

64

expect. (It would be interesting to pursue the philosophical issue of whether there are compelling reasons for preferring the [Linnaean] methodology of establishing a hierarchy of kinds on the basis of a single diagnostic character, and the historical question of whether certain individuals – Baron Cuvier comes to mind – were so impressed with Aristotle because of their own tendency to approach the biological realm with an eye to understanding differentiation rather than with the aim of organizing its species hierarchically.)

Aristotle's guiding question in his zoology seems to be, Why do all and only these animals have this feature? His answer seems to require starting with the *differentiae* and asking how widely a given *differentia* extends in relation to others – that is, he seeks to identify groups relative to some difference and not to identify the difference relative to a pre-established group. This method succeeds in identifying animals with commensurately universal *differentiae,* the first step toward causal accounts in the explanatory model proposed in the *Posterior Analytics.*

The distinction between factual and causal-explanatory inquiries in *APo.* II.1 and 2 is general and theoretical. I hope that my study of Aristotle's method in the *Historia animalium* and of its relationships to its companion studies of animals has given reasons for thinking that his zoological treatises took this distinction seriously. At the same time, I also hope that my remarks have deepened our understanding of 'that-it-is' investigations beyond the facile notion that they 'collect the facts'.

ACKNOWLEDGMENTS

The original publication of this paper benefited from comments on three occasions. At the conference on Aristotle's Biology held by Cambridge and Trenton State Universities (July 1985), the comments of David Balme, Alan Code, Geoffrey Lloyd, Allan Gotthelf, Martha Nussbaum, and Malcolm Schofield were especially useful. At the APA Pacific Division's symposium, Classification and Explanation in Aristotle's Biology (March 1986), the contributions of David Charles and Allan Gotthelf helped to clarify many issues relevant to the topic of this paper, and pointed out a number of weaknesses in its predecessor. During the conference, The Interaction of Science and Philosophy in Fifth and Fourth Century Greece, held by the Institute for Research in Classical Philosophy and Science (May 30–June 1, 1986), Charles Kahn, the late Joan Kung, Geoffrey Lloyd (yet again), Father Joseph Owens, and Robert Turnbull offered helpful suggestions. In addition,

Michael Ferejohn and Geoffrey Lloyd provided helpful written comments. This paper, as with all my work on Aristotle's biology, is pervasively indebted to discussions on the subject with Allan Gotthelf.

NOTES

1. Le Blond (1945, 19) and Balme (1987a, 13–17) draw precisely opposed conclusions from this fact.
2. Actually, Aristotle sometimes refers simply to "the histories" (*PA* 646a11; *Juv.* 478b1; *GA* 719a10, 740a23, 746a15) or to "the natural histories" (*PA* 639a12, 650a31–2; *IA* 704b10): the majority of the references in *PA,* however, are to specifically animal histories (660a9, 660b2, 668b30, 674b16–17, 680a1, 684b4–5, 689a18). It is presumably from these references that some ancient editor (probably Andronicus: cf. Keaney 1963, 57–58) derived the title that has come down to us.
3. I will focus on the extent to which the theory of finding middles relative to specific problems and the theory of inquiry in *APo.* II can give us a purchase on Aristotle's concept of an *historia.* In chapter 1, I concentrated on the remarks in the *APo.* about the move from incidental to unqualified understanding, about how locating predications at the commensurately universal level is a crucial part of this move, and about how the *HA* is consistently concerned to locate such predications. Some of the results of this study are directly relevant to this chapter's theme, and (in a somewhat developed and modified form) will be presented in section 3 of this chapter. I intend to leave open questions of the chronological order of the research reported in, and of the composition of, the various 'zoological' treatises and the *Analytics.* That the *Analytics* may serve to shed light on the biological works, or *vice versa,* is consistent both with the view that the *Analytics* is a result of reflecting on scientific research and explanation done or in progress, and with the view that it is a sort of model for the presentation of such research and explanation.
4. Previous theoretical uses of the term do not take one very far. Herodotus (*Hist.* I.1) announces his work as an *epideixis historiēs,* but the context suggests that he means little more than a presentation of reliable information. There is a consistent implication that *historia* is a basis for knowledge (cf. *Hist.* I.44, II.118, 119), though often this basis is a reliable report rather than something Herodotus has directly observed. In the Hippocratic medical treatise *De vet. med.* 21, the author tells us that by *historia* he means knowledge of what man is and of the causes of his coming to be. In *Phaedrus* 244c8, Plato conjoins *historia* with *nous,* where it seems to have the force of information acquired on the basis of signs. And at *Phaedo* 96a8–10, Socrates tells us that those who have investigated natural coming-to-be and passing away describe their wisdom as 'concerned with natural history'. The wider context suggests that they meant their wisdom to include an understanding of the causes of natural things. I draw attention to three points. First, while it is common to translate *historia* as 'inquiry' or 'investigation', the term more often designates the report or result of inquiry. Second, if Aristotle is restricting it to the report of a pre-causal inquiry, he is legislating this usage, which is not reflected in the above passages. Third, Herodotus *does* use the term to refer to reports and information which serve as the basis for knowledge, rather than to the state of knowing or the report of knowledge itself. Cf. Louis 1955.

5. There are a number of recent correctives to this: cf. Ackrill 1981b, Bolton 1987, Ferejohn 1982/3, and chapter 1 above.

6. Jonathan Barnes' two papers on this subject (1975b, 1981), for example, are limited almost entirely to the theory of demonstration in the *APo.*

7. Theophrastus introduces *De causis plantarum* by way of a similar contrast:

   That the modes of generation of the plants are numerous and how many they are and of what sort was said previously in the histories; but since not all modes of generation occur in all the plants, it is appropriate to distinguish which belong to each plant and through which causes, making use of principles in accord with their proper being. . . .

   On the overall relationship between the methods of Aristotle's *HA* and the *Historia Plantarum* by Theophrastus, cf. Gotthelf 1988 in Fortenbaugh 1988.

8. All translations from *PA* I are from Balme 1972 unless otherwise indicated.

9. The first question Aristotle raises is discussed in chapter 1, pp. 29–32.

10. It is of historical interest that the treatise by Euclid which comes to us under this title is highly mathematical in character.

11. Though, as Allan Gotthelf has reminded me, after making the distinction Aristotle goes on to say that the investigation will proceed "by separating on their own the things that were said in the *histories*" (*PA* 646a11–12). Since *PA* gives us all the data to be explained, and since at least some of it is inconsistent with *HA,* Gotthelf suggests that this passage instructs us to "put aside the information reported in *HA.*" I prefer the sense of Ogle's translation (1912, *ad* 646a12 n1), which takes Aristotle to distinguish these causal inquiries from the sort of report found in *HA.*

12. For detailed discussion of this passage, besides chapter 1 above, see also Kullmann 1974, 196–202; Lloyd 1979, 137–138 (and n64), 212; Gotthelf 1987b; Balme 1987b, 79–80.

13. For other passages indicating that biological explanation is to be demonstrative in character, see *GA* 742b18–743a1; *PA* 639a14, 640a2–9, 645a1–2; *IA* 704b12–705a2. Gotthelf (1987a, 170–172, 197–198) makes a strong case for a technical sense of 'demonstration' in these passages.

14. Translations of the *APo.* are by Barnes (1975a), unless otherwise noted. The expansions in square brackets are my own. (See now Barnes 1993 for more elegant, but more interpretive, translations.)

15. Cf. Alexander, *in APr.* 333.19ff. It is worth noting that *DL* V.25 lists a *Dissections* in eight books and one book of *Selections from the Dissections;* regrettably neither work survives, and to my knowledge there is nothing in the doxographical tradition which even hints at the form they might have taken.

16. My remarks on this problem owe a great deal to Allan Gotthelf's contribution to the Symposium on Classification and Explanation in Aristotle's Biology at the APA Pacific Division meetings in 1986.

    Partial demonstrations raise two central questions, one having to do with their form and the other with their explanatory force. I suggest that the form of such explanations is that the middle term will refer to the kind to which the referent of the minor term belongs as a sub-kind (see also chapter 1 above). For example,

    Having a heart belongs to all blooded animals
    Being blooded belongs to all birds
    _____
    Having a heart belongs to all birds

The virtues of this model are three in number. It appears to underlie numerous passages in the *APo.* (e.g., 73a16–20, 74a1–3, 74a25–32, 85b4–15); it is a pattern of reasoning that is found regularly in the biology (cf. chapter 1 above, pp. 25–29); and the major premise is the sort of proposition one would expect to find as the conclusion of the more basic explanations of commensurately universal predications, thereby giving a means of logical transition between universal and partial demonstrations. This is clearly the way in which Philoponus (*in APo.* 417.26–28) reads the passage: "Since these things follow animal, you will prove that perception or motion belong to humans and the rest through the middle, animal."

But this reading also has certain drawbacks. The discussion of demonstration and its relationship to definition in *APo.* II.8–10 generally treats the middle term as giving an account of the minor term, and that seems quite unlike the model just suggested. There, progress in understanding comes through acquiring more basic middle terms of the same logical sort, thus giving us better accounts of the minor term. It is not clear where this leaves us, however, for that discussion is not concerned with the move from knowing that a sub-kind has a feature to knowing that it has the feature in virtue of being the kind of thing that has the feature in itself.

This brings us to the issue of the explanatory force of partial demonstrations. Professor Gotthelf has urged that the appearance of the kind in the position of the middle term may simply be shorthand for saying "the sub-kind has the feature for the very reason that the kind does" (cf. Ackrill 1981b, 380), and there are passages which do suggest this gloss (e.g., *APo.* 91a2–5).

At the very least, Gotthelf has convinced me that formulations in Lennox 1987a and earlier drafts of the present work which suggest that partial demonstrations (which I term A-type explanations in chapter 1) could be demonstrative prior to understanding why the more primitive predication holds are wrong, for the reasons given in the text above.

17. A scholium to Euclid, *Elements* V attributes the discovery of the general theory of proportion to Eudoxus (cf. Heiberg and Stamatis 1977, I. 213.1–12; Heath 1956, II. 112–113).

18. See also the careful analysis of this passage in Gotthelf 1987a, 179–185.

19. Not true names, as Allan Gotthelf reminds me (cf. Balme 1962b, 90). The *APo.* 93b29–32 allows for definitions of 'name-like phrases', as does *Top.* 102a1–5. While I suspect that terms in the biology such as 'the horn-bearers', 'the pottery-skins', 'the live-bearing among the four-legged', are among the sort of name-like phrase he has in mind, we need to know more about the significance of 'name-like' here, specifically, about how name-like terms differ from terms which cannot be defined (see Bolton 1985, for some suggestions).

20. See *PA* 663b35–664a3, 674b7–17, where the production of horns (for self-defense) leaves less earthy material for teeth (accounting for the lack of two rows of teeth) and, thus, indirectly necessitates the possession of more stomachs (for the digestion of the poorly chewed nutrients). In this way, an animal's possession of horns accounts for these features.

21. Cf. *APo.* 78a28–b13. These are, of course, the passages at the basis of the distinction between *demonstratio quia* and *demonstratio propter quid.*

22. Broadly speaking, this interpretation of the passage has defenders from Philoponus (*in APo.* 429.32–430.7) to Ross 1949, 671.

23. On the developments in Aristotle's theory of division and its role in the biology, see Balme 1987b, 74–89.

24. I am here passing over a set of very difficult questions about what Aristotle would call a kind and why. In a classic article, Balme (1962b) first distinguished *differentia*-classes and true kinds. The former lack actual nominal identifications and collect animals for convenience on the basis of some shared feature or other. Pellegrin (1982; 1985, 103–106, 112; 1987, 334–337) argues that this is in fact typical of the way Aristotle identifies kinds – indeed Pellegrin goes further and argues that typically kinds are *parts* in Aristotle's biology, that his biology is more properly said to be a *moriology* than a zoology. (For my reservations on this score, see Lennox 1984, which is a review of Pellegrin 1982). And Allan Gotthelf (1985b, 1987a) has raised a number of questions related to this topic in the process of his work on the concepts of substance, essence, and definition in *PA*. Two points from our discussion of the *Analytics* are relevant to this issue. First, Aristotle clearly allows for definitions of things which have name-like phrases along with those having actual names. Of course, this merely pushes the question back one step to the question, What will count as a name-like phrase, i.e., How does Aristotle go about screening out those 'names' of merely accidental unities? Second, at least part of *APo.* II.14 presses us to admit into our scientific vocabulary phrases which are certainly not names (or even name-like), and provides us with a rationale for doing so. These phrases are similar in character to those Aristotle uses to designate previously unnamed extensive kinds in the biological works: translations usually mask this fact, 'oviparous quadruped' typically rendering a phrase which might be translated literally as 'the ones among the four-footed that lay eggs'. Indeed, Aristotle will list the two *differentiae* in either order (though I do not mean to imply that he does so randomly).

    This is clearly an issue that requires fresh and thorough re-investigation. At this stage, I am prepared to say that Aristotle is aware of the need to use certain mechanisms to extend the vocabulary of science in the direction of identifying unnamed groups that are unified in some way or other, that a syllogistic model of the logic underlying science requires that most of its terms be expressible as either subjects or predicates and be from diverse categories, and that Aristotle's zoological terminology is dominated by phrases that identify groups of animals as 'the ones that are (or have, or do) *X*', where *X* is a peculiar feature of those animals. These are among the 'phenomena' which an account of Aristotle's theory of scientific kinds will have to explain.

25. The extent to which elements of these ideas are reflected in Aristotle's zoology is explored in chapter 1 and in Kullmann 1974; Gotthelf 1987a, 1987b; Bolton 1987.

26. For other products of this collaboration, see Gotthelf 1987b and chapter 1 above. Gotthelf 1987b points out ways in which this research is consistent with the late David Balme's most recent work on the *HA*, work that was in progress as part of the preparation of a new edition of this text as well as a translation and commentary.

27. See Gotthelf 1987a, 192–193.
28. This contrast is dealt with in much more detail and more systematically at the beginning of Aristotle's discussion of differences in manner of life: see *HA* 589a10–590a18.
29. Serpents are denied the status of extensive kind, but they are discussed at length. The cetacea, however, are listed as one of the blooded extensive kinds (*HA* 490b9) and yet are not discussed in the review of external parts. Given their extreme peculiarity, which Aristotle stresses elsewhere (*HA* 588a31–b2), this seems doubly odd: cetacea are mentioned in Books 2 and 3, but only by way of contrast. The most extensive account of the cetacea occurs in the sections of *HA* dealing with *differentiae* of activity (reproduction, 566b2–27; respiration, 589a27ff.) and with dispositions (631a9–b4).
30. The *HA* makes frequent use of 'doubly quantified' expressions of the form 'As many as are *X,* all have *Y*'. Gotthelf (1988) suggests that the preponderance of this otherwise rare form of expression may signal Aristotle's concern in *HA* to identify primitively universal predications.
31. Notice that this provides a means of identifying animals without lungs while leaving the extension of the group open-ended. No reason is here given as to why animals with gills will not possess lungs, though one is provided in *Juv.* 476a6–15.
32. Earlier, in his review of the parts found in humans, Aristotle (*HA* 495a18–22) states that "the so-called esophagus (so named due to its length and narrowness) and the windpipe are within the neck; but the windpipe is positioned in front of the esophagus in all those animals which have it – and all have it which also have a lung." Indeed, the entire passage, 495a18–495b23, is taken for granted by the discussion in Book 2.
33. In his discussion of the ways in which the parts can differ, Aristotle (*HA* 486a25–487a1) makes a broad distinction between the ways in which parts can vary by 'excess and defect' (i.e. in degree, which includes variations in the quality, size or number of a structure), by analogy, and by the position of the part. This passage uses that distinction carefully, though the study of the relative positions of the windpipe and esophagus and of the differences in form of the lung at *HA* I.16 (see note 32) is taken for granted and not elaborated.
34. See Gotthelf 1985b, 31 and Balme 1962b, 90. As Gotthelf points out, while *PA* III.6 explains why certain animals possess a lung, it concludes by saying that having a lung is in their being. This raises a host of questions about the sorts of feature to be specified in the account of an animal's being, issues which lie beyond my present concerns.
35. I emphasize that this is merely an allusion to such an explanation, since all of *PA* III.6 is devoted to this task, a task which must involve a physiological account of breathing that can handle the fact that not all blooded animals breathe (by which Aristotle meant 'take in and expel environmental air'), and that some water animals do: see chapter 1, 25–26, for a brief discussion.
36. See Gotthelf 1988 and chapter 1, 25–32.
37. The importance of this point was brought home to me in discussion with Allan Gotthelf.

38. See Gotthelf 1985b, 41–42 for a discussion of this passage and its relationship to the patterns of explanation outlined in *PA* 640a33–b1. Gotthelf stresses the apparent difficulty for strong functionalist readings of Aristotle's biology posed by Aristotle's inclusion of such features as the dimensions of a part in the account of the being of an animal.

3

Aristotelian Problems

I INTRODUCTION

Posterior Analytics II.14 opens with the admonition to grasp problems (*problēmata*) by selecting in a certain way from dissections and divisions. This is not a prescription for which we have been prepared. What does Aristotle suppose we are grasping when we grasp a *problem?* What stage is thus achieved in the epistemic quest? The centrality of this concept in the *Organon,* the references to things said 'in the problems' scattered throughout Aristotle's works,[1] and the fact that entire books of '*Problems*' have come down to us as part of the Aristotelian corpus,[2] suggest that having an answer to this question would be worthwhile. From a wider perspective, the concept plays a crucial role in the history of Greek mathematical analysis, so that a better understanding of what Aristotle meant by this term might have broader historical significance.[3] This chapter provides an answer to the question regarding what Aristotle thinks we have achieved by grasping a problem; it also suggests some of the difficulties in the way of connecting the Aristotelian concept to the broader historical context.

II THE SOCRATIC BACKGROUND

The Aristotelian concept of a problem arises naturally from the Socratic world of question and inquiry. The 'logic' of Socratic elenchus is erotetic, not assertoric: the questions asked determine its trajectory. In the context of the gnomic revelations of Parmenides, Heraclitus, Anaxagoras or Empedocles, this is a decisive philosophical turn. It focuses attention on the *process* of coming to know. In particular it suggests that knowledge is the end point of inquiry and that inquiry involves the posing, and answering, of certain types of questions.

In this respect, Aristotle is at least the match of Plato as heir to the Socratic legacy. The predicables of the *Categories* appear as a list of different sorts of questions that may be asked and answered about any object (*Cat.* 4 1b25–27).[4] Knowing, Aristotle insists, is having answers to the question 'on account of what?', the types of answers being as many as the types of causes (*Ph.* II.3 194b18–20; II.7 198a14–17). The four forms of inquiry distinguished at the beginning of *APo.* II (about which, more later) are also characterized by reference to the different but related questions they seek to answer (89b23–35).

The prominence of the concept of a 'problem' in his account of reasoned inquiry is a further instance of Aristotle's inquisitional epistemology. But problems play a central role in both dialectical and scientific contexts. We are thus faced with the question whether the role of problems in these contexts is the same or different. In order to tackle this question, it is useful to distinguish two distinct forms that an answer might take.

(i) *Conceptual shift.* The term 'problem' in each context refers to a different form of proposition.

(ii) *Contextual shift.* The role of the propositions referred to as problems is different in each context, though the form of proposition is the same.

Ultimately, it shall be insisted that these are not exclusive alternatives, and that in the case at hand, subtle but important changes in reference are a consequence of the distinct roles played by problems in the dialectical and scientific contexts. But in any case, making the above distinction provides a way of representing differences in Aristotle's use of a concept in different but related contexts which is independent of theses about the development of Aristotle's epistemological views. It will be argued here that there is a conceptual shift in the concept of a 'problem', and that this is fully accounted for as a consequence of differences that Aristotle himself identifies between dialectical and demonstrative contexts. Such a conclusion is compatible with all developmental hypotheses that allow for the consistency of these two contexts.

III PROBLEMS IN THE *TOPICS*

The *Topics* opens with the following statement of purpose: "to discover a method by which we will be able to reason about every *problem* laid before us from reputable beliefs" (*Top.* I.1 100a18–20). According to the *Topics* a deductive argument consists of a statement (or collection of statements) in which certain things being laid down, other things follow from them from necessity. *Dialectical* argument is distinguished from demonstrative, not by its form, but by the epistemic character of the things laid down. It is quite generally true that argument derives *from* premises; and is *about,* or concerned with, problems.

> Those things the statements are derived from and those things the deductions are about, are equal and the same in number. But it is the problems the deductions are about. And every premise and problem makes manifest either a kind, property, or accident; for even the difference, insofar as it is kind-like, should be ranged with the kind. (*Top.* I.4 101b14–19)

Premises and problems differ from each other by virtue of their place in dialectical practice. Both arise within the question/answer context of dialectic. *Premises* arise from direct questions — "Is the definition of human two-footed land animal?" Such questions concern states of affairs about which all people, most people, or wise people agree, and thus expect a positive answer (104a7–15). *Problematic questions* concern alternatives — "Is two-footed land animal the definition of human, or is it not?" (101b26–37) "Are perception and knowledge the same or different?" (102a7) "Is every pleasure a good?" (108b35)[6] Problems thus require the dialectical respondent to adopt one or the other side of an alternative about which there are either no firmly held beliefs at all; or about which the many and the wise either hold contrary beliefs, or disagree among themselves (104b1–5). The reply to a problematic question, then, cannot help but be controversial — ". . . no one with a mind would propose as a problem what is apparent to everyone or most people; for these present no *aporia* . . ." (104a5–7). In answering, the respondent takes a position on a subject about which there is a puzzle for dialectical debate: a *problem.*

In the *Topics,* Aristotle uses the terms 'problem' and 'premise' sometimes to denote questions, and sometimes assertions.[7] This ambiguity is all but inevitable in the dialectical context. For within the rules of dialectic, that about which one reasons is *presented* in the form of an interrogative alternative: "Is it the case that S is P, or is it not?" "Is S P or not-P?" The next step, however, is for a respondent to adopt one of the alternatives, which is then challenged and defended. The above description of problems is formal enough to embrace both the question/alternative, and the accepted alternative.

Perhaps we should conclude from this, then, that the term 'problem' or 'premise' refers to that which is common to the question and answer:

namely, the statement which the question is about. On this view, if the question form is "Is it, or is it not, the case that p?", the problem is 'p'.[8] But this suggestion ignores the regular use of these terms to refer to interrogatives. How then should we deal with this ambiguity?

An alternative approach, adopted here, is to think of the interrogative context of dialectic as altering the epistemological status of a statement, whether proffered as an alternative or asserted as an answer. Aristotle first introduces the distinction between premise and problem by reference to the expectation that the asserted answer will be, if a premise, an *endoxon* to be argued *from,* or, if a problem, an *aporia* to be argued *about.* Studying the assertion in isolation, therefore, will not reveal whether it is a premise or a problem. The *sentential* content will be the same – it is the type of question the statement responds to that determines its dialectical status. This means that the epistemic attitude adopted toward it depends on the erotetic context. In particular, a *problem* begins as an alternative about which there is no general agreement. Consequently, to adopt one alternative rather than the other is not to assume its truth, but to put it out for debate.

IV DIALECTICAL PROBLEMS AND DIVISION

The very form of a problematic question recalls the divisional procedures of the *Statesman* and *Sophist.* Division proceeds by posing alternatives like that initially asked of Theaetetus about the angler: "Tell me whether we will put it that he is artistic, or artless, though having another capacity?" (*Sophist* 219a). The respondent must select an alternative, and then another such question is proposed, often though not invariably assuming the previous answer as stipulating the kind to be divided. For example: "Since acquisitive and productive encompass all the arts, in which, Theaetetus, shall we put the angler?" (*Sophist* 219d) At the end of the process one collects the answers to the various questions together as an account of A, as if it were an answer to the 'what is A?' question. The questioning process, since it poses an alternative to the interlocutor, is equivalent to Aristotle's posing of a problem. But the attitude adopted by the interlocutor is very different; akin to that appropriate to dialectical premises.[9]

In the *Topics,* then, questions of the above sorts generate either premises or problems. Certain methods for detecting confusions or errors in the answers adopted are discussed, and here again a good deal of what is learned in this part of dialectical training is the use of division for the testing of problems. As one example, *Topics* II.2 109b13–29 recommends that division will be useful in the solution of a *universal problem,* since it will aid us in seeing all the possible forms of the universal claim, and thus in locating a counter-example. His example is helpful. If the universal problem is "The same science studies opposites," one can divide 'Opposition' into its four types, and each of them into their types, and see whether there is general agreement, type by type, that one science studies both. To discover an instance where it is generally believed that opposites are studied by different sciences is to find grounds for challenging the universal problem. Thus while there may be genuine lack of agreement regarding the universal statement, if there is agreement on a counter-example, this can be used to challenge the person who affirms the general proposition.[10]

Aristotle is critical, of course, of claims emanating from the Academy that the procedure of division is *demonstrative*.[11] There is, he insists, no justification for selecting one or the other alternative provided by division – the method used in the later Platonic dialogues depends crucially on simply accepting the selection that is made.

V   FROM DIALECTICAL TO DEMONSTRATIVE PROBLEMS

In the *Analytics,* the concept of a problem plays a central role in two closely related discussions – *APr.* I.26–31 (and somewhat less frequently in the early part of *APr.* II.) and *APo.* II.14–17. In other contexts, the term is rarely used.[12] The fact that these chapters constitute discussions of certain *pre*-demonstrative activities suggests that the role of problems in demonstrative science is in what Michael Ferejohn refers to as its 'framing' stage.[13]

Problems in the Prior Analytics

As noted, except for a brief summative statement, *APr.* makes no use of the concept of a 'problem' until I.26, after which, especially in chapters 27–30, the term occurs regularly. In these chapters, Aristotle turns to a discussion of the *preliminaries* of proof, which likely accounts for the centrality of the concept 'problem'. That which, in *APr.* I.1–23, would be viewed as the necessary consequence of things already laid down – that is, as a *conclusion* – is, in the 'framing' context, viewed as something *proposed to be proved.* This is the demonstrative analogue to the alternative taken up for defense and attack in a dialectical context.[14] In the dialectical context of the *Topics,* a statement might be referred to either as a premise or as a problem, depending on whether it was a commonly accepted assumption of debate or a statement being debated. In the proof context of the *Analytics,* a statement might be referred to as either a conclusion or a problem depending on whether it was derived from premises or is in need of such derivation. Yet what is claimed at the opening of the *Topics* is true of both contexts – premises are the 'from which' of reasoning, problems are its 'about which'. The primary change is that the mathematically derived term *sumperasma* is introduced to indicate the results of deductive proof. Prior to such derivation, the same statement is a problem. *APr.* I.26 describes such statements as problems because it is concerned with the relative ease or difficulty of *establishing or refuting* the four possible forms of proposition (universal or particular affirmative and negative). *APr.* I.27–30 provide detailed recommendations for selecting premises relative to a proposed problem – in particular, for finding, among divisions, a middle term which 'follows' the problem's major term and which its minor term follows. In chapter 28, for example, the procedures outlined are ways of finding premises which can be used in establishing that some predicate does (or does not) belong to all (or some) of some subject. This presupposes that one of these four types of problems is posed prior to implementing the procedures involved in hunting down the premises. After the review of these procedures, Aristotle states: "It is evident, then, that, concerning each problem, one must look to the things previously said about each figure; for all the deductions are through these [figures]" (44a36–37).

Problems in the Posterior Analytics

Analogously, in *APo.* II.14–17, the focus of the discussion is on certain propositions as *objects of inquiry,* and in part on identifying these objects so that they can become the *explananda* of demonstrations. However, though there is this clear connection between these two texts, a decisive turn is taken in *APo.* II regarding problems which is not simply a consequence of their role in the 'framing stage' context. Problems are interpreted specifically as objects of *causal* inquiry (98a27–8, 31–4).[15] This step places problems firmly within the framework of scientific explanation – and historically, the why-question becomes, from this point on, the standard way of identifying a scientific problem. The *Problems* and *Mechanical Problems,* for example, introduce virtually every new topic by asking 'On account of what . . . ?'

In Alexander's comment on *Topics* 103b2–20, in fact, he argues that problems arising from questions about why X is Y or what the nature of X is do not seem to fit the characterization of 'problem' in the *Topics.* He then goes on:

So since the things stated this way at the start – 'why is this the case?' or 'what is this?' – are not ⟨dialectical⟩ premises, neither would problems ⟨stated this way⟩ be dialectical; still it is possible that the things proposed in this manner are natural problems, as has been said in the work *On Problems;* for natural things, the causes of which are unknown, are natural problems. To be sure, dialectical problems also arise concerning natural subjects, just as they do as well concerning ethical and logical ones; but the statements of the sort discussed earlier are *dialectical* problems, those which are of the sort just discussed are natural. And all the dialectical problems might be reducible to the 'that it is' and 'whether it is' inquiries, which are two of the four discussed at the beginning of the second book of the *Posterior Analytics;* for the 'why it is' and 'what it is' are not dialectical problems. (*in Top.* 63.9–19)[16]

Alexander is here claiming a relationship between the theory of problems in the *Topics* and the taxonomy of scientific inquiry which opens *APo.* II. He notes that the problems that fill the treatises *On Problems* – questions regarding 'why A is B' or 'what A is' – do not appear in *Topics* I. This cannot be due to subject matter: we are told later in *Topics* I that dialectical problems are either natural, ethical or logical (I.14 105a19–29), and many of the questions in *On Problems* concern natural phenomena. In Alexander's view, certain natural problems are dialectical – namely, those formulated as questions regarding 'whether p or not-p' – and certain dialectical problems can be natural – namely, those concerned with natural states of affairs.

Alexander then points out that all dialectical problems may thus be 're-duced' to the first two types of inquiry mentioned at *APo.* II.1, inquiries whether something is the case or not.[17]

What, concretely, does Alexander have in mind by this claim? Why, we might ask, cannot *any* problem have a place in either dialectic or natural science, depending solely on whether one's aim is to answer it 'according to the truth' or 'according to reputable opinion'? Or to put it slightly differently, why should the form of the question make a difference to whether a problem may be dialectical or not?

Alexander's answer to this question is not difficult to grasp. What is more, it is an answer that is grounded in Aristotle's account of dialectic, and thus may serve as a guide to answering our question about the origins of the scientific form of a problem, that is, as a why-question. The content of 'premises' and 'problems', according to the *Topics* doctrine, is the same – what distinguishes them is the nature of the question that gives rise to them, which in turn depends on the epistemic condition of the discussants regarding the question. As Alexander puts it:

> Dialectical problems are equal and the same ⟨in number⟩ as the premises; but these are questions regarding contradictory alternatives; . . . Hence just as 'What is man?' is not a dialectical interrogation, so neither is 'What is the nature of the prophetic waters?' a dialectical problem. (*in Top.* 63.5–9)

The claim that premises are questions regarding contradictory alternatives recalls the opening of the *Prior Analytics:*

> A demonstrative premise is different from a dialectical one, in that a demonstrative premise is an assumption of one or the other part of a contradiction (for the person who is demonstrating does not ask, but assumes), whereas a dialectical premise is a question regarding a contradictory alternative. (*APr.* I.1 24a22–25)

This passage concerns premises rather than problems, but in dialectic, as we have seen, this is a merely sociological distinction. Alexander is making an inference to what can count as a legitimate dialectical problem from the claim that dialectical premises are contradictory alternatives between which one is free to choose, and the correspondence between premises and problems. These introductory comments in *APr.* support Alexander's first premise. On the basis of this argument, he concludes that questions of the form 'What is X?' or 'Why does P belong to every S?' are of the wrong *form*. They do not permit one to answer with either contradictory alternative. They are questions which proceed from the already established truth of one half of a contradictory alternative.

Further, this conclusion is directly supported by remarks Aristotle himself makes in *Top.* VIII.2.

> But not every universal seems to be a dialectical premise, e.g. 'What is man?', or 'In how many ways is the good said?'. For a dialectical premise is that in response to which it is possible to answer yes or no; but this is not possible in response to these questions. (158a14–18)

The point concerns the logic of certain question/answer pairs: 'What is it?/ Why is it?' questions demand certain sorts of response – 'yes' and 'no' are not among them.

The results of our dialogue with Alexander may be represented schematically by distinguishing two levels of questions and answers:

Part I. Inquiry and Explanation

Level 1. *Common to Dialectical and Scientific Contexts*

Questions: Are there Ss (Ps) or not?   Is every S a P or not?

Answers: There are Ss (Ps).   Every S is a P.

Level 2. *Restricted to Scientific Contexts*

Questions: What is S (P)?   Why is every S a P?

Answers: S (P) is Ψ.   Every S is P because Φ.

On the interpretation provided by Alexander, dialectic provides the resources to move from the questions to the answers of Level 1. At Level 1, the differences between a dialectical and demonstrative problem are not logical, but epistemic. They are differences in the evidentiary basis of the answers. This seems to be the force of Alexander's statement that dialectical problems can be 'reduced to' the first two inquiries mentioned at the beginning of *APo.* II. And it suggests, perhaps, that the sense of *anagein eis* here is 'to raise or elevate to', rather than 'to reduce to'. Logically, the questions and answers of Level 1 look the same; but in the context of demonstrative inquiry, answers are evaluated according to a higher standard than in that of dialectic, that of causal demonstration, via 'middles'. Recall that *APo.* II.2 treats even 'whether it is' inquiries as disguised 'whether there is a middle' questions.

At Level 2, on the other hand, a new type of question emerges, one which assumes a choice between contradictory alternatives has been made. That is, Level 2 questions assume true answers to the questions of Level 1 have been established. The form of the question has changed, and affirmations and denials are no longer the right *types* of answer.

Still, is it not arbitrary to say such questions are not legitimate subjects of dialectical debate? To raise the question in a particularly compelling way; does not the *Topics* characterize certain dialectical problems as generated by questions concerning definition and kind, and will these not signify 'what it is' (*Top.* I.9 103b24–27)?

This worry misunderstands the interpretation being offered. That interpretation is based on the claim that what differs in the dialectical and demonstrative context is not the subject matter, but the form, of the problem. Definitional problems in dialectic start with a question about *whether* D is the definition of S *or not* – they do not start with, nor do they ever move to, the question, 'What is S?'

But then, if it is merely a matter of logical form, could we not re-frame such questions so as to fit a dialectical context? Aristotle says as much in *Top.* VIII.2. But in the process of changing the form of the question, you would have eliminated the causal/essential inquiries of Level 2. At the close of my discussion I want to raise the possibility that the exclusion of such inquiries from the dialectical context was intentional. It is better addressed after considering *APo.* II, on the connections between these two inquiries.

We noted that Alexander saw a relationship between the types of inquiry mentioned in the opening paragraph of *APo.* II and the problems of the *Topics.* Philoponus glosses the opening paragraph in a manner which shows that he too sees problems lurking in the background:

> The problems we inquire into are, Aristotle says, just as many as the ways in which we understand and know. . . . For just as the problem and the conclusion are in subject the same but in account different (e.g. 'The soul is immortal' is both a problem and a conclusion; for when it is proposed in an inquiry and we desire to demonstrate this, it is called a problem; but when it has been demonstrated, it is called a conclusion), thus also the things we seek and the things we know are the same in subject but different in account. (*in APo.* 336.4–11)

The word 'problem' does not occur, of course, in Aristotle's text. But Philoponus rightly sees it as the proper term to describe the objects of inquiry. Problems are, after all, that about which we reason. Philoponus nicely characterizes the difference between problems and conclusions in line with our understanding of the distinction in the *Prior Analytics.* It is not one of content, but of cognitive attitude – if a statement identifies an object of debate or inquiry, it is a problem; if the very same statement is the necessary consequence of a proof, it is a conclusion.[18] So in a general discussion of prologues in the *Rhetoric,* Aristotle refers to a problem as a proposition *proposed for the sake of* demonstration (*Rh.* III.13 1414a29–37).

The results of the discussion to this point can be reviewed by working through Philoponus' example. To view "The soul is immortal" as a dialectical problem is to assume a lack of agreement (and perhaps simple puzzlement, 104b12–17) over the question "Is the soul immortal or not (i.e. or mortal)?" This question could be elevated to an existential scientific inquiry, by moving it out of the realm of conflicting reputable opinions and into the realm of inductively grounded truth. But unqualified knowledge regarding whether the soul is immortal or not is secured only when we know the reason why, which involves another level of inquiry, concerned with such questions as "What is soul?" "What is immortality?" and "Why is the soul immortal?" Questions of this sort have no place in the dialectical context. Nor, since it cannot ground answers to its problems on anything other than reputable opinion, can dialectic by itself ever proceed to such questions. Thus there is a new question context into which the claim that the soul is immortal may be imbedded – the context of causal inquiry leading to understanding. But as we are about to see, this contextual shift will transform the meaning of Level 1 questions. Shift in context brings with it a shift in conceptual content.

The questions Aristotle uses as examples of first level inquiry are *formally* indistinguishable from the problems of the *Topics*.

> For when we seek whether this or that is the case, putting it in a number, for example *whether the sun is in eclipse or not,* we are seeking 'the that'. A sign of this is that upon discovering that it is in eclipse, we stop; and if we know from the start that it is in eclipse, we do not seek *whether.* But when we know 'the that', we seek 'the reason why'. (89b25–29; emphasis added)

But in *APo.* II.2, Aristotle begins to move in a direction which subtly interweaves the different inquiries at each level. For it turns out that when we inquire *whether S is P* we are inquiring *whether there is or is not a middle for S-P* (b37–38); and when we seek *the reason why S is P* we also seek *what the middle for S-P is* (89b39–90a7). Thus, even the question about *whether* a predication holds is a question about its causal base; and every inquiry *into* that causal base involves a search for the true nature of the cause.[19]

What then happens to the concept of a *problem* given the above move? The hypothesis that will be argued for here has two components.

(i) The extension of the concept is widened by Aristotle to include the objects of why-questions, themselves interwoven with 'What is it?' questions. The two-level picture of inquiry is maintained, but both levels become relentlessly causal. And this must have a significant impact on what will count as a scientific or demonstrable problem.

(ii) There is also a *restriction* on the notion of a problem which follows from the demonstrative context. In the *Topics* and *Prior Analytics,* problems (like premises and conclusions) could be universal or particular (cf. esp. *APr.* 47b9–14, *Top.* 108b37, 119a32–3, 120a1–4). But demonstration is of what holds always or for the most part, and necessarily. If problems are to be questions which identify the objects of scientific inquiry, then they must be universal affirmations and negations.

VI   *APo.* II.14–17: PROBLEMS IN DEMONSTRATIVE INQUIRY

APo. II.14–17 is in part concerned with the development of a research strategy which, borrowing from ancient geometric analysis, I will term 'problem reduction'. It is a strategy designed to aid in the eventual explanatory unification of a subject.

APo. II.14 argues that, in grasping problems one must select methodically from divisions.[20] Having done so, one will see that the same *logos* will explain that a property belongs to a number of different kinds because they are all forms of a more general kind to which the property belongs primitively. By proceeding in this way, a more fundamental problem – why the predicate belongs to this primitive kind – is identified. Whatever explains *that* will explain why it belongs to the forms of that kind.[21] For example, whatever explains why a lung belongs to every breather will explain why it belongs to every bird, horse or human, *qua* breather (98a4–11). The argument of this chapter goes on to extend these recommendations for 'grasping problems' to instances where the wider kind is unnamed (98a13–19), and even to cases where the different problems concern properties which are only analogous (98a20–23).

APo. *II.15: Two Reconstructions*

A natural question to pursue, given the conclusions of *APo.* II.14, is under what conditions are a series of particular problems appropriately thought of as 'the same problem'. Are analogous problems, such as 'Why do birds have bones?', 'Why do fish have fish-spine?' and 'Why do sepiae have sepions?' really the same problem, as that discussion suggests? This is the concern of II.15. (Parenthetically, it should be noted that the goal of identifying apparently distinct problems as explanatorily related is nearly always the reason for the use of the term 'problem' in Aristotle's 'scientific' works.)[22]

A common reading of this brief chapter, taken for granted by Philoponus, Ross, Tredennick and Barnes,[23] is, I am convinced, wrong. I will offer a minimally interpreted translation, and then defend an interpretation that is preferable to the common reading on both textual and philosophical grounds. (The paragraphs are numbered to facilitate comparison of the alternative interpretations.)

1. Problems are the same, some by virtue of having the same middle, for example because all are 'mutual replacement'. But some of these problems are the same in kind, as many as have differences by being of other things, or otherwise, as for instance 'due to what is it echoing?', or 'due to what is it mirroring?' and 'due to what is there a rainbow?'; for all these are the same problem in kind (for every one is a reflection), but are different problems in form.

2. But some of the problems differ by one middle being under the other middle, for example, 'Due to what does the Nile flow more when the month is

ending? Because there are more storms when the month is ending'. 'But due to what are there more storms when the month is ending? Because the moon is waning'. For these problems are related to one another in this manner. (*APo.* II.15 98a24–34)

The standard interpretation of this chapter sees the entire discussion as concerned with showing different ways in which apparently different problems are the same. It understands the numbered paragraphs above as follows.

1. Certain *problems* are the same because they have the same middle; some of these *middles* are the same in kind, but differ in form.[24]
2. Other *problems* are the same by having different middles related so that one is under another.

The reading I propose agrees that the uniting thread of the chapter is identifying cases where apparently distinct problems are related. But it disagrees on how these relationships between problems are to be understood. The paragraphs are thus to be interpreted as follows.

1. Certain problems are the same:
   (a) some because they have *the same middle,*
   (b) others because they are *the same problem* in kind, though different in form.
2. Certain problems, while related, differ from each other because the middle which explains problem A *is* problem B, which is explained by another middle.

The latter interpretation is preferable on philosophical grounds. This can be seen by looking at the examples in each of the two paragraphs.

On the reading just proposed, the first paragraph distinguishes two ways in which distinct problems may be the same. The first is not expanded upon, but the example suggests problems which are not simply *generically* the same, but the same without qualification.[25]

The traditional reading of the second case takes the parenthetical clause to tell us what the identical *middle* is – reflection is the generic name for distinct, specifically different, *causes*. On the recommended alternative, it refers to the generic *problem,* echoing, mirroring and rainbows being all forms of reflection. All the problems concern reflection, but they are differentiated nonetheless.

This difference, over how one should take the reflection example, is of no great moment, however. Problem identity in this case is of the sort discussed in the previous chapter: Certain things will be true of rainbows, mirror reflections and echoes, *because* they are all (forms of) reflection. Thus, modelled on II.14, 'reflection' *is* the (preliminary) middle relative to its sub-kinds, though this is simply a way of saying that the account of reflection (whatever it turns out to be) will also apply to them, in so far as they are all forms of reflection.

The enigmatic nature of the examples renders the choice between the alternative interpretations more difficult than it might otherwise have been. Aristotle states them by means of two quasi-impersonal verbs referring to processes, and one noun subject without a verb. For these examples to be of any value, they need to be interpreted. I shall expand them along the lines of Aristotle himself, in expanding on the principal examples of II.1–10, "why is it thundering?" and "why is it eclipsing?" which for the sake of demonstrative form convert to "why is there a certain noise in the clouds?" and "why is the moon suffering eclipse?" Such expansion can be done in ways favoring either interpretation. At this point it is only important that it be shown that an interpretive expansion consistent with the reading being defended here is possible. Below I give the expanded versions, followed by the unexpanded.

Problem 1. Why does echoing belong to (certain) sounds?
(Why is there echoing?)
Problem 2. Why do reflected images belong to (certain) surfaces?
(Why is there mirroring?)
Problem 3. Why do multicolored arcs (sometimes) appear in the sky?
(Why are there rainbows?)
Generic Problem – What is the cause of reflection?
(Why is there reflection?)

Evidence in support of this expansion can be found in the discussion of four related natural reflective phenomena discussed in *Meteorology* III.2–6. Of those mentioned in *APo.* II.15, only the rainbow is discussed. Furthermore, all the examples involve 'atmospheric' phenomena and are thus prima facie more alike that those mentioned there. Nevertheless, the way Aristotle approaches the relationship among them suggests that reflection is the name for the generic *explanandum,* not, as the standard interpretation claims, for the fundamental cause.[26]

Mete. III.2–6 forms a continuous discussion, which begins with the following announcement:

We must now discuss haloes, rainbows, mock suns and rods, what each is and due to what cause each comes about; and indeed all these come about due to the same causes as one another. But first we must grasp the affections and the attributes about each of them. (371b18–22)

The next few paragraphs do lay out "the attributes and affections" for all four phenomena, at the conclusion of which Aristotle makes a transition to the discussion of their causal explanation.

> So then, these are the attributes regarding each of these; but the cause of all of them is the same; for every one of these is a reflection. However they differ in their manner ⟨of reflection⟩ and from which things ⟨there is reflection⟩, and in as much as it turns out that the reflection comes about relative to the sun or to some other of the sources of light. (372a17–21)

It will be noted that the first sentence parallels with some precision the reflection example in *APo.* II.15. That chapter's enigmatic "of other things, or otherwise" can be taken to refer to differences in the origin and the manner of reflection, as here. For example, we are told that we must take as proven "from the proofs regarding the optical ray . . . that the optical ray is reflected just as much *from air and all smooth surfaces as from water* and the reason why *some reflectors reflect shapes while others only colors*" (372a30–34; emphasis added).

What lessons can we take away from the *Meteorology* discussion for the examples of problems alike in kind in *APo.* II.15? First of all, we gain insight into what Aristotle means by saying that they are all due to the same cause. Identifying them all as *reflective* phenomena means that all involve some relationship between a source of light, sight and a reflective body. That, in turn, means that geometrical optics is available as an explanatory device.[27] But most of the explanations in these chapters are concerned with showing how the *different* attributes associated with each *distinct* form of reflection flow from specific forms of optical causation. What, then, does the claim that they are all due to the same causes amount to?

The identification of the generic sameness of these problems makes possible the *unification through explanation* of a number of formally different phenomena. Aristotle is stressing the formal differences among these generically identical problems. There may be a common explanation of reflection, but its application (to return to the *APo.* II.15 examples) to sound and to visual phenomena will introduce qualitatively different processes.

Let us now turn to the example used to exemplify the third class of related problems – the periodic increase in the flow of the Nile, which became a classic problem generating a rich and continuous set of solutions throughout the Hellenistic period. Is Aristotle claiming that we have here another sort of case in which apparently different problems are in fact the same, as the interpretation of Philoponus and Ross suggests? I think not. It will help to begin with a simple reconstruction of the example in 'problematic' format.

> *Problem I*
> Why does increased flow occur at month's end?
> **Because storms occur at month's end.**
>
> *Problem II*
> **Why do storms occur at month's end?**
> Because waning of the moon occurs at month's end.[28]

Cast in this problematic format, the example is reasonably clear. Nile flow increases near the end of the month because storms increase then. You have provided a causal middle linking increased Nile flow and the end of the month. But you have also created a new problem – why do storms increase at month's end? The causal middle for this problem should be more fundamental. And indeed, the claim that the moon wanes at month's end is both a definitional identification – that is what it is to be month's end – and the more fundamental cause of increased Nile flow at that time, since it is the cause of the increased storm activity. We have, in this example, moved from a preliminary explanation of a problem to a more fundamental explanation, which simultaneously provides a definition of the preliminary explanatory middle term.[29]

In the above pattern, the statement of the content of a *problem* is followed by a statement of its causal explanation. This is a common explanatory format in scientific works such as the *Parts of Animals* and the *Meteorology.* Since they combine the requisite three terms, such problem/solution pairs can quite easily be reconfigured as syllogisms, but this is not an exercise Aristotle often indulged in.[30] Second, it is also the formulaic pattern of the books of *Problems* that have come down to us – in these, however, a number of *tentative* solutions are suggested as hypotheses. Third, it shows how problematic inquiry works *toward* first principles – in later terminology, is an act of *analysis,* rather than *synthesis.*

Chapter 15, then, provides us with two quite different ways of getting back to more basic scientific (i.e. demonstrable) problems. The first is a matter of identifying the generic problem common to a number of distinct cases (grasping something of what it is, in the language of *APo.* II.8); the second leads the investigator from a less to a more fundamental causal investigation of the same problem. Identifying more basic problems for pursuit thus seems to be one aim of this brief chapter. Another is to uncover the different ways in which *explanatory unification*[31] can be achieved, a fundamental goal for the model of science of the *Analytics.*

Recognition that such distinctive phenomena as echoes, mirror images and rainbows are 'of a kind' can lead to a variety of theoretical advances. The discovery of a plausible explanation of one of these suggests the possibility of an analogical transfer of that explanation, perhaps in a modified form, to the others. Alternatively, one might determine that certain properties, at first assumed to belong primitively to one or the other phenomenon, actually belong primitively to the kind. Inquiry would thus be led to focus on the more abstract level.

In either case, inquiry into problems is, in the demonstrative context, inquiry into the *reason why* the problematic predication holds, and at the core of such inquiries is *problem reduction* of various kinds. Such problems are beyond the scope of dialectic, since they are fully embedded within the context of a search for causes.

Problems and Causes: APo. II.16, 17

Chapters 16–18 are structured around a basic *aporia:* must causes and effects be convertible? The question is a natural one to ask, given the process of problem reduction now in place – for that process seems designed to generate explanations in which the 'explanatory middle' extends more widely than its *explananda.* The *aporia* is followed by a number of objections, the answers to which further clarify Aristotle's concept of a scientific problem.

Objection one: If cause and consquent were convertible, could not either one be proved through the other (and thus how could one distinguish cause and consequent)?

Answer (already provided in *APo.* I.13): Only one of the convertible items provides the reason why the other belongs to its subject. Predication relations may convert; causal relations do not.

Objection two: But could there not be a situation where some attribute A belongs to two different subjects D and E, yet B causes the A-D connection while C causes the A-E connection? Thus the same attribute is produced in different subjects by a different cause. There must be some cause of these predications, but it could be a distinct cause in each case.

Answer (in Aristotle's own words):

> Rather, if *the problem* is always universal, is not the cause also some whole and that of which it is the cause universal? Leaf loss, being delimited to a certain

whole, even if there are forms of it, is determined to these universally, either to plants or to such and such plants; so that the middle as well must be equal to these and that of which it is the cause, and must convert. (98b32–37)

How strongly are we to take the claim that problems are "always universal"? Ross, without argument, gives it the strongest possible reading. "A problem, i.e. a proposition such as science seeks to establish, is always universal, in the sense explained in I.4, viz. that the predicate is true of the subject *kata pantos, kath'auto,* and *hi auto* . . ."[32] We have seen that this is certainly not generally the case for problems. Ross is drawn to this reading of 'universal' here because it provides a natural interpretation for the final claim Aristotle makes about problems in this discussion: In demonstrations in accordance with signs or accidents, in which the same attribute can belong to different subjects due to a different cause, the objects of investigation do not seem to be problems (99a5–7). One reason this might be true is that cause and consequent are not convertible in sign relations. But this does *not* imply that genuine problems *must* be convertible. For the problem is simply the potential *explanandum.* Aristotle certainly thinks that inquiry should aim toward convertible problems, but that does not commit him to the stronger thesis that no non-convertible proposition is a genuine problem.

Neither of these passages, then, demands a stronger reading of 'universal' than that embodied in the earlier restriction imposed by the demonstrative context; namely, that problems for demonstration must be universal affirmations or negations. Nothing here implies that they *must* also be fully convertible propositions.

A more modest interpretation is suggested by the example Aristotle uses. It follows the problem format noted above precisely, beginning with a statement of the universal problem:

Due to what do the trees lose their leaves? If it is due to coagulation of moisture, then, if a tree loses its leaves, coagulation must belong, and if coagulation belongs, not to anything but to a tree, it must lose its leaves. (98b37–39)

Clearly the only conversion Aristotle is concerned with here is that between coagulation and leaf loss. Prior to this, he makes the point that coagulation must be equal in *extent* to the various *forms* of things which lose their leaves, but that has no implication for the convertibility of the problem one begins with.

Notwithstanding the fact that these chapters are clearly discussing ways of *reducing* problems to those in which subject and predicate *are* convertible, there is no evidence that Aristotle would want to restrict the concept to such predications. And neither the evidence from Aristotle's own scientific works, nor from the extant texts of problems, suggests such a restriction. This is, after all, as it should be. Problems represent the objects of demonstrative investigation, *not* the conclusions of demonstrations.

This brings us back to a question raised at the close of the discussion of the *Topics:* Does Aristotle have general philosophical reasons for formulating problematic questions in that work so as to exclude the inquiries central to demonstrative science, the causal inquiries referred to as problems in *Posterior Analytics* II? As a way of providing an answer, let me review the two aspects of dialectic that seem most akin to demonstrative inquiry.

First among these are problems formulated as questions about whether some state of affairs is or is not the case. Even here, as we saw, the similarity is only skin deep. Since in *APo.* II.2, Aristotle reconfigures 'whether it is' inquiries as attempts to determine whether there is a causal middle relative to some predication, even these questions, which Alexander sees as the common property of dialectic and demonstration, are only superficially so. The various means described in the *Topics* for testing problematic responses do not include the steps that take one to the point where one grasps that there is a middle, though not what the middle is. The studied avoidance of certain sorts of problematic questions in the *Topics* reflects Aristotle's views on the limitations of dialectic, and on the differences between the demonstrative and dialectical contexts.

Another case of apparent overlap between dialectical and demonstrative problems would appear to be definitional inquiries. Since dialectic involves the formulation and testing of definitions (cf. *Topics* I, VI, VII), and since scientific definitions are true answers to 'what is it?' questions, why shouldn't these dialectical problems be formulated as 'what is it?' questions, the answers to which could then be starting points of demonstrations? This question has too many aspects to be fully answered here, but the outlines of an answer can be provided as follows.

First, the dialectical context precludes direct 'what is it?' inquiry. Within dialectic, one begins with a proposed definition of something. If a *problem* is to be formulated regarding it, it must be a definition about which there is not common agreement. The problem, then, is the question of whether the purported definition is actually the definition of the subject or not. The dialectician, then, faced with the assertion that a statement S counts as a definition of something X, asks "Is S the definition of X or not?" There are a rich set of dialectical procedures for testing the adequacy of such definitions, but nothing is said about the procedures to be followed in moving empirically toward the formulation of such a definition. The natural scientist, on the other hand, asks "What is man?" about the objects so designated, and follows a rich set of empirical practices in moving toward formulating an answer. Dialectical problems could certainly be formulated about proposed answers to these questions, and serve as the starting points of debate. But how the definition was arrived at, and what its empirical status is, are not questions that arise in formulating such problems.

Second, the *APo.* II account of demonstrative definitions ties them so intimately to fundamental answers to scientific problems – i.e. to causal why questions – that the inability of dialectic to formulate and answer inquiries into causes precludes it from establishing scientifically fundamental definitions.[33]

VII   ARISTOTLE AND THE MATHEMATICIANS

An alternative to the above explanation for the differences in the form of a problem in the *Topics* and the *Posterior Analytics* would be one which claimed the differences reflect a development in Aristotle's thinking about problems.

One reason for doubting this developmental explanation for the differences in dialectical and scientific problem sets is that the problem form discussed in *APo.* II is already mentioned in Plato. It is clear from a reference in the *Republic* that there was some notion of a problem already playing a role in pre-Euclidian mathematics. In *Republic* VI, Socrates suggests that astronomy and harmonics should be pursued in a manner akin to geometry or arithmetic, ascending to *problems* – investigating such things as which numbers are consonant and which not *and why* (531c1–4; cf. 530b6–c1). The problems being investigated here include both whether and why questions, and the context is scientific inquiry.

Our primary sources for the history of geometry – especially Pappus and Proclus – include extensive discussions of disagreements among earlier geometers regarding how to distinguish theorems and problems.[34] Aristotle provides the only extensive discussion of the concept which is close to the source of this debate – a frustrating prospect, of course, given his notorious lack of concern for the needs of twentieth century historians of ideas. Nevertheless, by looking now at the use of specifically mathematical examples in the 'problematic' chapters of *APo.,* light may be shed on a key concept in ancient mathematics.

In *APo.* II.17, Aristotle returns to a number of geometric examples which were central to his presentation of ideal demonstrative understanding in *APo.* I. But here they are used as examples of *problem reduction.*

For example, he argues that problems may be similar in kind, analogous or homonymous. The causal middles needed to answer these problems, he claims, will be related to each other in the same manner as their problems are.

> For example, why do proportionals alternate? For the cause in the case of lines and of numbers is different and yet the same, *qua* line ⟨the cause in the case of lines is⟩ different, but *qua* having such an increase ⟨the cause is⟩ the same. Thus it is in all ⟨proportional⟩.

> But in the case of color being like color and figure being like figure ⟨the middle⟩ is different in different cases. For the term 'like' is a homonym in these cases; for in the latter perhaps it is having proportionate lines and equal angles,[35] while in the case of colors it is that there is a single perception or some other such. And the problems analogously the same will have the middle analogously ⟨the same⟩. (99a8–16)

The first example here was used at *APo.* I.5 74a17–25 to note the way in which a science can progress from incidental to unqualified understanding of a subject, from a number of partial proofs to a single universal one. The point there stressed was that alternating proportionality is in fact a property primarily of magnitude, and only consequently of lines, planes and so on. In this later discussion, the example is used to indicate conditions under which different problems can be treated as forms of a single, generic one.[36]

This is quite similar to Proclus' description of the 'reduction' of the problem of doubling the cube to that of finding two mean proportionals, which he attributes to Hippocrates of Chios.[37]

> Reduction is the transfer from one problem or theorem to another, from the knowledge or construction of which the one set before us will also be manifest; as for example when those inquiring into the doubling of the cube transferred the inquiry to another, which this follows, the discovery of the two means, and henceforward sought how, given two straight lines, two mean proportionals might be discovered. (Proclus, *in Eucl.* [Friedlein 212.24–213.11])

The distinction between problem and theorem here is likely a matter of whether construction is involved in the solution or not, and is probably not found prior to Speusippus.[38] But it is clear that the context is inquiry aimed at discovering how a certain figure is produced. The reduction of the problem is *not* its solution, but simply the realization that its solution follows directly from the solution of a more basic problem.

Sometimes, however, it is important to recognize that problems are different. Pappus (*Collection* IV 36.57–59) distinguishes plane, solid and linear problems, in the context of showing that "the ancients" were puzzled about how to solve the problem of the trisection of a given rectilineal angle, because they attempted to do so by means of planes "when it is by nature a solid problem; for the sectioning of cones was not yet familiar to them." Here the problem continues to be the *object of inquiry.* Further, as in Aristotle's examples, advance on a particular problem involved finding out the *kind* of problem it was, for this was crucial to identifying the appropriate starting points for its solution.

These are just the sorts of norms of inquiry which Aristotle is developing in *APo.* II.14–18. I can imagine three possible reasons for this general similarity. First, the mathematicians of Aristotle's day might have been pursuing problems in the ways Proclus and Pappus describe, and Aristotle was developing some of his thoughts regarding inquiry by reflection on their research. Second, Proclus and Pappus may have reconstructed the history of mathematics in accordance with the model of the *Posterior Analytics.*[39] Third, whatever the geometers before Aristotle were doing, those who come after him describe geometric activity in ways that owe much to the *Analytics,* and Proclus and Pappus base their accounts primarily on these post-Aristotelian sources.

There is evidence which supports each of these three statements, but very little evidence independent of these late sources on the character of Greek mathematics before Aristotle. However, one can derive a minimal description of early geometric activity from the evidence which suggests the following.

1. The concept of a problem was used to identify an object of inquiry, rather than an object of a proof.
2. The concept of a problem was restricted to descriptions of the analytic, rather than the synthetic, procedures of the geometer.
3. Often, problems were stated in the form of questions posing alternatives.
4. There is, throughout the problem tradition, a focus on the reduction of related, particular problems to a single, more basic one.[40]

This is sufficient to justify consideration of the possibility that Aristotle, in his discussion of inquiry no less than of demonstration, was attempting to develop a general program for scientific inquiry which was modelled on the geometric analysis of problems.[41]

NOTES

1. Cf. Bonitz 1870, 103b17–30.
2. Cf. pp. 847–968 of I. Bekker's edition of 1831.
3. Discussed in section 6, below.
4. Cf. Ackrill 1963, 79.
5. Without becoming embroiled in the controversy over the meaning of *sullogismos* in the *Topics*, something must be done to preserve the division made in I.12, 105a10–13, of dialectical argument into 'syllogism' and 'induction'. The discussion in I.12 refers back to this passage for an account of 'syllogism', and thus I take it the distinction applies here as well.
6. Compare Plato *Philebus* 65d4–7: Socrates: "Well then after this examine in the same manner 'measuredness', whether pleasure possesses more of it than prudence, or prudence more than pleasure? Protarchus: The inquiry you have posed as a question is again an easy one . . ."
7. Contrast the problems (= theses) of 104b18–28 with the problems (= questions) of 105b19–29; similarly the premise (= question) of 104a8 with the *selected* premise (= assertion) of 105a34–b18.
8. This might seem to be suggested by Robin Smith's claim that problems are 'categorical sentences types', though he also says that in the *Topics* there is an implication that "a problem is a two-sided question" (Smith 1989, 148).
9. Indeed, in *APo.* II.5, Aristotle criticizes the Platonic use of division on grounds that proper demonstrative conclusions should not depend on the respondent conceding to the questioner, but must follow necessarily even if the respondent rejects it.
10. Cf. *Topics* III.6 120a32–38.
11. See the arguments against division as a method of proof in *APr.* I.31 and *APo.* II.5.
12. To be precise, there are 19 occurrences in *APr.*, all but one of which occur after 42b; there are seven occurrences in *APo.*, six of which occur between 98a and 99a. The other (at 88a12) is a reference to 'the Problems'.
13. Cf. Ferejohn 1982/3, ch. 1, and Ferejohn 1982, 375–395.
14. This interpretation finds support in Philoponus, *in APo.* 336.4–11.
15. Barnes (1975a, 239–241; 1993, 250–251) simply imports the account of a problem from *Topics* I into his discussion of *APo.* II.14–15, rendering 'why' questions as "Is S P or not?." W. D. Ross (1949, 662–25), following the Greek commentators and Zabarella closely, is sensitive to the difference.
16. Part of the passage is given as fragment 112 from the lost work *On Problems* in V. Rose 1886, 106. Rose quotes 62.30–63.4 and 63.11–19, omitting 63.5–10 without explanation. Ross follows suit in the Oxford translation (Ross 1952, 104).
17. Alexander is referring to the initial distinction between seeking what something is and whether it is. But Aristotle quickly qualifies this distinction in two respects. Both inquiries are seeking whether there is or is not a *middle* (89b38), and both can be construed as 'whether it is' inquiries, of a qualified (predicative) or an un-qualified (existential) nature (89b37–39, 90a2–6). Thus in a demonstrative con-

text even these 'pre-demonstrative' inquiries have implicitly causal/definitional intent.

18. Smith (1989, 114, 26b26–33n; 148, 42b27n) claims that 'problem' in *APr.* invariably has the sense 'type of categorical sentence'. The argument of this chapter is that, while problems in dialectical, deductive and demonstrative contexts all 'involve' a 'categorical sentence', that sentence will be a *problem* only so long as it is the subject of debate or inquiry. They will thus be either questions or answers, but never conclusions, though the same categorical statement might well be. This is also the way Philoponus reads Aristotle.

19. For different accounts of inquiries aimed at knowing 'that there is a cause', cf. Bolton 1987, 133–140; Charles 1990, 148–154; Lennox 1990, 169–175. The view I defend is this: 'knowing that there is a middle' is having reliable information that the predication being investigated is non-incidental. The inquiry aimed at such 'knowing' involves organizing information about a domain in divisions, and selecting terms from those divisions out of which premises for a problem may be constructed. This method is described formally in *APr.* I.27–30, and as a preliminary to demonstration in *APo.* II.14. Thus it will include the investigation of problems. Cf. Ferejohn 1990, 9–12, 119–123; chapter 2 above.

20. Chapter 1 above; Ferejohn 1990, 29–30.

21. This interpretation of the 'problem grasping' methodology of *APo.* II.14 is a slight modification of that presented in chapter 2 above. The deductions in that chapter, termed 'A-type explanations', use the primitive kind-term as their middle term. I now suppose that in such cases, the kind-term is a mere place holder for whatever is the cause of the predicate belonging primitively to that kind. In footnote 16 in chapter 2 I note that Allan Gotthelf had suggested this reading, but at the time I wrote that essay I was not convinced. The argument in Charles 1990, 151–152 has not convinced me to reject another aspect of my interpretation of *APo.* II.14, however – namely, that the 'selection from divisions' described is a means of fixing the predications for which causes are to be sought, preliminary to grasping what the middle is. Further, while I do not deny that questions about non-commensurate universal predications can be problems, one aim of II.14–17, as I see it, is to specify conditions under which such non-commensurate universal predications are to be treated as instances of a more general, commensurate universal predication – thus identifying the predications for which *unqualified* demonstration is possible.

22. Cf. *GA* 724a7, 746a16; *Probl.* 874a34, 885b4.

23. Barnes' *translation,* however, recommends the interpretation I favor, as will be seen.

24. Ross (1949, 665) goes so far as to say that this will be true of all cases where the middle is the same, but gives no reason for this.

25. The example could be interpreted in a variety of ways. Ross (1949, 664–65) notes Aristotle's use of 'mutual displacement' in a number of very different explanatory contexts, which has the effect of erasing the difference between the first and second cases. On any interpretation, however, these should be contrasted, so this interpretation seems wrong.

26. The following is not meant as a serious attempt to forge links between the two discussions, such as is found in Freeland 1990, 287–320.

27. The reference to the science of the rainbow as subordinate to geometric optics in *APo.* I.13, the remarkable fit of methodology between *APo.* II.15–18 and *Mete.* III.2–6, and the latter discussion's general agreement with the philosophy of mathematics we find in *Ph.* II.2, *APo.* and *Metaphysics* M lead me respectfully to dissent from Wilbur Knorr's argument (Knorr 1986, 102–107) that *Mete.* III.2–6 are not by Aristotle. (Cf. Lennox 1985, 29–53.) This does not imply, however, that Aristotle was responsible for the geometric analysis in the *Meteorology,* nor even for the recognition of its applicability to the rainbow. That he adopts openly extromissionist descriptions of the 'optical ray' here is a general puzzle about geometric optics – Aristotle is not alone among intromissionists in adopting this mode of description when geometry is in use.
28. Barnes' brief synopsis (1975a, 241; 1993, 251) is: "Suppose two problems, *Is S P or not?* and *Is S′ P′ or not?* These may be connected . . . if P = P′ and S′ is the middle term in the proof of S is P." This summary cannot be derived from the text.
29. Note the similarity of the pattern of inquiry here to that outlined in *APo.* II.8, on which see Bolton 1987, 137.
30. And of course *APr.* I.35 tells us it will seldom be the case in actual science that such reconfiguration will be of value.
31. I am using this term in precisely the way that Philip Kitcher does in his essay by the same name (see Kitcher 1981, 507–531). This is achieved by the application of the same explanatory pattern to a wide variety of problems. Its modern historical roots are in Newton's method for establishing a *vera causa.* Cf. Herschel 1830, ch. 6.
32. Ross 1949, 667 and 98b32–8n.
33. Bolton 1993 has an illuminating discussion of this point.
34. It is worth noting that Aristotle's formal definition of a 'dialectical problem' is "a *theorem* aiming toward choice and avoidance or toward truth and knowledge . . ." (*Topics* 104b1–2). 'Theorem' is thus the genus of such problems. But Aristotle makes no further use of this distinction. This allows us to extend Alan Bowen's claim about its origins – not only is it post-Platonic; it is apparently post-Aristotelian.
35. Euclid, *Elements* VI, Def. 1.
36. Presented in Euclid, *Elements* V; Def. 12 defines alternation. The general theory of proportion is characterized by Aristotle as a recent discovery, and is usually attributed to Eudoxus.
37. Cf. Euclid II.14 and VI.13; Theon 2.3–12; the various solutions described by Eutocius (in Heiberg 1910–15, vol. III, 54.26–58.14, 66.8–70.5, 78.13–80.24, 84.12–90.13); and Pappus (in Hultsch 1876–78, 62.26–64, 242.13–250.25).
38. As Bowen 1983, 12–29, argues. This paper has been of enormous help in the development of this section of the discussion, though Bowen's argument focuses on the context of proof rather than discovery. At this point I am agnostic about his claim that 'problem' has two quite different meanings in these two contexts, though Aristotle provides no support for it.
39. An alternative taken very seriously in Knorr 1986, 356–357.
40. On all points, cf. the discussion in Knorr 1986, 15–48, 348–370; all four points are made explicit in Proclus' commentary on the first book of the *Elements* (Friedlein 79.11–81.4).

41. Many thanks to Alan Bowen, Allan Gotthelf and Mary Louise Gill, for fruitful discussions of a preliminary draft of this chapter. Many thanks as well to my fellow participants in the conference *Logic, Dialectic, and Science in Aristotle* at Kansas State University, where my original paper was first aired, especially to Robert Bolton, Michael Ferejohn, Cynthia Freeland and Robin Smith. This chapter owes much both to their own contributions and to their questions concerning mine.

4

Putting Philosophy of Science to the Test:
The Case of Aristotle's Biology

I INTRODUCTION

Aristotle's *De Partibus Animalium* consists, at the broadest level, of four books. The first is devoted to articulating (I'm quoting its introduction) " . . . standards, by reference to which one will judge the manner of the demonstrations ⟨of natural inquiry⟩, apart from the question of how the truth has it, whether thus or otherwise" (639a12–15). Books II–IV, on the other hand, are introduced as attempts to provide causal explanations for the facts regarding the parts that belong to the various kinds of animals, facts systematically organized in the *Historia Animalium* (646a8–12). This means that Aristotle's *De Partibus* consists of an introductory book on the *philosophy of* biological science, and three books *of* biological science.

Such an arrangement provides the student of this great work an opportunity to explore one of the perennial issues in the history and philosophy of science: the connection between a scientist's theory of science, and his actual scientific practice. This issue has been a center of controversy regarding (among others) Galileo, Newton, Lyell, Darwin, and Einstein. So has it been with Aristotle.

II THE ISSUE IN ITS ARISTOTELIAN CONTEXT

Aristotle was the first Greek thinker to articulate self-consciously a taxonomy of intellectual pursuits. At the widest level, he distinguishes theoretical, practical and productive areas of knowledge, based on their fundamental aims and subject matter. Theoretical knowledge is sub-divided into mathematics, second philosophy (the study of natural objects) and first philosophy (the study of being *qua* being). In addition, the four books of *Analytics* present a theory of scientific knowledge, a remarkably rigorous and systematic account of what a body of propositions must be like in order to count as a theoretical science.

That theory, notoriously, seems to put insuperable barriers in the way of there being a natural science that could live up to its ideals. At best, it appears to be enigmatic on how its prescriptions would apply to a natural science (as opposed to mathematics) – it never mentions a distinction between matter and form (and therefore never raises the question of whether a proper definition of a natural object or its parts should include reference to its material nature), and never mentions conditional necessity, even in the brief discussion of natural processes occurring both for an end and of necessity (*APo.* II.11 94b27–95a9). Yet many of the examples in the *Posterior Analytics* are drawn from natural science (especially meteorology, botany and zoology), and are discussed side by side with mathematical examples – as if a rigorously formal demonstrative science of nature posed no special problems. Aristotle's biological practice, on the other hand, presents us with a relentlessly theoretical explanation of why animals have the parts they have, and develop and behave as they do. Yet, according to some, it looks nothing like the prescriptions for proper theoretical science in the *Analytics.* Where are the axioms, definitions, theorems, and proofs, they say? Why does it not look like Euclid's *Elements?* To quote one proponent of this view, G. E. R. Lloyd:

> It is not just that actual explanations set out in syllogistic form are difficult to find: the whole discourse of the practising natural scientist resists, one might say, being recast in the mould of the ideal formal language that the Organon desiderates. (Lloyd 1991, 394)

The existence of *PA* I offers an obvious approach to this issue. Suppose it does, as the *Posterior Analytics* does not, detail what a theoretical science of natural objects should be, but does so in a way that *specifies* and *builds on* the *Analytics* ideal, rather than abandoning it? There would be a philosophy of biology based on the *Analytics* philosophy of science. We may then ask whether the explanations of *PA* II–IV mirror the philosophy of biology provided in *PA* I.

The work I have done over the last six years has convinced me that *PA* I was intentionally written to answer the question of how the *Analytics* model of science is to apply to Aristotle's paradigm natural substances, animals; and that *PA* II–IV carries out the program of *PA* I (cf. chapters 1 and 2). In the sections that follow I will provide an outline of the arguments of *PA* I.1–5 viewed from this perspective, and provide a single, extended example of zoological explanation from *PA* III that shows Aristotle practicing what he preaches.

III   ARISTOTLE'S PHILOSOPHY OF BIOLOGY:
PARTS OF ANIMALS

I begin with a brief outline of *PA* I's five chapters.

Chapter 1. Biology's Explanatory Principles. Chapter 1 begins with a series of questions, all but the first of which being answered as they are raised. There is a gradual transition to a statement of principles and methods based on the results achieved by answering the initial questions. The following is an analytical outline of chapter 1:

(i) **639a16–b5.** At what level of generality should our investigation be organized, i.e. should we study the nature of each species independently, or should we focus first on general attributes and then on more specific ones as necessary?

(ii) **639b5–10.** Should the biologist follow the lead of astronomy, first establishing the facts of the domain, and only then studying the reason why and the causes of these facts?

(iii) **639b10–21.** Since natural processes are subject to both motive causation and goal causation, which should take priority in our study?

(iv) **639b22–640a9.** Do we find both unqualified and conditional necessity governing natural processes, and if so, what is the nature of conditional necessity, and of the demonstrations appealing to it?

(v) **640a 11–33.** Since animals come to be, should we attempt to understand the fully developed animal by reference to processes leading up to it (as most of our predecessors have), or vice versa?

(vi) **640a33–641a14.** Should we study only the material constituents of animal bodies, or their parts and especially the functional capacities specific to each of their parts?

(vii) **641a14–641b10.** Since studying an animal functionally is in effect to study the soul, should biologists study all aspects of the soul, or should reason be excluded?

(viii) **641b10–642a1.** Biologists cannot study their subject in abstraction from matter, since nature always acts for the sake of an end, which involves studying the relation of what is potentially something to its full realization.

At this point, a gathering up of the results is clearly apparent.

(ix) **642a1–24.** There are two sorts of causation, teleological and necessitarian, and they are related through the concept of conditional necessity. This is part of what prior natural philosophers lacked in their investigation of nature.

(x) **642a24–31.** The other thing they lacked was a clear notion of essence and definition. Socrates provided this, but in a moral and political context rather than in the study of nature.

(xi) **642a31–b4.** Finally, a closing passage provides a sketch demonstration explaining breathing, in which both sorts of necessity are operative.

The entire chapter shows an overarching concern with the issue of how to integrate the account of natural substance in terms of matter and form, and teleology and necessity, with the ideals of a demonstrative science laid out in the *Posterior Analytics.*

For example, listen to the text referred to in (ii) above:

> . . . ⟨should⟩ the natural scientist first study the appearances regarding animals and their parts, and only then state the reason why and the causes, just as the mathematicians demonstrate the astronomical appearances, or should he proceed in some other way? (639b5–10)

The question here is methodological: whether investigation should first establish the explananda of a domain – the 'appearances' – before articulating their causal explanation. It is answered in the affirmative at 640a13–15 (and see *HA* I.6 491a7–14, *PA* II.1 646a8–12, *IA* 1. 704b6–11 for the principle stated in biological practice). Aristotle here claims mathematical astronomy as his model for this methodology. Elsewhere, in the *Prior Analytics,* we find a detailed articulation of this position.

> Thus the principles are provided by experience in each case. I mean, for example, astronomical experience provides the principles of astronomical knowledge; for when the appearances had been grasped sufficiently, astronomical demonstrations were easily discovered. *And it is likewise with any other art and science.* So that if the predicates about each thing have been grasped, we will be well-prepared to exhibit their demonstrations. (46a20–24)

The *Posterior Analytics* adds that a demonstration involves knowing the cause(s) of the fact, and using that knowledge to construct proofs in which the cause is identified by the middle term of the proof, the term which is subject of one premise and predicate of the other (cf. 89b24–31, 90a5–7).

But only once, and only briefly, does the *Analytics* acknowledge that natural objects come to be both for the sake of something and from necessity

(94b27–95a9), and it doesn't respond at all to the obvious question: what would a demonstration look like in that case? Again, in *PA* I Aristotle faces this question directly, and forcefully.

> But the mode of demonstration and of necessity in the natural sciences is different from that in the other theoretical sciences. This has been discussed elsewhere. In the other sciences, the starting point is what is, in the natural sciences what is to be; for ⟨one would say⟩ 'since health or man is such and such, this must be or come to be', rather than 'since this is or has come to be, that from necessity is or will be'. (640a1–6)

The "other discussion" referred to here is almost certainly *Physics* II.9, in which such demonstrations are shown to be formally analogous to those in mathematics, but in content different in just the way sketched above (cf. Gotthelf 1987b, 197–198).

PA I.1 and *Physics* II.8–9 refine both the concepts of 'that for the sake of which' and of 'necessity' so that they become, through the notion of *conditional* (or hypothetical) necessity, intimately related explanatory tools for organic investigation. It seems clear from *APo.* II.11 (94b26–95a9) that Aristotle had not yet formulated, in the *Analytics,* the concept of conditional necessity – perhaps because the matter/form analysis, upon which it depends, was also not yet clearly formulated. The concept of conditional necessity is, however, highlighted in the outline of proper explanation provided in *PA* I.1. In two central passages (639b21–640a9, 642a1–17), the existence of certain materials and processes is said to be necessary *given that* certain goals are to be – given that there is to be an eagle, certain materials and processes *cannot not be.*

Chapters 2–3: The Reform of Division. The *Prior* and *Posterior Analytics* are both critical of the use made of logical division by Plato and other Academics, and yet give division an important place in the scientific enterprise (cf. *APr.* I.31, *APo.* II.5, 13). The second and third chapters of *PA* I develop a systematic reform of division which avoids a set of problems arising from the Platonic method (for a detailed account of the reforms, cf. Balme 1987b, 71–80). It would take a paper in itself to go through the entire package of reforms, but it is generally acknowledged that there are three principal ones that imply the others:

(i) Division must not be restricted to dichotomy, if its aim is to aid in grasping the real natures of things. (642b5–20)
(ii) There must be a method of pursuing many different divisions simultaneously with respect to the same kind. (643b10–13, 644a3–10)

(iii) The products of a division must be determinate forms of the general differentiae, to avoid 'accidental' division; e.g., if the general feature being divided is wing, the products of the division must be forms of wing.

Here again, we see Aristotle adopting both the criticisms of Platonic division presented in *APr.* I.31 and *APo.* II.5–6, 13, and, as there, insisting that, *properly revised,* it has an important place in science. But the *Analytics* once more gives little guidance as to how a suitably revised divisional method would work in the complex world of organisms. *De Partibus Animalium* I.2–3 provides the application, but only after further revising the method.

Chapter 4. Finding the Appropriate Kinds. Chapter four provides an answer to a question growing out of the revised method of division: Since the starting point of proper division is a multiply differentiated general kind, how are we to establish the appropriate kinds initially? Aristotle argues that one looks for a significant number of 'species' that are similar in overall bodily configuration, and have parts, habits and modes of activity the perceptible features of which differ only in degree on the same scale. If these criteria are met, as they are in the case of bird and fish, you have identified a natural kind. Aristotle ends up identifying seven or eight others that, it seems, were not so recognized prior to his work.

Chapter 5a. Exhortation to Biology. Perhaps the best known passage in the *Parts of Animals,* and certainly the most elegant, constitutes the first 48 lines of chapter five (in the Bekker text). The passage is a plea to put aside the natural distaste for dissection, in the interests of philosophy. To whom is Aristotle addressing this wonderful little bit of rhetoric? The answer, it seems clear, is to someone who wonders whether it is possible to have scientific knowledge of non-eternal composites of matter and form. He begins by dividing natural substances into those that are eternal and those that come to be and pass away, arguing that, while the former are more noble, the latter are more knowable, since "the perishable plants and animals live all around us" (644b22–645a4). He goes on to stress that the study of living things provides great pleasure "to those by nature philosophical and able to know the causes of things" (645a9–10). More than anywhere else, the good for the sake of which things exists is clear in the living world (645a23–26).

Finally, if you aren't yet convinced to cut open the next fetal pig, Aristotle reminds you that you too are flesh and blood, and if that fetal pig's guts offend you, you must find yourself offensive (645a26–30). He concludes by pointing out that this study must involve matter, since an end for the sake of which implies something which comes to be and exists for the sake of that end (645a30–36).

Chapter 5b. The Integration of Division and Teleological Explanation. The rest of the chapter distinguishes division of the attributes to be explained and division of the causes relative to those divisions, making rich use of the ideas on division of kinds in chapters 3 and 4, while integrating these lessons with two from chapter 1: the principle 'first grasp and organize the appearances, then study the causes', here becomes 'first apply division to the proper attributes, then attempt to divide off their causes'; but in order to apply this principle to biology, given the method of conditional necessity, there must be parallel divisions of parts and of functional activities, since the former are for the sake of the latter (645b18–646a4).

Aristotle's aim throughout is the application and development of a general model of scientific explanation set forth in the *Analytics* to matter/form composites, with their matter causally dependent on (and conditionally necessary for) their functionally defined form. It is a *tour de force* in the philosophy of biology. One can understand how, after reading it, Charles Darwin wrote (in one of his last letters) to its 19th century translator, Dr. Wm. Ogle, that it established that Linnaeus and Cuvier were "mere school boys" to old Aristotle. Since he mentions only having read a third of Ogle's volume, it is primarily *PA* I to which he was referring, and he is dead right. Neither of those 18th century geniuses provided anything quite so systematic and powerful in the philosophy of biology as this, and indeed on division and teleology both had learned their lessons from Aristotle.

IV   LESSONS ABOUT METHOD FROM PRACTICE

Suppose, then, that *PA* I provides a philosophy of biology that is an application of the *Analytics* model of science to the study of matter/form composites governed both by material and teleological principles. This does not guarantee that Aristotle's biological practice *conforms* to the ideal. Furthermore, *PA* I has not resolved one question regarding this 'application' – how will multi-causal demonstration conform to the demonstrative ideal? To deal fully with these issues, one would need to look in detail at the entire biological enterprise, since not only do specific explanations have to take a certain form, but so does the entire research program. I cannot begin to make the general case here, but the explanatory example I have chosen will at least hint at the overall structure of the *Parts of Animals*.

V   THE CHEMICAL MECHANICS OF FAT AND BLOOD

Aristotle concludes his discussion of blood and fat (in *PA* II.3–5) by declaring that he has stated "what each of these is and on account of which causes" (651b18–19). Later, in *PA* III, the possession of fat by the kidneys offers us an example of Aristotle's infamous 'dual explanations', in which he argues that a part belongs to the animals that possess it *both* from necessity *and* for the sake of something. By looking at each of these explanations, one begins to get a picture of the demonstrative structure of the *PA* as a whole, since the latter account is demonstratively dependent on the explanations in II.5 we are about to examine (cf. Gotthelf 1987a, 1997; Detel, 1997).

Soft and hard fat, or lard and suet, are both produced by concoction of excess blood, the excess being present because an animal is well nourished – not a bad theory, actually. The overarching explanatory goal of the chapter is why there is lard in certain blooded animals, but suet in others. Nevertheless, since both are products of excessive blood, Aristotle is able quickly to note that "because none of the bloodless animals have blood, none of them have soft or hard fat" (651a26–27). That is:

p1: Being blooded belongs of necessity to no bloodless animal.
p2: Having lard and suet belongs of necessity to all that are blooded.
 c: Having lard and suet belongs of necessity to no bloodless animal.

Not surprisingly, since he opens the discussion by noting that the two fats differ in accordance with the differentiation of blood, he proves that those animals with fibrous blood have suet and those with non-fibrous blood have lard. Thus, in accord with the recommendations of *PA* I.5, there is an explanatory dependence of the division of the fats upon a causally more basic division of blood-types. The form of explanation is in line with the *Analytics,* but the content is very much determined by the composite nature of the substances in question.

These explanations depend, in their turn, on a detailed theory of the differential effects of heating and cooling materials of different elemental constitutions. This theory is borrowed from *Mete.* IV, to which we are twice referred for a justification of premises in *PA* II (646a16, 649a32–33). Thus to understand fully the demonstrations regarding fat and its formation, we need to see how firmly established are Aristotle's views about the effects of hot and cold on various sorts of earth/water compounds.

VI *PA* III 7–9: KIDNEYS AND THEIR FAT

In the discussion of suet and lard in general, the natural scientist is focused almost entirely on understanding their *material* differences. The fat around the kidneys, on the other hand, is another matter, and introduces the problem of multi-causal demonstration.

PA I.1 640a33–b2 divides explanations in biology into those that establish the conditional necessity that a part belong to a certain kind of animal, and those that establish that a part, though not strictly necessary, is good for the life of the animal in question. The explanation for kidneys provides us with an example of the latter sort of explanation.

> The kidneys belong to animals that have them, not out of necessity, but for the sake of what is good and fine. For, in accord with their distinctive nature, they are present for the sake of the collection of residue in the bladder, in those animals in which such deposits are large, so that the bladder may better perform its function. (*PA* III.7 670b23–28)

That is, animals with excessive fluid residues could survive without kidneys – they have kidneys because that is better for them than the mere possession of a bladder.

The kidneys themselves are not present of necessity, then; but they *are* the fattest of viscera both of necessity *and* for the sake of an end. Kidney fat results from the process by which kidneys filter blood as it moves from the blood vessels surrounding the kidneys, through the kidney walls, to their hollow core, from which it flows into the ureters and bladder. As a result, a residue of blood, well-prepared for concoction, forms on the outside of the kidneys. The nature of that residue is either lard-like or suet-like, depending on the nature of the filtered blood. This explanation explicitly refers back to the explanation of the production of suet and lard (cf. *PA* III.9 672a12–13). Though it is thus a necessary by-product of this 'filtering' process, kidney fat also forms for the sake of an end.

> On the one hand lard arises of necessity – on account of the cause just given, that is, as a consequence of what happens of necessity in animals with kidneys. On the other hand, it comes to be for the sake of preservation, i.e. for the sake of preserving the hot nature of the kidneys. (*PA* III.9 672a13–16)

As we saw earlier, the *Posterior Analytics* acknowledges such explanations in natural science, but provides no answer to the question of how to integrate this idea with his views about the causes as the middle terms in demonstrations. *PA* I suggests a priority ranking of the causes, and briefly sketches the form a teleological explanation (with conditional necessity) will take, but still leaves the reader with many questions about the structure of dual explanations. By looking to his actual biological practice, we find answers to our questions.

There is, in this case, a *single fact* requiring explanation: In animals with kidneys, kidneys are the fattest of viscera. The explanation of this single fact, however, is complex. We are given both antecedent material conditions and processes which, being present, must produce one of two kinds of fat around the kidneys; and we are told that kidney fat arises for the sake of the contribution it makes to preserving the (hot) nature of the kidneys. This makes it crystal clear that goals are causes, and that in explanation they take priority.

The explanation offered in this chapter involves reference to (at least) two causes. And since the middle term in a demonstration identifies the cause of the predication to be explained, the questions raised at the beginning of this paper arise in a particularly compelling way here. What will a multi-causal demonstration look like in practice?

Very briefly, the explanation by reference to necessary consequences has the following demonstrative form.

p1:Being well-concocted residue of blood belongs of necessity to the product of kidney filtering.
p2:The product of kidney filtering belongs of necessity to all kidneys.
c1:Well-concocted residue of blood belongs of necessity to all kidneys.
p3:Being lard and suet belongs of necessity to well-concocted residue of blood. (*PA* II.5)
p4:Well-concocted residue of blood belongs of necessity to kidneys. (c1)
c2:Lard and suet belong of necessity to kidneys.

I have fully spelled this out as two syllogisms, though the reader can see that the second simply identifies lard and suet with well-concocted blood residue. Making this step explicit has the virtue of showing how this explanation borrows premises from previous results about (a) the nature of fat and (b) the product of kidney filtration.

What of the teleological explanation, and its integration with necessity? The logic can be construed in a variety of ways (one of which is offered below), but the essential points seem clear: all the viscera require heat to function; flesh protects the heat of the viscera around the heart; but kidneys, being located at the loins, lack this sort of protection, and fat is provided for this purpose instead.

It seems clear, however, that even if kidneys didn't need fat around them (imagine them located in a fleshy part of the body), fat would nevertheless arise by the necessary process described. It is an empirical question for Aristotle whether 'residues' produced by such a process also come to be for a function – bile arises in essentially the same manner, but serves no function for the organism (cf. IV.2 677a12–30). As he puts it:

> Sometimes the ⟨formal⟩ nature makes use even of residues for the benefit ⟨of the organism⟩; nevertheless it is not for this reason necessary to seek 'for the sake of what end?' in every case – but when some things are such and such, others often result from necessity because of these. (677a16–18)

Thus, it is necessary that fat appear as a necessary by-product of the proper function of the kidneys; whether the nature of the organism uses such by-products for an end is not. In the case of bile, Aristotle notes that it only arises as the by-product of *impure* blood, and thus has a universally negative effect on the animals that have it (677a19–29). Having determined this, he concludes, "[t]herefore it is apparent that bile is not for the sake of something, but dross, off-scouring" (677a29–30).

Fats derive from a similar process; but they derive from a pure blood ripe for being 'well-concocted'. Nothing precludes them serving a useful function then; furthermore, the presence of fat around the kidneys is necessary for their proper function.

To fully capture the structure of this 'dual explanation' for kidney fat, then, a minimum of three modal syllogisms would be required. Furthermore, the same fact is explained via two distinct sets of syllogistic inferences. By looking at the formal structure of these explanations, we learn a good deal about Aristotle's biological science.

VII PHILOSOPHY OF SCIENCE, PHILOSOPHY OF BIOLOGY AND BIOLOGY

Contemporary philosophy of biology was defined by the fact that logical positivism and logical empiricism focused almost exclusively on physics – whether one looks at Woodger, Sommerhoff, Nagel, Beckner, Schaffner, Ruse or Hull, the central questions were defined by that legacy. From the beginning, philosophically inclined biologists such as George Simpson and Ernst Mayr protested the importation of inappropriate philosophical standards from physics, and the philosophy of biology eventually listened.

For Aristotle, on the other hand, the paradigm natural substances were not the common material constituents of the universe, but the most active, complex and organized of bodies, the living ones. He might, one would then suppose, have been free to develop a philosophy of science with biology as its paradigm case.

But it was not to be. The *Posterior Analytics* shows a Platonic influence in at least one crucial respect – the paradigm case of scientific knowledge in that first great treatise of philosophy of science is mathematics. Thus Aristotle faced problems analogous to those of contemporary philosophy of biology. Most generally, those problems can be summed up in the question: to what extent can a general philosophy of science, initially defined by reference to mathematics, be applicable to a world of teleologically organized, complex living systems? This, I have argued, is the central question of *Parts of Animals* I, the first great treatise in the philosophy of biology. Aristotle's biological practice, I have additionally suggested, is a valuable source of insight into Aristotle's answer to that question.

ACKNOWLEDGMENTS

It is a pleasure to thank the Division of Research Programs of the National Endowment for the Humanities for their support during work on the translation and commentary of Aristotle's *De Partibus Animalium*. It was Professor Finocchiaro's wonderful idea to invite Daniel Jones of that division to chair the symposium in which this paper was presented, and I would like to thank him for agreeing to take part. Finally, I would like to thank Maurice Finocchiaro for organizing the symposium, and Ernan McMullin for his perceptive comments on my paper.

5

The Disappearance of Aristotle's Biology:
A Hellenistic Mystery

The mystery is this: the period after Aristotle's death saw unprecedented developments in astronomy, medicine, optics, arithmetic, geometry, and mechanics – the first flowering of the 'special sciences'. Aristotle and Theophrastus, rightly viewed as the originators, respectively, of the sciences of zoology and botany, should stand at the beginning of the flowering of biology. But they do not. I do not think it is an overstatement to say that the next work that can be called a biological treatise – one focused on a theoretical study of animals or plants as such – is Albertus Magnus' *De Animalibus,* in the twelfth century of the Christian era.

Zoology, as a special science, was the invention of Aristotle. Apparently, as eloquently as *Parts of Animals* I.5 sings of the immeasurable pleasures awaiting those who study animals philosophically, with an eye to causes (645a9–10), his rhetorical energies were wasted. Aristotle's zoological research program disappeared. This chapter outlines the evidence for this claim, and considers a pair of hypotheses that could account for it. The first, which I reject, is that the research program disappears because the works that lay it out disappear. The second, which I also reject, is that there weren't people around sufficiently talented to understand and carry on the tradition begun by Aristotle and Theophrastus. A close study of Galen and the Greek Commentators, who were familiar with the biology as we have it, casts doubt on both these hypotheses. I conclude by suggesting that the explanation must lie in deeper, philosophical, problems that later Greek philosophy had with Aristotle's research program.

I ARISTOTLE'S ZOOLOGICAL RESEARCH PROGRAM DEFINED

The first step is to specify what it is that is alleged to have disappeared. For there were certainly great developments in anatomy and physiology during the Hellenistic era, and various collections of stories, anecdotes and 'marvels' about animals. In order to make it clear that neither of these traditions were heir to any part of Aristotle's zoological science, that science needs to be clearly defined.

The question of whether a research tradition disappeared should be distinguished from a closely connected, but distinct question – whether the originals of the works that now make up pp. 486–789 of the Berlin Academy edition of Aristotle *also*, for all intents and purposes, disappeared. For there is an oft-repeated story that much of the Lyceum's library disappeared shortly after Theophrastus' death, only to reappear sometime in the first century BC, then to be edited for Roman distribution by Andronicus.[1] The view that emerges from the following study is that the absence of Aristotle's zoological writings is neither a necessary nor a sufficient explanation for the absence of a Hellenistic biological research tradition based on Aristotelian beginnings.

Over the last thirty years, there has been much progress in understanding Aristotle's biological treatises – by which I primarily mean the *Historia Animalium, De Partibus Animalium, De Incessu Animalium, De Motu Animalium* and *De Generatione Animalium* – as an integral part of his natural philosophy, and in seeing the 'zoology' as a comprehensive whole. This does not preclude attempting to trace its development – whole projects develop too – but it does preclude trying to understand each of the biological works in isolation from the others. To use the phrase I introduced in the last paragraph, it involves seeing these works as parts of a *research program*, and indeed a program embedded within Aristotle's broader philosophical outlook, reflected in the *Metaphysics, Physics* and *Organon*. A number of my papers have pressed the thesis that certain aspects of the method of the *Parts of Animals* and *History of Animals* derive from Aristotle's views on the nature of scientific inquiry and scientific demonstration in the *Analytics*.[2]

Let me put a hard edge on the rather fuzzy concept of a 'research program'. Minimally, a cognitive enterprise is a 'research program' if:

(1) It identifies a relatively *self-contained domain* as a single object of investigation.
(2) That investigation aims at *theoretical knowledge* of this domain.
(3) Basic *principles for that investigation* are explicitly identified.
(4) A set of *concepts and methods* is defined, at least partly domain specific, for answering questions within the domain.

Aristotle's zoology meets each of these criteria easily.

(1) **A Self-contained Domain.** The *De Anima* and *Parts of Animals* I clearly demarcate ensouled substances as a distinct set of objects with organs and tissues defined by, and existing for the sake of realizing, their instrumental capacities (cf. *de An.*II.1 412b4–11; 4 415b9–28; *PA* I.1 641a6–29; 5 645b15–20). Animals are set off from plants in having, besides the shared capacities of nutrition and reproduction, the distinctive capacities of perception and (in virtually all cases) locomotion (*de An.* II.3 414a29–b17; *PA* II.1 647a22–24).

(2) **Aim of Theoretical Knowledge.** *PA* I is clearly a philosophical introduction to the science of living nature and not simply the first book of *PA*. It makes regular reference to the aim of such a science – it is scientific understanding (*epistēmē*) – and explicitly defends the view that this aim includes establishing definitions, discovering causes and providing causal demonstrations of those non-primary predications that hold of necessity (see esp. 639a15, 639b8–9, 640a1–2, 644b32–645a15).

(3) **Investigative Principles.** *PA* I.1 and *Phy.* II.8–9 refine both the concepts of 'that for the sake of which' (final causality, if you like: I don't) and of 'necessity' so that they become, through the notion of *conditional* necessity, intimately related explanatory tools for organic investigation (*PA* I.1 639b12–640a9, 642a2–14; *Ph.* II.9 200a7–15). In fact it seems clear from looking at the discussion of necessity and teleology in *APo.* II.11 (94b26–95a9) that Aristotle had not yet formulated the concept of conditional necessity – perhaps because the matter/form analysis, upon which it depends, was not yet clearly formulated. The concept of conditional necessity is not mentioned in the *Analytics,* but is highlighted in the outline of proper explanation provided in *PA* I.1. In two central passages, the existence of certain materials and processes is said to be necessary *given that* certain goals are to be – these materials come into being and continue to be for the sake of some end. This is conditional *necessity* because, given the appropriate conditions, certain materials and processes *cannot not be* (cf. *Metaph.* Δ 5, 1015a20–22).

In addition to conditional necessity, *PA* I gradually moves us toward a *functional theory of form,* by criticizing simple structural theories of essence on grounds that a structure incapable of doing its own work (as defined by the life of the organism it serves) is that structure in name only (640b5–641a17). Similarly, *Meteorology* IV.12 makes it clear that, while we can give a generic account of all uniform bodies whether organic or inorganic (for example, ignoring the differences between bone and clay) a full definition of a uniform part of an animal will require that we get down to the business of

PA II, which is determining what those uniform parts are *for.* This is a set of principles for the study of living things, *qua* living.

These last two concepts – conditional necessity and form as living functional capacity – are *domain-specific* and intimately related. The development of the appropriate tissues and organs is not necessitated unconditionally by the existence of the relevant elements – rather, they emerge through the activation of a potential for *form,* as Allan Gotthelf has called it (Gotthelf 1987b). The completed organic structure is necessary, *given* the activities which constitute the animal's life. That is, it is *necessary for,* not *necessitated by.*

(4) **Proper Concepts and Methods.** These domain-specific principles imply certain methodological practices. This is not the place for a detailed analysis, but the following admonitions of *PA* I are worth mentioning:

(i) It recommends giving priority in investigation to goal causation over motive causation (639b11–21).

(ii) It is critical of Platonic division (642b5–643b8) and recommends a non-dichotomous, multiple differentiae method as the only way to use division successfully in zoology (643b9–26, 644a6–11).

(iii) A method of explanation is outlined whereby an animal's defining features, as well as those conditionally necessary given the defining features, those that make the animal's life better given those defining features, and the per se incidental features of animals, and of their development, are all objects of scientific knowledge (e.g., 640a33–b4, 642a31–b4).

(iv) It specifies conditions under which generic understanding of a feature is preferable to a case-by-case analysis (639a15–b3, 644a12–b15).

(v) It distinguishes between identity in kind, identity in form and identity by virtue of analogy, with the concept of variation by 'the more and the less' playing a crucial role in determining kind identity (644b1–15, 645a36–645b28).

(vi) The opening discussions of both *HA* I.1 and *PA* II.1 divide parts into simple and complex, defining simple parts as those which divide into parts like themselves and like the whole (uniform parts), and those which are instruments constructed out of tissues and thus combining many different capacities. (Note that *HA* I.1 draws the distinction without reference to capacities or goals, while *PA* II.1 explains the existence of each level of organization as *for the sake of* the next functional level – earth for flesh, flesh for hands, hands for touching and grasping.)

Many more such domain-specific conceptual and methodological tools could be mentioned, but enough has been said to provide a concrete idea of what is intended by this aspect of a 'research program'.

The above is sufficient, I think, to establish that Aristotle provided his students with a richly articulated zoological research program, and one rooted in his metaphysics and epistemology. But he did much more than this, of course. He was a practising biologist of considerable skill and insight. Now it might have been the case that the practice failed to mirror the above recommendations to such an extent that this by itself might explain the demise of the program. But in fact the practice does conform to these ideals, at least well enough that if you are familiar with them, and with chapter and verse of the zoological works, the relationship between the two is hard to miss.

II THE PROGRAM DISAPPEARS

It is not stretching matters, then, to say that Aristotle provided Greek science with a thoroughgoing biological research program, one deeply embedded in his views about the nature of scientific knowledge and the nature of composite substances. Were this not the case, the mystery which the history of Greco-Roman science presents to the student of Aristotle's zoology would be greatly diminished. Unlike mathematics and medicine, this research program seems to have fallen mostly on deaf ears. Only Theophrastus appears to have grasped what Aristotle was trying to do: the *Historia Plantarum* and *De Causis Plantarum* are related to each other in a manner carefully modelled on the relation of the *Historia Animalium* to (especially) *De Generatione Animalium.*[3] He was, however, part of this research program much more than its heir.

But after that, what is there? Pliny gives us a truly frightening list of vanished texts upon which he relied. Still, ponder the following:

1. The admittedly fragmentary evidence for the Lyceum itself indicates very little interest in biological subjects at all.[4]
2. Not one of the ancient commentators on Aristotle saw fit to comment on any of the major zoological treatises – the earliest, extremely spotty, zoological commentaries are those of Michael, probably in the tenth century AD – and even here, there is virtually no independent discussion of *HA*. Yet the commentators refer to the biological treatises (as Diogenes Laertius, for example, did not) by their post-Andronican titles, suggesting

(though not implying) that they had access to them if a commentary was desired. The relatively empirical character of the biology is no explanation for their being ignored by the Greek commentators – the *Meteorology,* for example, was regularly commented upon.

3. Pliny had not the slightest idea of (or interest in) an organized causal/ explanatory science of animals. In fact, a number of passages suggest he was actively opposed to such a project on skeptical grounds, as in the following from *Natural History* XI 8:

> denique existimatio sua cuique sit: nobis propositum est naturas rerum manifestas indicare, non causas indicare dubias. (Pliny 1967–71)

> [In the end, let everyone judge for himself; as for us, we propose to indicate the manifest natures of things, *not* their doubtful causes.]

Granted this is a period of epitomes, handbooks, and commentaries, the best of these in other sciences nevertheless understood their sources – none of the works in which we find passages cribbed from Aristotle or references to his work show the slightest grasp of what he was doing – leaving open, for the moment, the question of Galen. To put it rhetorically – where is Hellenistic/Alexandrian zoology?

4. Then there are the mysteries of the lists of Aristotle's works, and the various titles under which the various works seem to have been known.[5] To start with the most well-known and glaring problem: the list of DL V 25 refers to nine books *On Animals* and one book *On Failure to Reproduce,* which together are usually assumed to refer to our *HA* I–X. Other interesting references include eight books of *Dissections,* one of *Selections from Dissections,* one *About Composite Animals,* one *About Mythological Animals,* two *On Plants,* and one on *Physiognomy.*

Thus Diogenes Laertius seems to know of a collection which included *HA* I–IX, a separate treatise on failure to generate which at some later point is occasionally attached to *HA* (as Balme indicates [Balme 1991, 26–29], the ancient order of the texts encouraged this practice) and a number of other works, all of which are lost. There is nothing in the contemporary corpus of Aristotle answering, for example, to "the Dissections" and "the Selection from Dissections," but that there were such works is suggested by a passage in the *Posterior Analytics.* The opening of *APo.* II.14 98a1–2 refers to the necessity, in grasping problems, of using "selections from dissections and divisions." This chapter is focused entirely on zoological examples and problems, suggesting that there was such a work produced by Aristotle himself, subsequently lost. The fact that they did not make up part of the edition that Andronicus put together is yet another mysterious aspect of our story.

On the other hand, there is no mention in this list of any of the remaining treatises which make up the rest of the modern edition's 'biological' treatises – no *Parts, Generation, Movement* or *Locomotion of Animals,* for example. And, for reasons to be mentioned shortly, we needn't assume that Diogenes was actually familiar with *HA* either. Perhaps the most puzzling thing about Diogenes's failure to mention these works is that even on the earliest dating possible for him, he lived after the Andronicus edition of Aristotle was available. The evidence for this comes from Galen.

There is, however, another list of Aristotle's works, attributed to a Ptolemy in the Arabic tradition, which does include reference to these works. And the work of Paul Moraux, Ingemar Düring and John Keaney has provided us with a reasonable narrative about how this came about – as opposed to more speculative reconstructions of previous generations of scholars.

Many of the earliest references to Aristotle's zoological work – those of Apollonius, Aelian, Artemidorus – seem to have relied not directly on *HA,* but on an *epitome* by Aristophanes of Byzantium, which refers to itself as an *Epitome on Animals,* but which appears to have indiscriminately mixed summaries from *PA, GA* and *HA* on the same animal – it also, ominously, refers to itself as a *historia* of each single animal.[6] One aspect of Aristotle's research program which is richly exemplified in his practice is the idea of identifying differentiae at the most general level first, and then through division more specific differentiae when necessary. The organization of *HA* displays this elegantly. There are multi-book sections devoted to parts, activities, modes of life and habits; within the four books devoted to parts, those correlated with blooded animals are covered first, then those correlated with bloodless animals. Within blooded animals, the external non-uniform are discussed first, the internal next, and finally the uniform parts. For any particular part, what can be truly said of every such part is said first, then the different forms it can take are described, and so on. Very rarely, the divisions descend to the level of a particular species (the peculiar humps of the camels, or tongue of the wryneck, for example). If you want to know what Aristotle had to say about any particular animal kind, you would have to read sections from all over *HA,* since it is not organized animal by animal, but general differentia by general differentia, from widest to more narrow. And unless there is something quite peculiar about a species, it may never get mentioned at all. When particular animals *are* mentioned, it is typically as part of a list of examples of animals with some general feature.

Thus the talk of a *historia* of each animal suggests a fundamental misunderstanding of Aristotle's project, and one which appears systematically throughout our story, until a theory of *HA*'s purpose emerges as common currency in the Greek commentators – a theory which is not based on familiarity with that work itself. This epitome reads like a randomly arranged discussion of various interesting animals. It bears no evidence of the organization of the work it allegedly epitomized.

During the Alexandrian/Hellenistic period, references which can be clearly identified as to passages taken from *HA* refer to it most commonly simply as *On Animals,* as it is referred to in the list of Diogenes. There is some slight evidence that all or part of that treatise is occasionally referred to, perhaps based on its incipet, as *On Parts,* but the evidence for this is slight, and ambiguous. However, nowhere in these early references does *HA* seem to be referred to by name as *On Investigation of Animals,* though these works are occasionally described as giving a *historia* or *historiai* of various things.[7] There seems good reason, then, to think that the Alexandrian collection of Aristotle's writings did *not,* during the Hellenistic period, include the 'zoological writings', but only a general *epitome* of Aristotle's writings about animals. Furthermore, this work was organized, not as Aristotle would have wanted, but more like a modern 'natural history', information of all sorts gathered together animal by animal, rather than function by function or part by part.[8]

Is it then mere coincidence that the *On Animals* mentioned at Diogenes Laertius V 25 has the same number of books as the modern *HA?* Probably not. Nor, however, does it imply that Diogenes had *our HA* before him. It is likely that the epitomes would all mention the number of books summarized, and that would be all Diogenes would need for bibliographic purposes. Two lines of evidence support this. First, the two-book epitome of Aristophanes excerpts material from all ten of our books. Second, the Arabic "Tract comprising Excerpts from Aristotle's *Book of Animals,*" attributed to Maimonides, follows what was the original book order of *HA* (that is, with the modern VII and IX reversed), and, like Aristophanes, has what appear to be randomly selected excerpts from each book, including Book X, along with some material from *PA* and *GA,* mostly near the end. It too lacks any of the theoretical unity that holds the original together and ties it to the other biological works. The Cambridge editor and translator of the Arabic text operates on the unquestioned assumption that the work derives from Aristotle's *Historia Animalium.*[10] But the treatise on which the Arabic is based is entitled "*Excerpts* from . . . ," and the excerpts are said to be from "Aristotle's Book of Animals," a title Aristotle never uses for *HA* when he refers to

117

it in the other treatises. A more reasonable assumption is that Maimonides was working from an *epitome* of Aristotle's zoological studies, one that concealed all traces of Aristotle's research program. Indeed, the similarities between Aristophanes and the Arabic text suggest either that the archetype of one served as the source for the other, or that they derive from a common source.

If the names of works preserved by the list of Ptolemy which are not mentioned by Diogenes Laertius are based on the edition of Aristotle's works prepared by Andronicus, as seems likely, it appears to be *that* edition which first provides the learned world with the title *peri zôiôn historias,* and the fruits, more or less complete, of Aristotle's zoological research program. This is the standard manner of reference in the Greek commentators on Aristotle, by which they distinguish the *HA* from *PA, GA,* and the rest.

It is also, of course, the way the other biological works actually refer to *HA* when they direct us to what is said in 'the histories about animals', 'the animal histories', 'the natural histories', or 'the written histories', references quite often conjoined with references to pictorial representations known collectively as 'the dissections'. These dissections are also referred to in our *HA* on numerous occasions. This adds further support to the idea that none of these works was available in Hellenistic times – had they been, this method of reference to *HA* would have been the obvious one, being the one Aristotle himself used.

The fact that the *HA* never refers to the other biological works is also relevant to our mystery about what happened to Aristotle's research program. Not only does *HA* never make direct reference to other biological works, it offers no indication to the reader that the next stage in the research program outlined in *HA* I.6 – to attempt to discover the causes – had ever been carried out. Readers restricted to *HA,* or to works which epitomized the biology in ways that erased any distinction between the causal treatises and the *HA,* would have only the most gnomic of hints about Aristotle's research program. The distinction between a preliminary investigation establishing the phenomena and a later investigation of their causes is the heart and soul of Aristotle's biological research program – once lost, the point of *HA* being organized as it was would also be lost. An epitomiser of *HA* would, at that point, have every reason to arrange the material in a different manner, e.g., gathering information about specific animals together in one place. There would no longer be any theoretical justification for its complex and difficult structure.

The Greek commentators, though they never deign it worthy of commentary, use *The Investigation of the Animals* as the standard title for *HA.* And, since *HA* only once refers to itself as *historia,* and never names itself such, while in *PA, GA,* etc., various permutations of that title appear, used in the standard formula for book reference (*In the* . . .), this form of reference suggests that the entire zoological corpus was by then available for study. This evidence is only suggestive of such availability, however, since the few references to the biology in these commentaries don't indicate a first-hand acquaintance with their contents.

The same, however, cannot be said for Galen, to whom we now turn.

III   ARISTOTLE IN GALEN

Galen would seem to be in as good a position as an Ancient could be to understand the Aristotelian zoological research program. He clearly had access, as Diogenes apparently had not, to the Andronican edition of Aristotle's treatises. He directly quotes, as we will see, *PA* II and III, and *HA* I, and refers, by means of the titles that have come down to us, to material in *PA* I, *GC, Mete.* IV, *GA,* and the parts of *Parva Naturalia* on life and death, youth and old age, etc. That is to say, he apparently had access to the whole zoology, and he knows it in roughly the form we have come to know it. At some point in his life he must even have had it at hand for purposes of quotation – indeed, it is possible to form a reasonable hypothesis about which of our manuscripts are based on the tradition that Galen is reading. Further, he deeply respected Aristotle, and was, like Aristotle, a brilliant student of anatomy, physiology, logic, the theory of knowledge and metaphysics. He had the material to hand, and the intellect, to understand the Aristotelian project. Did he? The verdict we must reach, I believe, is that he did not.

Most of the more than 250 references to Aristotle in the Galenic corpus include him in lists of thinkers who all allegedly agree on some doctrine or other – Hippocrates, Plato, Theophrastus and Chryssipus being the most common bedfellows. There are, in addition, nearly two dozen references to Aristotle's zoological treatises by name, and many more where one can be quite sure of the passage Galen is relying on though it is not explicitly identified. And one can add to these a number of references to the *De Anima* II which show knowledge of Aristotle's views on the nature of the psychic capacities and their organic base, and to *GC* and *Mete.* IV on hot, cold, moist, and dry, earth, air, fire, and water, the nature of uniform bodies, the distinction between concoction and decay, and the passive capacities which distinguish uniform bodies from each other at the material level.

As Moraux has carefully discussed in his contribution to the Balme Fest-schrift,[11] Galen actually quotes *PA* II twice in *Quod animi mores corporis temperamenta sequantur* and *PA* III once in *De semine*. In addition, there are two passages in *Quod animi mores* where he quotes *HA* I. A careful look at these quotes is, for our purposes, both revealing and discouraging. Moraux compared Bekker's text of Aristotle's *PA* with Müller's of *Quod animi mores* for variants. He did not, however, take the next step of comparing Galen's 'quotations' with the manuscript variants of *PA*. Often, the existing manuscripts of Aristotle show no trace of the reading found in Galen.[12] In some cases, however, Galen's quotation of Aristotle varies in scientifically irrelevant respects, but is correlated with the same group of Aristotelian manuscripts. We are thus able to identify with some assurance those among our manuscripts that derive from the family of manuscripts with which Galen was familiar.

When one focuses on the scientifically relevant differences between Galen's quotations and our manuscripts, the comparisons are disturbing. Moraux's comment that "the differences between the text of Galen and the text of Aristotle are minimal" is very misleading. For example, at *PA* II.4 (650b14–651a19), Galen reads "the nature of *the body* is the cause of many things with respect both to the *form* and perception in the animals," while all our extant Aristotle manuscripts read "the nature of the *blood* is the cause of many things with respect both to the *character* and perception in the animals." This is hardly a minimal difference – and there are many such substantial differences. Unless Galen was quoting from memory – and the slight variants that agree with one group of manuscripts rather than another tell against this – these variants, unmatched in any modern manuscripts, raise the possibility of manuscripts of quite ancient authority with substantial differences of doctrine compared to ours.

That rather gloomy reflection aside, what do these references tell us about the Hellenistic understanding of Aristotle's zoological works? Amazingly little. Moraux's goal was to establish the influence of Aristotle's *de Partibus Animalium* on Galen. He discusses a number of general philosophical principles that Galen acknowledges as Aristotelian, and then a number of specific aspects of Aristotle's anatomy and physiology which Galen discusses, usually in order to disagree.

My focus will be somewhat different. I am particularly interested in whether Galen's references and discussions of Aristotle indicate a particular understanding of Aristotle's research program. There are some key passages which can help us in this respect, and it is now time to look at them.

Perhaps the most abstract question, given the picture of Aristotle's zoological research program drawn at the outset, is whether Galen understood Aristotle's distinction between a preliminary inquiry aimed at establishing *that* X is the case, followed by an attempt to understand *why* X is the case. That is, would he have been able to understand the differences, and connections, between the aims and methods of the *HA* and those of the causal/ demonstrative treatises, such as *PA* and *GA?* The answer is unambiguously yes, as the following passage from book VI of *De Usu Partium Corporis Humani* shows:

> For it is now proposed to show not that this comes about with respect to the animal's body, but why; but since "the that" from necessity precedes "the why," just as Aristotle in fact said, it is impossible to explain the uses until the activities are recalled. (3 495.18–496.4 [Kühn edition])

Even without Galen's explicit acknowledgment of debt to Aristotle, the reliance on Aristotle's distinction between two stages of theoretical inquiry is impossible to miss.[13]

The next question, then, is whether Galen recognises that Aristotle's zoological works are organised so as to reflect this distinction? The answer, as far as I can see, is that he does not. And here one has to be very careful with the evidence. The May translation of Galen's *De Usu Partium* may serve as an example, the more so because it is of such high quality.[14] The footnotes to this edition contain many references to *HA*, reflecting May's knowledge of the *HA*'s contents. When Galen refers to some detail about this or that animal, part, or behavior, the notes point us to Aristotle's comment on the same detail in *HA*. However, as David Balme has exhaustively established (1991, 21–25), the *HA* consists to a significant extent (roughly 40%) of material reported both in *HA* and in the other zoological treatises. Most of the actual references in *De Usu* are not to *HA*, but to one of the causal treatises. Indeed, in the Galenic Corpus as a whole the explicit references to *HA* are rare, and always in the interest of supporting a view already established by reference to *PA, GA* or some other treatise. Furthermore, the references are never to the theoretically most central aspects of *HA*, and show no understanding of how the treatise is organized or what its overall purpose is. A number of details indicate this.

1. While Galen refers constantly to Aristotle as support for his view of the constitution of the uniform or simple parts, and to Aristotle's distinction between uniform and non-uniform parts or organs, he nowhere refers to the most important theoretical account of these distinctions, the opening page of *HA* I.

2. While fascinated with the relationship between differences of character and differences in bodily make-up, Galen nowhere discusses Aristotle's fourfold list of differentiae in *HA* I.2, around which that entire treatise is organised – that is, differences of parts, actions, lives, and characters. Rather, the bulk of his *HA* quotations and references are from the brief 'physiognomic' remarks in *HA* I.8–10. Here Aristotle remarks on relationships between differences in facial characteristics and differences in character, the former being (fallible) indicators of the latter.

3. Interestingly, given this apparent familiarity with *HA* I, Galen never comments on the theoretical discussions of the first six chapters.

4. To be specific, Galen passes over in silence the passage in *HA* I.6, in which Aristotle refers to the forthcoming study of the differences and the attributes of all the animals as a *historia,* and says that after it he will attempt to go on and discover the causes in each case – for *historia* allows one to distinguish the 'subjects about which' of demonstration from the 'principles from which' (491a7–14). There then follows a description and theoretical justification of the order of presentation to be followed in *HA*. Again, Galen never mentions it. Yet all of this theoretically rich discussion is less than a 'Bekker' page prior to the passages Galen quotes extensively, on face and character.

All of this, and much more, appears lost on Galen – at the very least, it doesn't interest him. The *HA*, like the *PA* and *GA*, is referred to or quoted as ancient support for specific claims Galen seeks to establish. It has little further significance for him.

All this deepens the mystery with which I began. Galen was, of course, a philosophical physician. Even his most theoretical treatises should be viewed as theory required for medical practice, not pursued as a general study of animals pursued for itself. At some stage of his long career he apparently had access to all of the physical treatises of Aristotle that we have, and he appears to have read them carefully.

He does not, however, seem to have read them in order to determine what Aristotle, the student of animal nature *qua* animal, was doing – when he came across a zoological claim which he thought was correct, or which supported his own position, he adopted it, often with thanks. When he strongly disagreed with such claims, usually on specific matters of anatomy and physiology, he also usually said so. If all one had of Aristotle was what is to be found in Galen, there would be no hint at all of the nature of his zoological research program. Our look at Galen's use of Aristotle suggests renewed caution in using him as a source for Hellenistic medicine. For even when he quotes, he may not be quoting what his source took to be important, and he may be quoting in a way that distorts his source's work. This is true even when, as with Aristotle, he admired his source.

IV   UNSATISFACTORY POSTSCRIPT

So, then, where does this leave us? A study of Galen's discussion of Aristotle's biology shows us that, while familiarity with the founding texts is possibly a *necessary* condition for carrying on Aristotle's zoological research program, it is not *sufficient*. In addition, there need to be active and intelligent researchers who share the theoretical aims and methods articulated in those founding texts, and who are capable of extending the program of investigation beyond where it was left by its founder. After 300 BC, there seem to be no such people.

The mystery we began with was 'Why?' The two hypotheses most often adduced to answer it do not suffice. I would like to suggest a third. The research program presented in *PA* I asks a lot of its practitioner. In particular, he must believe that things that come to be and pass away are legitimate objects of theoretical, demonstrative science. During the Hellenistic period, those who were not skeptical of claims to knowledge generally were liable to skepticism about such a science for one of two reasons. They might reject the theoretical study of living things while studying certain plants and animals as an adjunct to various arts, such as medicine, for example. Or they might reject the demonstrative ideal, either because of rejection of Aristotle's causal realism, or because of doubts about whether there could be a demonstrative science of such variable and contingent things as animals and plants.

The impassioned plea for the philosophical, causally oriented study of the "animals and plants around us" in *Parts of Animals* I.5 was, I suspect, in part motivated by fear that those most inclined to such a study would turn their talents to less messy, more noble subjects. Aristotle hoped to attract to its investigation those who saw the living world as he did – as the ideal place to study being *qua* being.

Albertus Magnus, the mentor of Thomas Aquinas, was such a person (Tkacz 1993). As far as I know, he was the first such person in the West after Theophrastus. But many were to follow – echoes of Aristotle's research program can be heard in the works of Gesner, Caesalpinus, Fabricius, Harvey, Ray, St. Hilaire, and Cuvier, all of whom studied Aristotle's biological works carefully. Charles Darwin said more than he perhaps intended when he remarked, in a letter thanking William Ogle for a copy of his translation of the *Parts of Animals,* that his gods, Cuvier and Linnaeus, were "mere schoolboys" to old Aristotle. For both of these members of Darwin's pantheon read Aristotle's biological works not as antiquaria, but as sources of methodological guidance. It was not until the nineteenth century, with the advent of evolutionary, experimental and mathematical modes of thinking in biology, that theoretical biology developed a fundamentally different research program (or programs). Twenty-two hundred years is a long wait. But when one realizes that the Aristotelian program lay dormant for roughly fifteen hundred of those twenty-two hundred years, one acquires a different perspective.

NOTES

1. There continues to be disagreement over whether the Hellenistic evidence supports or undermines this story. According to most reliable sources, Theophrastus willed his library, which included versions of Aristotle's works, to a Peripatetic named Neleus, who removed them from Athens to Skepsis, where they were kept in a cellar. In the first century BC they were returned to Athens. At some point during the next one hundred years, Sulla carried copies to Rome, where Andronicus of Rhodes was commissioned to produce the edition of Aristotle's writings upon which all our mss. appear to be based. All of the ancient sources for this story can be found in Düring 1958, 337–425; for an overview, see Chroust 1962, 50–67.
2. See chapters 1 and 2, above; Lennox 1990; Gotthelf 1985; Charles 1990; also the pioneering study of Kullmann 1974.
3. See Gotthelf 1988.
4. The only possible exceptions are Phainias of Eresos (fr. 36–50 Wehrli) and Eudemus of Rhodes (fr. 125–132 Wehrli). The references to Phainias of Eresos suggest a work on plants in five books. The references in Aelian, *De nat. an.* III–V to Eudemus do not suggest systematic biological research.
5. For valuable discussions of the ancient lists and titles of Aristotle's works, see Düring 1950, 1956, 1957; Moraux 1951; Keaney 1963; Balme 1991.
6. Edited by S. P. Lambros, *CAG: Supplementum Aristotelicum* 1.1 (1885). The manuscript includes passages attributed to a long list of authors besides Aristotle, and includes passages which correspond to those found in works of Aristotle's other than *HA,* in the same manner as the Arabic work mentioned later. For someone with both Greek and Arabic, a careful comparison of their corresponding passages might turn up a shared source.
7. The titles of the modern manuscripts are by no means uniform: see Bekker's apparatus at 486a5 for the variants.
8. Keaney 1963, 55, disputes this, but I find his arguments unconvincing.
9. For a list of correspondences, see *CAG: Supplementum Aristotelicum* 1.1: 266–270.
10. Matlock 1967.
11. Moraux 1985, 327–344.

12. For example, in Galen's quotation of *PA* III 666b13–16 at *De placitis* I.8.3–4 (Kühn 1965), a hiatus is present which is removed by elision in all our manuscripts, and in two cases *te* is added to strengthen conjunctions.

13. Cf. *De semine* 4.512.7–513.8, 620.4–13 (Kühn 1965) for similar methodological acknowledgments. A good discussion of Galen's debts to Aristotle's philosophy of science can be found in Hankinson 1991, esp. xxiv, 100–106 (on division), 109–124 (on demonstration).

14. May 1968.

II

Matter, Form, and Kind

Those animals that, though the same, differ in degree, are those which are the same in kind. By 'kind' here I mean, e.g., bird and fish. Each of these is different in kind, and there are many forms of fishes and birds.

Aristotle, *History of Animals* I.1 486a22–25

Some of the birds are long-legged. This is because such birds live the life of a marsh-dweller, and nature makes the organ for the function, not the function for the organ.

Aristotle, *Parts of Animals* IV.12 694b12–15

The chapters in this section all show a concern to correct common misconceptions among biologists and historians and philosophers of biology about Aristotle's concepts of 'form' and 'kind', traditionally (and, I argue, misleadingly) translated 'species' and 'genus'.

These papers, along with the work of David Balme and Pierre Pellegrin, have made a difference. The most extensive re-conceptualization of Aristotle in light of this work is to be seen in the recent history of evolutionary biology by David Depew and Bruce Weber, *Darwinism Evolving*. But the myth of Aristotle as a "typological essentialist" who simply "brought Platonic forms down to earth" is still alive. It stems from a single ideological driving force, Neoplatonism. I can put the point no better than David Balme did more than a decade ago:

The extraordinary later misinterpretations of Aristotle, the magical entelechies and real specific forms, must be largely due to these imported concepts – Species, Essentia, Substantia – which presided like three witches over his rebirth in the Middle Ages, but should be banished to haunt the neoplatonism from which they came. (Balme 1987c, 212)

Amen.

The four chapters in this section all concern Aristotle's conception of biological form. Everyone who has taken an introductory class in Greek philosophy learns that Aristotle's substantial beings are composites of matter, the stuff from which they are made, and form, the organization of that stuff. In application to living things, however, this apparently simple idea is anything but simple.

In the first place, the form of a living thing is its soul, and Aristotle considers soul to be a unified set of goal-oriented capacities – nutritive, reproductive, locomotive, and cognitive. Thus one is disabused of the idea that form is mere structure and shape.

In the second place, the same Greek word – *eidos* – that in some contexts is translated 'form' is, in others, translated 'species'. Indeed, there are translators who will translate this word in both ways in the same chapter of the same work. They do this because Aristotle on some occasions uses his term *eidos* to refer to the 'formal', rather than the material, aspect of substances, on other occasions to refer to a determinate form of some kind (his term is *genos*) of animal or other. To confuse matters yet more, Aristotle occasionally speaks of *genos* as matter, and of *eidos* as the 'final' or 'complete' differentia of the *genos*. At this point one may begin to develop sympathy with the positivist inclination to dismiss metaphysics as nonsense. But however one may feel generally about that inclination, it is not to be followed if one's goal is to understand Aristotle's conception of living beings.

The first two chapters in this section aim to locate Aristotle correctly on two crucial issues in the history and philosophy of biology – essentialism about natural kinds, and the permanence or impermanence of species. These chapters advocate abandoning 'species' as a translation for *eidos* and 'genus' as a translation for *genos,* arguing that 'form' and 'kind' are adequate for all contexts. "Are Aristotelian Species Eternal?" examines the key passages that are most often cited to establish that Aristotle was an eternalist about species. What those passages actually say, I insist, is that reproduction is a biological capacity for maintaining the form of the reproductive agents. The view Aristotle defends is *not* that species are eternal, but that through reproduction, the *individual* partakes in the eternal – it becomes 'eternal in form'.

One consequence of understanding reproduction this way is that every act of reproduction presupposes a prior parent with the form to be reproduced, whose existence presupposes a prior parent, and so on. I also look at those passages which are sometimes cited to indicate that Aristotle is not in principle opposed to the "indefinite modification of species," to use Charles Lyell's phrase. None suggest the slightest hint of an evolutionary perspective. I insist that in the absence of the compelling fossil and biogeographical evidence that emerged in the 18th century, and given Aristotle's theory of formal reproduction, there was no good reason for him to take 'evolutionary' views seriously. But this is not because he believes either in 'typological essentialism' or in 'the eternality of species'.

What, then, *is* his view about animals that are 'one in form', or 'one in kind'? Negatively, Aristotle does *not* think that individuals that are one in form 'share the same essence' and are differentiated by their 'matter' or 'accidents'; nor does he think that a 'genus' is a collection of species so defined. Birds constitute a typical Aristotelian kind. They are a true kind because at a very general level the many forms of bird are the same with respect to a number of traits – or as he would say, differences. They have wings, a certain sort of two-leggedness, feathers, and beaks; they are flyers, and egg-layers. Understanding the *forms* of this kind involves understanding the ways in which these characteristics of the kind are differentiated, the primary task of Aristotelian division (see "Divide and Explain" and "Between Data and Demonstration" in Part I).

Ultimately, the explanation for these differences is teleological – the peculiar legs and beak of the spoonbill or the curlew (see Figures 7.1 and 7.2, pp. 163–164) develop for the sake of their ways of life, their *bioi*. Forms of kinds differ, Aristotle consistently argues, by degree or by 'the more and less'. There is, then, no reason in principle why two different forms of bird shouldn't have members with legs of the same length or beaks of the same shape, or feathers of the same color. And yet these are the very features which one takes it are 'essential' to being one sort of bird or another. It thus seems as if Aristotle is saying that two different kinds of bird could be identical with respect to a feature that is supposed to distinguish them.

My solution to this paradox is that the 'package' which constitutes a distinct bird form precisely adapts it to its way of life – specifying 'what it is to be a Spoonbill' is explaining how its differentiae are precisely those needed to live as it does. (There is no word in Greek which is correctly rendered 'essence' in my view, but the Greek phrase *to ti ên einai*, precisely translated 'what it is to be', is often translated 'the essence'.)

Note that I said all of this without specifying taxonomic levels or categories. Understanding Aristotle's biology demands this. The above paragraph describes a way of thinking that can be applied to any domain in which similarities and differences at more specific levels can be seen as differentiations along various dimensions of similarities at a more general level. To approach him with the Linnean hierarchy in hand is a sure-fire way to miss most of what is interesting in his thinking about biological kinds.

The last two chapters in this section are two recent and closely related papers. Here the focus is on Aristotle's distinction between form and matter – but more particularly on his view that the task of the natural scientist is to study both the formal and the material nature of his subject matter. These chapters both examine what that actually means in practice in biology: how does the zoologist study the matter of animals, how does he study the form of animals, and how are these two investigations related to each other? "Material and Formal Natures" argues that a good deal of biological explanation appeals fundamentally to the details of the material differences found in different animals; and that the active 'nature' appealed to so often in Aristotle's biology is not some cosmic Dame Nature, but the formal natures – viewed as inherent sources of reproduction and self-maintenance – of the animals under investigation.

The most likely source of skepticism about this latter claim are the many passages in which Aristotle asserts that "nature does nothing in vain," a phrase that echoes loudly through the history of science from Newton to Darwin and beyond. The last chapter in this section examines these passages in Aristotle (and some echoes of them in William Harvey's *De Motu Cordis et Sangunis*) and argues that even here, it is to the formal natures of individual animals that Aristotle refers. This chapter is in certain ways the unifying center of this collection. In defending the view that this principle is in fact a zoological 'first principle' such as the *Posterior Analytics* would insist that zoology should have, it is related to the project of Part I. But since this 'first principle' in fact underwrites the appeal to teleological explanation, it also serves as an introduction to Part 3. And finally, it represents my most recent efforts toward understanding the active, causal role of the formal nature of animals in organizing and structuring their material nature for life-maintaining ends.

6

Are Aristotelian Species Eternal?

For a human being generates a human being.

Aristotle

INTRODUCTION

It is often claimed that Aristotelian species are eternal.[1] But this *must* be a difficult claim to establish because of the inherent problems in understanding Aristotle's concept of *eidos* where the Greek may be translated variously as 'form' and 'species'. The best evidence for the claim is of two sorts:

a. A number of passages in the central books of the *Metaphysics* argue that the form does not come to be.[2]
b. Three passages (one each from *De anima*, *De generatione animalium*, and *De generatione et corruptione*) say something about reproduction and being eternal – although just what it is they say is by no means clear.[3]

My thesis is that this evidence, when properly understood, will support the claim that there is an eternal generation of organisms which are one in form and that, as a consequence of this, kinds including species, are eternal. The evidence does not, however, support the claim that Aristotelian forms are eternal. In addition, it will give us a better understanding of Aristotle's reasons for placing sexually reproductive organisms at the very heart of his ontology.

I

A central motivation of much pre-Aristotelian Greek philosophy was the search for some aspect of the world which is changeless, eternally just what it is.[4] Basic to this search was the conviction, whether implicit or stated explicitly as it is at the opening of Parmenides' *Way of Truth,* that if something exists it cannot not be (DK 28 B2). Parmenides assumes that to suppose something comes to be or passes away, one must also hold that it *was not* or *will not be.* And this, he takes it, contradicts the assumption that what is cannot not be. To commit oneself to something's existence is thus to deny that it has a temporally limited career, that it came to exist in the past and will cease to exist in the future. Parmenides tells us there are many signs which point in this direction.

> There is only one way left to be spoken of, namely, that it is. And on this way are many signs that what is is unborn and imperishable; for it is entire, immovable, and without end. It was not in the past, nor shall it be, since it is now, all at once, one, continuous; for what birth of it will you look for? In what way and from whence did it grow? Nor will I allow you to say or think 'from that which is not', for it is not to be said or thought that it is not. How could what is thereafter perish? And how could it come into being? For if it came into being, it is not, nor if it is going to be in the future. So coming into being is extinguished and perishing unimagined. (DK 28 B8)

Plato similarly argues that Forms are to be distinguished from their participants or copies partly in virtue of their being timelessly (or always) F, while participants or copies are F at one time and not-F at another.[5] Other things might come to be F and cease to be F, but the Form is F eternally.

Aristotle carefully distinguishes generation and perishing from other sorts of change (*Ph.* V.1–2). He also defines time as a measure or number of change (*Ph.* IV.11 220a3–4, a24–25; 12 220b17, 221a1–9, 221b23). Something can thus be said to *exist* eternally and yet be capable of undergoing a variety of changes. A planet which *was* at place p_0 at time t_0, is *now* at p_1, and which *will* be at p_2, at t_2, is nonetheless everlastingly just what it is – a planet. In addition, while Aristotle insists that something's being *eternal* implies that it cannot not be and, therefore, that it cannot come to be or pass away, he denies that simply claiming something exists commits one to any claims about its necessarily existing.

This paper will explore an extended sense of 'being eternally' developed by Aristotle and applied to a special class of natural substances. I shall not here speculate on his motive for developing this notion; but one of the consequences of his doing so is that living things are capable of being subjects of scientific knowledge, though each and every one of them comes to be and passes away.

II

I shall translate the Greek *aidios* as 'eternal'. While a popular translation, it is somewhat misleading because it need not imply anything more than existence at every moment in time. If it were not co-opted for *to aei,* 'everlasting' might be preferable. Indeed, in Aristotle these terms often seem to be synonyms. We can avoid the misleading aspects of the current translation by holding Aristotle's account of its meaning and its typical referents in mind. Aristotle attributes unqualified eternity to objects which exist necessarily and which are thus incapable of coming to be and passing away.[6] The sun and moon, as well as the cosmos as a whole, are said to be eternal things.[7] He also argues that natural forms do not come to be or pass away (see note 1 above), arguments we will look at carefully later. Before doing so, however, I shall examine the passages listed in (b) above. The passage from *GA* II.1 will serve as the centerpiece of the discussion; the other passages, as support for the interpretation offered.

As David Balme (1972, 155) indicates, *GA* II.1 731b21ff. is a continuation of the argument of *GA* I which seeks an explanation for why the male and female are found in distinct organisms in some cases. This topic has been a central theme of Book I, but comes especially into focus once Aristotle begins to consider the relative contributions of each to the production of a new organism (*GA* I.17 721a30ff.). Near the end of *GA* I, Aristotle remarks on why the sexual capacities are combined in one organism in plants, yet found in different organisms which are the same in form (730b35) in sexually reproducing animals (*GA* I.23). He argues that if organisms have no function other than the production of seed (731a25–26), the separation of the sexual capacities into distinct organisms is pointless and claims that plants have only this one function or activity.[8] And while he goes on to note that higher animals have other (cognitive) functions, he does *not* tell us what advantage higher animals gain by having the sexual functions in separate organisms.

The opening argument of *GA* II is intended to answer this question. How the interactions of efficient causes and materials produce males and females will be considered as part of the general discussion of organic development (cf. especially *GA* IV.1–3). But he intends to tell us immediately, ". . . how it is due to the better and the cause for the sake of something" (*GA* II.1 731b23–24).

The argument actually consists of three hierarchically related teleological explanations. First, Aristotle wishes to establish why the various kinds of organisms there are always exist (731b24–732a1). Next, he asserts that the male and the female exist for the sake of there being those various kinds of organisms (732a3–4). Finally, he argues that it is better, where possible, that the male and female capacities be separated (732a4–7). And while I am ultimately only concerned with the first argument, it is important to know that it is a preliminary step toward another conclusion. Balme (1972) translates the passage as follows:

> For since some existing things are eternal and divine, while the others are capable both of being and not being, and since the good and the divine is always according to its own nature a cause of the better in things that are capable, while the non-eternal is capable both of being [(and not being)] and of partaking in both the worse and the better, and since soul is a better thing than body, and the ensouled than the soulless because of the soul, and being than not being, and living than not living, – for these reasons there is generation of animals. For, since the nature of such a kind cannot be eternal, that which comes into being is eternal in the way that is possible for it. Now it is not possible in number (for the being of existing things is in the particular, and if this were such it would be eternal) but it is possible in form. That is why there is always a kind – of men and of animals and of plants. (731b24–732a1)

Aristotle begins this discussion by distinguishing between existing things which are eternal and those which are capable of both being and not being. This capability, we are told elsewhere, is inherent in objects which are constituted of the ordinary material elements of our world (*GC* II.9 335a33–34).

The next few lines (b26–31) argue that the soul, which is the form of an organism, is responsible for an organism's being alive and therefore being in a better rather than a worse state. The organism's life is assumed to constitute its good, and it is asserted that an organism is capable of being and not being, that is, of possessing its good and losing it. The relevant premises have not been provided to establish what Aristotle now concludes, namely, that "for these reasons there is a generation of animals." He recognizes this, as the 'for' opening the next line shows. What he now wishes to argue is that *if* animals reproduce,[9] that which comes to be can be eternal – not numerically eternal, but eternal in form. Generation takes place *for this*.[10]

I should like to focus attention on three aspects of the argument which is intended to establish this claim. First, Aristotle asserts that it is individual organisms that are able to be eternal in form, not kinds. This is the point of his denying that what comes to be can be eternal numerically, a denial that would be pointless were kinds under discussion. And notice that he says the *nature* of generated things cannot be eternal, but the generated things can be eternal in a way. In section 4 I shall explore the metaphysical motives behind this careful distinction.

Second, the conclusion of this subsidiary argument – that there is always a kind of men, animals, and plants – is derived from the claim that things which come to be can be eternal in form by means of reproduction. The use of *genos* in the conclusion is, I think, intended to reinforce this point. The context indicates that *genos* is being used in its etymologically primitive sense, defined by Aristotle in *Metaph.* Δ.28 as 'a continuous generation of things which have the same form . . .'; ". . . for example, we say 'as long as the human *genos* exists', which means 'as long as the generation of men continues'" (1024a29–31). When Aristotle concludes that *therefore* there is always a *genos,* or kind, of men, animals, and plants, the force of this remark is that there is a continuous reproduction of organisms of various kinds which are one in form.[11]

Finally, there appears to be an argument which shows why generated things cannot be eternal numerically, that is, *as particulars.* This is a crucial step in the argument to show that the act of formal reproduction which animals can perform takes place because it achieves some good end. The goodness of eternal being or living is taken for granted. Now if living things could be numerically and individually eternal, reproduction would not achieve anything the reproductive agent did not already have. It is only on the assumption that individual immortality is not possible that reproduction makes teleological sense. Let us recall Balme's translation of the crucial lines:

> For since the nature of such a kind cannot be eternal, that which comes into being is eternal in the way that is possible for it. Now it is not possible in number (for the being of existing things is in the particular, and if this were such it would be eternal) but it is possible in form.

Balme (1972, 156) suggests that the argument can be filled out as follows:

1. All beings either are eternal or come into being;
2. What a thing is, it is as a particular;
   Therefore if a thing is eternal in number, i.e., as a particular, it is eternal.
3. But things that come into being are not eternals.
   Therefore things that come into being are not eternal in number.

The basic difficulties are in the portion of the translation Balme puts in parentheses. It is unclear what is the subject of the verbs in the protasis and apodosis of the conditional clause, what is the referent of 'such', and in what way the remark about the being of things being in the particular contributes to an explanation why generated things cannot be numerically eternal. One might have expected that Aristotle had no need of an *argument* to show that things which manifestly come into being are not eternal. And in fact, as Balme reconstructs the argument, one might ask why we cannot directly infer the conclusion from premise 1.

This very fact suggests an interpretation. Balme's expansion of the argument in his note renders the second clause of this parenthetical remark as an assertoric conclusion, rather than as the contrasted counterfactual it is. I wish to suggest that it is no part of the argument to show that the things which come to be cannot be numerically eternal; rather it shows that a Platonic solution will not work. I would reconstruct the argument as follows:

Generated things cannot be eternal in number, for being of all things is in the particular.

Aristotle here expects us to see that since the particular under consideration is a generated thing and since its being is not apart from it, it cannot be eternal in number. He is assuming, however, that if the being were not in particulars another suggestion might be made, namely, that what the generated thing is, its essence, could be numerically eternal even though the generated thing was not. This is ruled out by the fact that a particular thing's being is not apart from it.

Now he considers – counterfactually – the Platonic alternative: "while if the being were a such, *it* (not the generated thing) would be eternal." So, even if we were to suppose an essence or being separated from the generated thing, this would not provide grounds for ascribing eternality to the generated thing which participates in it.[12] A Platonist may respond that it is the *form* that is numerically eternal, and this is unaffected by the temporary participation in it of perceptible individuals. But Aristotle wants to make it clear that on *his* account of the relationship between a generated thing and its being, this answer will not work. It is for this reason that he must either deny eternality to generated things *tout court* or else claim there is a qualified sort of eternality. Typically, he develops the second option.

An implication of this interpretation is that the nature or being of a generated thing is no more numerically eternal than it is. Thus, if Aristotle holds that a species can be eternal, it is not because the being of species-members is eternal in number. This interpretation will gain indirect support when we see him insisting on the same point against Plato in *Metaph.* Z.8.

The assumption that lies at the heart of Aristotle's complex argument is that eternal being is preferable to temporary being. This, combined with a fundamental hypothesis about living nature, which is stated eloquently in the following passage, leads to his claim that he has given us an explanation of why organisms reproduce themselves.

> One of the beginnings of investigation we posit and which we are accustomed to use often in the study of nature is that nature never produces in vain, but always produces the best among the possibilities for the being of each kind of animal. (*IA* 2 704b14–17)

The best *possible* state of affairs for organisms, given that each of them cannot exist eternally, is that each of them be a member of an everlasting, continuous series of organisms which are the same in form if not in number. Reproduction occurs *because* it procures the best possible state of affairs for the reproducing organism.

But now what does it come to, to be eternal in form? The answer to this question is slightly more explicit in *De anima* II.4 than in *GA*. There, producing a likeness[13] is said to be a most natural organic function. It allows the parent "to partake of the everlasting and divine so far as is possible." This it accomplishes because while "the reproducer does not persist, what is like him does; what persists is not one in number, but is one in form" (415b5–6).[14] Every act of reproduction involves the preservation of the form which the parent has beyond his life span. Presuming the reproduced organism also reproduces, and so on, the form characteristic of the kind should persist indefinitely.

But if we now recall Aristotle's account of what it is to be eternal, a serious problem arises. To say that anything is eternal is to claim that it *necessarily* exists. Yet it now appears that the existence of the form characteristic of members of a kind is contingent on members of that species doing what comes naturally. If an organism can be eternal in form only if its form necessarily exists it looks as if the organism is out of luck.

Aristotle crafts an answer to this problem at the end of *Generation and Corruption* II. The arguments of chapters 9–11 of that work are complex and difficult, and the argument I focus on depends on a number of these earlier arguments. I will content myself with picking up two conclusions Aristotle thinks he has already established and will then proceed to the passage which is of direct relevance. Those two conclusions are:

1. It is possible that, even in the domain of generated things, some sorts of generation take place *of necessity.* (337b8–14)
2. This is possible only if (a) there is no limit to the series of generations being considered, (b) there is nevertheless an origin for the series, and (c)

the generations which make up the series are cyclical, involving some sort of 'return'. (338a4–12)

Point 2(c) can actually be inferred from 2(a) and 2(b), given that the only processes in an Aristotelian universe which can go on without limit and still have a 'source' are those which keep returning to that source at periodic intervals. An unlimited linear series, with no source and no goal, is literally a non-starter for Aristotle.

Near the close of *GC* II.11, then, Aristotle raises a question about animals, 'Why do men and animals not return to themselves so that the same individual is regenerated?' (338b8–10). On the face of it, an odd question to ask. But I believe the problem of the necessity of continuous reproduction of things which are the same in form is at issue here. He is first pointing out that we have no direct analogue to the periodic recurrence of a heavenly body at a similar relative position in the heavens. One could imagine a world in which organisms periodically went through a decay, after which time those very same materials were reorganized into that very same organism again. But this is not such a world. Perhaps, then, any given instance of biological generation is merely hypothetically necessary. "For it is not necessary, if the father was generated, that you be, though it is necessary that if you were, he was" (338b11–12). Thus it seems that at every point along a reproducing kind, one can legitimately worry about the future of the kind!

But this is not Aristotle's considered opinion, and there are indications in the wording that already give this away. The argument is formulated in the language of 'hypothetical' necessity, but it is not an example which illustrates hypothetical necessity at all. Typically, it is the materials, including organic parts, required for the production of a natural object that are said to be necessary on the hypothesis that there is to be such an object.[15] This should make us suspicious that the above remarks are only preliminary to Aristotle's solution. In fact, Aristotle is setting us up for an argument to show that biological generation meets the requirements previously established for going on eternally.

> But again, a starting point of the investigation is this: Do all things return to themselves in the same way or not, but some do so numerically, others only in form? Now it is apparent that those things which are indestructible (though their being is such as to be in motion) will be numerically the same (for the change follows on the thing changed). But those things which are destructible, must return in form, but not in number. Therefore, the water which comes from air and the air which comes from water are the same in form, not in number. But even if these things are the same in number, this is not the case with those

things the substance of which comes to be, and which is the sort of thing capable of not being (*GC* 338b12–19)

In the case of the generation of the solstices, the same indestructible being returns over and over again to the same place. Here one and the same being exists throughout the change of place. The 'recurrence' is a recurrence of one and the same substance at the same location: literally a return. Next Aristotle considers the regular recurrences of the elements in a rain/evaporation cycle. He first supposes that the water which comes to be from air is not *numerically* the same water which did so previously, and uses this as an example of a *formal* recurrence, that is, what is now coming to be through condensation is water, though not the same body of water which evaporated previously. The case is problematic and he realizes it. Supposing a reasonably closed system, there is no reason why the very same continuous body of air could not first cool sufficiently to become water and then become air again – that is, the very same body of air.[16] Aristotle goes on to consider the non-problematic cases of "formal, but not numerical, recurrence" – organisms.[17] If there is recurrence in these cases at all, it is clearly not numerically the same substance which recurs, but something of the same kind. This is the answer to his original, strange question. Men and animals do "return to themselves," not by the same individual composite regenerating itself but by its generating something the same in form. The claim is that biological generation is the kind of change which is capable of going on eternally. It is not an unlimited linear series, but a continuous reproduction of the form of the kind. The form, in fact, can be viewed as the ever present beginning and end of the process of generation.

But it now appears we have leapt to the other horn of a dilemma. Having established biological generation's credentials as an eternal feature of the world, Aristotle would appear to have removed it from the realm of the hypothetically necessitated. A rather fundamental Aristotelian distinction between the world of generated things and the domain of the eternal, ungenerated things seems to have broken down.

This worry rests, I believe, on a misunderstanding of this distinction. Hypothetical necessity is still present in the organic world, and only in that world. Nothing said so far implies that materials which are necessary for an organism's generation will *by themselves* generate that organism. And if not, they are only necessary on the hypothesis that an organism of some kind is to be. And that in turn requires reference to the form of that kind. Indeed, this is what we are constantly reminded of (see Joachim 1922, 271 on 337a34–338b19).

For in house-building too it is more the case that *these* things take place because the form of the house is such than that the house is such because it comes to be in this way. For coming-to-be is for the sake of being, not being for the sake of coming-to-be. (*PA* II.1 640a16–19).

Having said that organisms of a kind exhibit the proper sort of generation to be capable of being eternal in form does not, however, give us an explanation of why they are so. But the explanation is now not hard to find. In the first place, Aristotle is insistent that any case of the generation of an organism of high complexity requires the pre-existence of an actual organism of that kind as its source of generation (e.g., *Metaph.* Z 1032b30–32, 1034b13–19; *GA* I.1 715b8–16). This implies an everlasting series of such organisms in the past, for every pre-existent parent will require its own pre-existent parent (see Waterlow 1982, 152–158).

But what of the necessity which insures the future recurrence of organisms with the same form? Here Aristotle insists that one of the basic features of every 'perfect' organism's soul is a natural impulse to reproduce (see above). The *Politics* insists that the natural basis of social organization is that male and female must be brought together "for the sake of generation," but that

this is not due to forethought but just as in both the other animals and plants the striving to leave behind another such as oneself is natural. (1252a28–30)

Just as the presently existent organism of a kind implies an everlasting series of previous such organisms, so does it imply continuous generation in the future. Each organism has, as part of its natural make-up, a natural disposition to make a copy of itself – and an aspect of that copy is the possession of the same disposition. An occasional 'imperfect' species member may be produced, but given the number of organisms around in any generation, that this natural disposition is present 'for the most part' is all that is required to ensure the future recurrence of organisms the same in form.

Aristotle, then, has a carefully developed theory to the effect that organisms of various kinds, though incapable of individual immortality, are eternal in form. To support this theory he argued that the process of coming to be of an organism of a specific kind and then its eventual passing away was to be viewed as one 'cycle' in an everlasting, continuous preservation of the form of that kind. This model of biological reproduction possesses the requisite biconditional necessity to allow the concept of eternity to apply to it, and yet leaves unaffected the applicability of hypothetical necessity to natural generation. The requisite materials for an organism do not absolutely necessitate its existence, but are nonetheless necessary *if* such an organism is to be. Hypothetical necessity is possessed by materials or organic parts in virtue of their being necessary for the life of an organism without by themselves necessitating an organism.

This everlasting, recurrent production of organisms which are all one in form is what Aristotle is talking about when he concludes, in *GA* II.1, that there is always a kind of men, animals, and plants. The natural way to take 'kind' here is in the way defined in the opening lines of *Metaph.* Δ.28 – a continuous generation of individuals which are the same in form. If one thinks of a species in this genetic, historical manner, then it is tolerably clear that Aristotle held that species were eternal. He is just as inclined, however, to speak of the individual of a kind as eternal, not in the unqualified sense (as the kind is) but by virtue of being the same in form with every other instance of the kind.

III

A kind is unqualifiedly eternal and the generated members of the kind are eternal in form, because every member of the kind is the same in that respect. Is the *form* of the kind, then, eternal? I have already stated that Aristotle was inclined to say 'No.' To say of something that it is unqualifiedly eternal, it must be numerically one, and I will now argue that Aristotle rejected the idea that the form of a kind is numerically one. On the other hand, one might suppose he thought that a form is eternal in virtue of the form it has. To make either claim after hearing the *Parmenides* and composing *On Ideas,* appears out of the question. The first raises problems of how a thing numerically one can be participated in as a whole by many individuals; the second seems to imply self-predication.

And yet Aristotle does insist in the passages cited under (a) above (p. 67) that the form does not come to be or pass away, and this is the basic mark of what is eternal. *Metaph.* Z.8 shows Aristotle clearly aware of the difficulties he is in. Having argued that it is not the form characteristic of spheres or humans that comes to be, but spheres or humans (which have those forms), he (without obvious provocation) launches into an attack on the classical theory of forms. I imagine an *agent provocateur,* perhaps in the front row, wondering about these forms that do not come to be or pass away, while those things which have them do – wondering if the lecturer was not much less of a renegade in Platonic circles than he imagined.

A good route into this web of issues is briefly to trace out the dialectic of *Metaph.* B.4 999b1ff. Aristotle's worry is that if only perishable perceptibles exist, nothing will be eternal. If this is true, he insists, generation will not be possible. Why? Because every generation requires a 'from which' and a 'by which' (identified respectively as matter and form) and, given that it is impossible for something to come from nothing or be due to nothing, the matter and form must pre-exist any case of generation. He provisionally concludes that generation requires an ungenerated form over and above the perishable perceptibles.

The next move is to consider this form's status. He decides that, whether one says the form of all men is one or *many and different,* the results are equally absurd. Either all men end up being one or else, strictly speaking, not *men* at all. He also asks whether the first principles are one in form or one in number and again decides neither result will do. Either the principle in virtue of which the many things are one is itself only one in form in each case and loses its ability to be the principle of identity among many distinct individuals,[18] or it is one in number and becomes an individual by itself, not something common to many things.

The resolution he does not consider is to apply the predicates 'one in number' and 'one in form' *only* to composite individuals, and to insist that there are numerically many individuals which are one in form. This avoids the paradoxes involved in imagining that the form is either one in number or one in form. At the same time it allows one to hold both that the being of each individual is proper to it (*Metaph.* Z.13 1038b8–10) and that, in certain respects, all men are indeed one – not numerically, but formally, i.e. in virtue of their form. The payoffs, if this position is defensible, are important. It provides a form which is ungenerated, for there are always human beings. On the other hand, it avoids the apparent concomitant of this, that there is numerically one form. In fact, Aristotle appears to want to restrict the primary reference of the notion of unity to composite substances and groups of substances. If this is so, forms cannot be said to be numerically one or one in any other way. Rather, as we shall see, forms are what make natural and artificial objects one, *both* numerically and formally.

In developing the distinction between numerical and formal unity in *Metaph.* Δ.6 and I.1–3, this is the position Aristotle seeks to articulate. The various sorts of unity there discussed are related in such a way that being one in the less extensive way implies being one in the next more extensive way. "But the posterior follows from the prior, e.g., whatever is one in number is also one in form, but not everything which is one in form is one in number" (1016b35–1017a1). Thus something is a numerical unit by virtue of being one of a kind; therefore, if it is numerically one, it is one in form. On the other hand, it is not the case that all things one in form are numerically one; that is, it is possible that numerically many things are one in form.

Aristotle has what might be called minimal and robust criteria for being a numerical unit. As he puts it in *Metaph.* Δ.6, "Again, in a way, we say anything is one if it is a continuous quantity, but in a way not, unless it is some whole, but this means unless the form it has is one" (1017b12–14). In *Metaph.* I.1[19] the unity an individual has in virtue of its form is explicated causally as either due to the activity of a craftsman imposing organization on a heap of materials or (especially) due to nature, i.e. due to the individual "having in itself the cause of its continuity" (1052a25). Notice that being a continuous quantity is still crucial, but the sort of continuity involved in organized, functional bodies is given pride of place. The reason for this lies in the central position *indivisibility* plays in Aristotle's account of unity at all levels.

If one defines unity as continuous quantity, then every unit is potentially many, e.g., each 'unit' of water or mud is potentially as many units as you like.[20] Its 'unity' consists simply in the accident of currently not being divided.

Aristotle, however, insists that 'that which is individually indivisible is indivisible in number, while that which is indivisible in understanding and knowledge is indivisible in form' (1052a32–33). He concludes:

> The one then is said in these many ways, the naturally continuous, the whole, the individual, and the universal; but all these are one by being indivisible, in some cases the change being indivisible, in others the cognition or account being indivisible. (1052a35–36)

Unlike a continuous 'heap' of uniform stuff, an organized body, especially one which must be organized in a certain way to function,[21] is not potentially many. Insofar as it is a sandal or porpoise, an artificially or naturally organized body is "one by being indivisible."[22]

Aristotle has, then, a notion of form as the source of the numerical unity of a parcel of material – "the cause of unity for a substance" (1052a34). But he also insists that numerically many individuals may be one in form. He gives a variety of explications of this formal unity possessed by many individuals: having one account (1016b33, 1052a36), or being indivisible in conception (1052a32,35; 1016b2) or in account (1052a33). He expands on these notions of indivisibility of conception and account in an identical manner:

143

> Again, things are called one the account of the essence of which is indivisible
> relative to another which reveals the thing's essence. For just by itself every
> account is divisible. (1016b33–35: cf. Δ.6 1016b1–4, I.8 1058a18–19)

This passage indicates that it is not having just any account in common
that makes many things one in form – the account must also reveal what it is
to be any of the objects in question. Now an account of the genus would do
this, but it is *divisible* (into accounts of various species), so Aristotle insists
that, among essence-revealing accounts, it is the one which is not further
divisible that is common to the many things one in form (cf. *PA* I 642b7–20,
643a8–12, 644a24–26 and *Metaph.* 1034a5, 1038a16). One more detail in
this passage is revealing: he notes that by itself every account is divisible.
'By itself' is in opposition to 'relative to another which reveals the thing's
essence'. In any division, a point will be reached beyond which further
division will not further reveal features essential to the kind in question.

Can we provide a concrete specification of that point? I think not. Two
principles of Aristotle's approach to definition militate against doing so. The
first is his willingness to seek definitions for kinds at various levels of *our*
taxonomic hierarchy (see Balme 1987b; Pellegrin 1986; Gotthelf 1985b).
What is incidental to being a bird, however, may be *crucial* to being a wading
bird, a crane, or a sandhill crane. Thus, a feature is 'essential' or 'incidental'
only relative to a kind, not in itself. Take legs, for example. Birds are by
nature bipeds of a peculiar sort: their legs bend at the "knee" in the opposite
direction to that characteristic of the other true bipeds, human beings. Some
birds have long, others short, legs – having long legs is not essential to being
a bird, and to divide further than forward bending biped would reveal nothing more about what it is to be a bird.

In this example, the division has stopped short of quantifiable differences
in degree and one might think this is precisely the point at which division
ceases to reveal anything essential. But two related considerations rule out
this suggestion. First, Aristotle treats the forms of his greatest kinds as
differing, for the most part, only by 'the more and the less'.

> Among the birds, the differentiation relative to each other is in the excess and
> deficiency of the parts, that is, according to the more and the less. For some are
> long-legged, some short-legged, some have a broad and some a narrow tongue;
> and similarly as well in the other parts. (*PA* IV 692b3–6)

If such distinctions were in some absolute way incidental, Aristotle would be
unable to define anything below the level of the nine greatest kinds he
recognized.

A second consideration which makes this suggestion implausible is that these sorts of differences are in fact *crucial* to the being of all kinds of creatures. If a crane is to survive and flourish, it *must* have, not simply "long" legs, but legs of a certain length defined relative to its body, neck, environment, and so on.

> Whichever birds live in swamps and eat plants have a broad beak, this sort being useful for digging and for the uprooting and cropping of their food. (*PA* IV 693a15–17)

> Certain of the birds are long-legged. And a cause of this is that the life of such birds is marsh-bound; and nature makes organs relative to their function, not the function relative to the organ. (694b12–15)

Aristotle usually does not consider lower levels of fine-tuning to the environment than this (though he does occasionally). But this, and dozens of similar passages, are enough to show that differences of degree in the affections of organic structures will often be part of an account of the being of a kind.

In *GA* V Aristotle discusses incidental properties in just this manner. He is reviewing the affections by which the parts of animals are differentiated. He has in mind such things as color differences in the eyes, hair, or feathers of organisms which in other respects appear identical. He insists that if the feature is neither a part of the kind's nature nor proper to each kind, then it neither came to be nor is *for* anything – nor should it be considered part of that kind's defining characteristics.[23] He makes it quite clear, however, that what sort of feature fits the above description cannot be specified out of context: "For an eye is for something, while being blue is not, *unless* this affection is a property of the kind" (*GA* V 778a33–34). Aristotle seems to have used the fact that a feature is a property of the kind as evidence for that feature's being one that is present for the sake of something. If a species were universally blue-eyed, being blue-eyed in all likelihood would be for something. Similarly, in most species of lobster, Aristotle argued that the right claw is always larger than the left and he willingly gives a functional explanation for this fact. However, when the members of a species do not have the larger claws always or usually on one side, Aristotle concludes that having one larger than the other is a matter of chance, and not for the sake of anything (*PA* IV 684a25–32).

Every individual of a kind will have its organs and tissues completely differentiated – they will have a precise hardness or softness, texture, hue, temperature, viscosity, width, length, and so on. But there will only be a *range* of these precisely determined qualities relevant to the organism's life.

Individuals which differ *within* this range do not differ essentially or in form, but only 'incidentally'. Two individuals may be one in form while having every qualitative property instantiated to a different degree within the appropriate range. But these individuating differences will not be relevant to a functional account of the organism's nature. Thus, when *PA* I.5 makes recommendations for the study of the activities that constitute an organism's life, it issues the following directive:

> Therefore we must first state the activities, both those common to all and those that are generic and those that are specific. (I call them common when they belong to all animals, generic when they belong to animals whose differences among each other are seen to be in degree. For example, I speak generically of 'bird' but specifically of 'man' and of every animal that has no differentia in respect of its general definition. What they have in common some have by analogy, some generically, some specifically.) (645b21–27; trans. from Balme 1972)

The addition, "in respect of its general definition," stresses the point that the actions of races of men, or of individual men, may differ but not in ways relevant to a *formal* account which is general for all men.

Let me now restate the sense in which an organism can be eternal in respect to form. An individual organism is eternal in form if it is the product of, and in turn has a natural disposition to produce, an individual the general, essence-revealing, account of which is identical with its own.

If to be one in form is to be indivisible in account in the way specified, then numerically many individuals will be indistinguishable with respect to that account. If that account names only teleological features, those individuals will be identical with respect to those features required for each of them to exist. Those features, however, will vary from individual to individual in ways which do not entail altering their functions in ways crucial to the organism's life. Such variation between individuals is compatible with their being one in form. To be eternal in form, on the other hand, is to be a member of a necessarily reproducing series, each member of which is one in form. An organism is eternal in form if it is disposed to make a copy of itself which is indistinguishable from it with respect to those characteristics which contribute directly to its life.

That being one in form, as we now understand it, provides a solution to those aporetic problems from *Metaph.* B.4 with which we began this section, seems clear from *Metaph.* Z.14. Aristotle there contrasts a view which makes the form numerically one and the same (as you are the same as yourself [1039b1]) with one which makes the account revealing different things one

(1039a24, b8). The former view seems to force one toward a notion of forms as separate, *thises,* and substances (1039a33–34), a view of forms of natural kinds Aristotle rejects (1039b18–19).

One of the advantages of such forms, however, is that, being separate from perceptible individuals, they stand outside the realm of generation and destruction. Aristotle sees the existence of generation as dependent on the form characteristic of the kind not coming to be or passing away (*Metaph.* Z.15 1039b20–1040a8, B.4). But can he defend this dependence while insisting that the forms of natural kinds are organized capacities of numerically many temporally limited bodies? This is the problem of *Metaph.* Z.8.

IV

Metaph. Z.8 divides into two distinct sections. The first (1033a24–1033b19) argues that a proper analysis of generation requires us to assume that neither the matter nor the form of what comes to be itself comes to be (cf. B.4 999b6–17). The second (1033b19–1034a8) involves a puzzling critique of the middle dialogue theory of Forms. Ross (1924, II 188) has stated the connection clearly:

> Aristotle passes now to consider a doctrine which might seem to follow from his denial of the creation of form, viz. the Platonic doctrine that Form exists eternal and independently.

The chapter begins with a principle established in the previous chapter: that which comes to be does so *by* something (identified as the source of the beginning of generation) *from* something (the matter), and comes to be something, e.g., a sphere or a circle (Z.8 1033a24–28). Aristotle immediately draws a conclusion from this, namely, that just as matter is not generated (argued for at 7 1032b15–1033a2), neither is form.

G.E.L. Owen (1978–1979, 20–21; Burnyeat et al. 1981, 45) has pointed out the importance of a small grammatical distinction in this chapter which goes a long way toward sorting out some of its confusions.[24] For Aristotle says:

a. *A* sphere or a circle is what comes to be; and
b. *The* sphere does not come to be.

The reference in (a) is to embodied instances of the form as indicated by the lack of a definite article. But it is to such instances as *instances of a kind,* not as unique individuals, which Aristotle refers by the use of the demonstrative pronoun.[25] Thus, he does not argue that the form does not come to be in Socrates or Callias or that instantiations of the human soul do not, but only that the form characteristic of human beings does not come to be. This suggestion finds confirmation in a later chapter of *Metaph.* Z where Aristotle says parenthetically, "For the being for a house is not generated, but only the being for *this* house" (15 1039b24–25).[26]

Following this remark, Aristotle goes on to argue that demonstration and definition cannot be of the perceptible composite.[27] This passage fulfills a promise made at the end of chapter 13, that the manner in which a substance can be defined will be discussed later (1039a20–24). Aristotle's claim that the form does not come to be is the claim that there is always something around with that form. Does that imply the existence of numerically one thing, the form of man? It cannot: in the case of natural things, it is difficult even to imagine the form apart from its appropriate materials, and it certainly cannot exist in separation. And yet Z.15 opens with a distinction between the formula bound up with the matter and the formula generally,[28] and (again) between a house's being and *this* house's being.

Building on our earlier discussion, I would suggest the way to understand this distinction is in terms of different levels of differentiation (and correspondingly different accounts). Being a house does not entail being one precise size, shape, or having bricks of a certain precise density or dimension. The *general* account will leave all this out. But something can be this house I am sitting in only if it has these various precise features. The form of man does not come to be, since there are always functioning humans about; but the precise way that I fulfill those functions will differ in many incidental ways from the way you (or my daughter) does.

Aristotle introduces the notion of being one in form in order to sort out an ambiguity in the claim that Socrates and David Balme have 'the same form'. Aristotle is concerned that this phrase not be taken to mean that there is numerically one thing, the form, in which David Balme and Socrates share. Having the same form might also mean David Balme and Socrates are the same in the organized set of capacities their distinct bodies possess. That is, the reference of 'the same' may be directed to composite individuals, not to the form. The phrase 'X and Y are one in form' does not possess the same ambiguities, for it treats individuals as one *in respect of* form, rather than suggesting there is some one thing they are all related to. At the same time, Aristotle's conception of natural substances as *unities* drives him to insist that the precise way my organs are structured and function is the way the human soul is for me. Or so I am claiming. The test of these claims, apart from textual support (which it is admittedly possible to interpret otherwise), is in their ability to make sense of otherwise paradoxical arguments in *Metaphysics* Z.

In *Metaph.* Z.8, Aristotle demonstrates that the form is not generated by showing that if it *were,* since every generation is for something, generation would never take place.[29] But he also says that *this* form is generated *in another* (1033a34, b10: cf. *GA* I 730b15). Again he sums up his previous argument against the generation of the form by saying, "If there is a generation of the being for a sphere generally, there will be a something from something" (1033b11). All of these texts present a consistent theory if there is a generation of this sphere here (i.e. the blue marble one three inches in diameter in the corner of my room) but not a generation of what it is to be a sphere (for there are always solids with surfaces every point of which is equidistant from a center).

The constant reference throughout this argument to the generation of this form *in another* looks toward the characteristic feature of biological generation – the making of another like the producer. It is this feature which Aristotle thinks allows us to dispense with separate, substantial forms, the point of the second half of *Metaph.* Z.8.

Is the form characteristic of the kind, e.g., the sphere as opposed to spheres, separate from these spheres here? This is the question which opens the chapter's second half, and it is a reasonable question, given the first half of the chapter. For Aristotle has been distinguishing between the sphere and the composite bronze and wooden spheres here: one might easily lapse into speaking of the sphere itself and the many spherical things. If the position he is arguing for is to be clearly distinguishable from Plato's, it will be in virtue of how he answers this question. Two responses are offered:

1. Generation would never take place if 'a sphere' signified a this, that is, some definite thing. Generation requires 'a sphere' to signify the sort of thing which comes to be. (1033b21–27)
2. Platonic forms are of no use in accounting for organic generation, and this is a severe difficulty because living things are most of all substances. (1033b27–1034a8)

The argument for (1) is sometimes taken as an argument against separate and independent forms, but I believe the argument is more general than that and has to do as much with a clarification of Aristotle's own position as with criticizing Plato's. The question Aristotle is responding to is – is there a sphere besides these spheres here? One way of taking this question is as a request for clarification of the claims that things which come to be come to be (e.g.) spheres or humans, and yet the form of sphere or human does not come to be. Might this mean that things which come to be spheres come to participate in the form, which itself does not come to be?

The request is perfectly reasonable. Aristotle tells us that spheres (1033a27–29, 1033b2), men and plants (1032a18), bronze spheres (1033b8–9, 1033b10, 1034b11), the form in another (1033a33, 1033b10), and the whole named in virtue of the form (1033b17) come to be. He *denies* that the bronze, the sphere, the circle, the form, the essence, the shape in the sensible, the being for a sphere generally or the thing spoken of as form or being, come to be (1033a29, 32–34, 1033b5, 6, 11, 16; 1034b10). His language is fluid enough to be contradictory if contexts are dropped. In chapter 9 *a* sphere and bronze are said *not* to come to be (1034b10); in chapter 8 *a* sphere *is* said to come to be (1033a28, 1033b2).

The reason for this fluidity is made clear in Z.10–11. General terms like 'man' and particular terms like 'Socrates' seem to refer sometimes to general or particular composites, and sometimes to the form of a kind or some individual's form (1035a5–6, 20–21; 1035b1–3; 1036a1–2; 1037a5–10, 1043b2–4). My view is that Aristotle would prefer to legislate the latter use out of existence, but in fact simply notes the ambiguity. As Loux (1979) points out, this approach is a descendent of Aristotle's recognition (*Ph.* I.7–8) that phrases like 'the white' can refer to the property a thing has after a change (whiteness) or to the thing itself described with respect to that property (the white thing). So 'human being' can refer to the human soul, to what it is to be a man, or to human beings described generally – to use Z.8's terminology (1033b17).

The problem is that the same term now may designate both what *cannot* come to be and what *does* come to be, precisely the distinction Aristotle wishes clearly to demarcate. The question opening this section of the chapter can be expanded to reveal its motivation as follows. Is there a sphere or house, that does not come to be and is apart from the material ones [which do come to be]? That is, does this distinction between spheres which are generated and the form of a sphere which is not force us to accept the existence of separate forms?

The rest of the chapter offers two closely related negative answers to this question. The first (1033b21–28) is that, given the properties separate entities must have, a sphere apart from the material ones would be quite useless in explaining their generation. The second (1033b29–1034a8) is that, given the analysis of generation just offered, biological reproduction has the requisite properties to account for natural generation and, therefore, it is unnecessary to posit forms as paradigms. The basic conceptual link between these two responses is the distinction between terms which refer to matter/form composites as *thises,* and those which refer to them as *suches.* "Now the complete this, Callias or Socrates, is just like the bronze sphere here, while the human and the animal are just like a bronze sphere generally" (1033b24–26). Terms like 'house', 'sphere', 'animal', and 'human' do not refer to other thises distinct from these ones here – they refer to the sort of thing these ones here are. But sortal reference already includes a 'material' component. Aristotle has yet to give his strongest argument for this way of conceiving terms for natural kinds, but the conclusion of those arguments is already quietly assumed.

The role of sortal referring expressions in these arguments is all-pervasive. In the first answer, Aristotle insists that terms like 'house' and 'sphere' do not refer to 'thises and definite things' (1033b22) but to suches (b23), and that the generator generates a such from a this, the result being a 'this such' (b23–24). And after the passage quoted above, he concludes:

> It is therefore apparent that the cause in the sense of the forms as some are in the habit of speaking of the forms, if they are something apart from the particulars, are of no use in relation to the generations and substances. (1033b26–28)

Why will *Phaedo*-type explanations not be of any use? He treats this as a conclusion which follows from what he has said in response to the initial question. The implicit answer is that terms like 'man' and 'sphere' refer to what comes to be, but refer to them as sorts or kinds of things. What must be explained is how a sort of thing comes to be. If one takes sortals to refer to already definite thises which are apart from individuals, the central problem of generation is left without an explanation. The *Phaedo*-type theory posits paradigm instances of kinds as the primary reference for sortal expressions. Aristotle insists to the contrary that their reference is directed to individuals but as instances of kinds.

But then if the fact that individuals can be properly signified by sortal expressions is not explained by those individuals' being somehow related to a paradigmatic primary reference, how is it to be explained? Aristotle's answer comprises the second response to his original question. Again the notion of individuals which are suches is crucial. Individuals come to be sorts of things by becoming such as the generator: and this is to become one in form with it (1033b31–33, 1034a7–8). "The generator is sufficient to produce and be the cause of the form in the matter" (1034a4–5). This claim must be read correctly. The generator does not make the form in the matter –

he makes the-form-in-the-matter, a material thing of a certain kind. Aristotle strains to explain the point in this chapter's well-known closing lines:

> But the complete thing, 'the sortal form' in these flesh and these bones here, is Callias or Socrates, different due to their matter (for it is different), but the same in form, for the form is indivisible. (1034a6–8)

The account which captures the being of Callias will also do for Socrates; any more precise account which might distinguish them will not capture their being. For many things to be one in form is for them to be not distinguishable at the level of differentiation which is described in a scientific account of what they are. The account will nonetheless be of what each of them is, not an account of a paradigmatic instance which each more or less resembles, or of a separately created form in which each 'partakes'. Forms are not produced; individuals of various kinds are. The explanation for there occurring so regularly individuals of kinds is a naturalistic one. Organisms have a natural disposition to make copies of themselves and can only come to be as the result of the actualization of that disposition. For a more detailed story of how this takes place, one must engage in a scientific study of organic generation. *That* it takes place is a given of common experience. If this argument is to have force, it must show that biological reproduction solves the same problems separate forms were intended to. I shall now indicate that it does just that.

According to the *Phaedo,* if I were to have asked Socrates in his last hours why this stuff came to be a horse, he would reply that nothing other than the Horse itself was responsible; this stuff became a horse by participation in the Horse itself. Aristotle's conception of biological reproduction allows him to adopt a causal model not far removed from this, which nonetheless renders separate and independent forms otiose. Something becomes a horse by coming to have the form characteristic of horses. Yet the form itself does not come to be, for it already is (in the sire). It persists while the various steeds come to be and pass away. Secretariat's being will pass away; but the sort of being Secretariat is, the equine soul, will not. And the form is responsible for this stuff becoming a horse both as the goal of the process and as its efficient cause. The materials necessary *if* Secretariat is to be will never necessitate Secretariat in an unqualified way. The agency of the equine form (in the male sire) is required to make the material *material for a horse* in an unqualified way.[30]

Biological reproduction, then, answers just those questions the *Phaedo* wants answered and makes the form of F responsible for what comes to be F without requiring that the Form be separated. In addition, it shows why the materialistic explanations Socrates was attracted to in his youth will never be adequate (*Phaedo* 96a6–99d5) and affords equal status to formal and teleological explanations, a goal hoped for but left unfulfilled in the *Phaedo* (97d5–99d1). As a response to the account of explanation in the *Phaedo*, this is a *tour de force*.[31]

One might respond that this analysis works fine for one small class of sensible objects, but that is all. However, Aristotle's account of the way artefacts are produced is self-consciously parallel to his model of biological generation (*Metaph.* Z.7 1032a20ff.; *GA* I.22, II.1 734b21ff.), and it is not clear that, among generated things, anything else counts as a substance.[32] Thus, this model has far greater generality than might at first be supposed, at least as an account of substantial generation.

Aristotle's argument in Z.8 moves from the presentation of a theory that the form characteristic of a kind does not come to be, to blocking a possible inference that might be drawn from this, namely, that forms are separated from instances of the kinds that come to be and pass away. He can block this inference, he believes, by noting that the best candidates for substance are able to make things like themselves in form.

Further, we have seen that, for sexually reproducing organisms, this replication process is necessarily eternal, which allows Aristotle to speak of an eternal reproductive kind, and to speak of instances of the kind as 'participating' in the eternal, and as being eternal in form. Is he willing then to say that the form of a biological kind is eternal?

There is, as far as I know, only one text which may be so construed, but at the very least it is tentative and ambiguous.[33] It is more striking how often he goes so far as to say the specific form is neither generated nor destroyed, and yet does *not* say it is eternal or everlasting. I believe he has good reasons for this.

For an individual to be eternal, as we have seen, either it must persist at any and every time, or it must be a member of a continuous series of individuals which are the same in form such that for each moment of time at least one member of that series exists at that moment. Now consider each alternative with the idea in mind that it is the form characteristic of the kind that exists at any and every moment of time. First, there is no one form characteristic of the kind if by this is meant numerically one thing. For Aristotle, to speak of 'one form' or 'the same form' is to speak of a respect in which many matter/form composites are indistinguishable. Thus, he would not wish to predicate eternality of a species-form in this way. Perhaps, then, a species-form is eternal because it is one in form in all the individuals which exist at any and every moment of time. This, however, is equally impossible.

The notion of being one in form was coined to characterize a manner or respect in which composite substances are one or the same. That manner was by virtue of their form. If forms of individuals are now said to be one in form, it must be in virtue of something the same in all of them, and so we get a regress analogous to the Third Man (but see note 22). I conclude, then, that the form characteristic of a species is not a candidate for eternity.

CONCLUSION

The appropriate answer to the question, Are Aristotelian species eternal?, depends on what one takes an Aristotelian species to be. If one supposes any group of organisms which are the same in form to be an Aristotelian species, then they are not necessarily eternal. Many species discussed in detail in Aristotle's biology lack the appropriate sort of causation to be eternal; moreover, Aristotle conceived of hybrids produced by individuals similar but not the same in kind. On the other hand, some organisms by their very nature reproduce their form in another, and *these* organisms constitute an everlasting branching series which Aristotle termed a *genos,* a kind. Now, it may be thought that, as these 'kinds' are eternal in virtue of each member being one in form, the form would be eternal. And it seemed prima facie plausible that Aristotle would hold this view, because he argued that the form characteristic of a species does not come to be or pass away. Nonetheless, he did not claim this, and for good reason. The form is not the sort of thing that can be eternal, even though it is the basis for predicating eternality of a natural kind and of its members.

My discussion suggests that the motives behind Aristotle's argument for the eternality of certain kinds went far beyond those suggested some years ago by David Balme (1972, 97):

> There is nothing in Aristotle's theory to prevent an 'evolution of species', i.e., a continuous modification of the kinds being transmitted. But he had no evidence of evolution . . .

Balme cites three types of evidence to support his claim that Aristotle was unconcerned about the 'fixity of species':

(i) new kinds arising from fertile hybrids (*GA* II 738b28–34, 746a30ff).
(ii) 'Dualizing' organisms, e.g., seals, bats, ostriches, cetacea.[34]
(iii) Aristotle's theory of heredity in *GA* IV.3, which explains the various differences one can see between parent and offspring.

154

But (i) is somewhat weakened by the fact that Aristotle in an earlier and more theoretical passage says,

> And for this reason [because the male supplies the soul for certain sorts of body], whenever a female and a male of similar kind copulate (which happens when the times are equal, gestation periods similar, and the sizes of their bodies are not very different), the first comes to be according to the likeness common to both,[35] but as time goes on and one thing comes to be after another, it ends up like the female form. (738b28–34)

New kinds arise, but not naturally (746a29), and they have a natural tendency to revert to the female sort, although exactly why is not clear. The evidence provided by dualizing kinds is equally unhelpful, for it is not as if any of these are not perfectly determinate kinds with their own form passed on from one generation to the next. The problem posed by these organisms is that they run afoul of certain popular divisions. But there is no suggestion in Aristotle that porpoises might have evolved from other air-breathing land animals or that they ever do anything but perfectly replicate their form.

Finally, the discussion in *GA* IV.3 of heredity shows simply that children often only resemble their parents in kind and not in nonessential details – a view that is perfectly consistent with the form of the kind being everlastingly reproduced.

Balme opened his discussion of this issue by making two claims: that continuance of species does not entail fixity, and that theological motivations for believing in the fixity of species are not relevant to Aristotle. On the first claim I cannot agree. If to continue a species is to continue replicating its form, it does entail fixity. With the latter proposition, however, I emphatically agree. Nonetheless, I hope I have demonstrated that certain metaphysical principles interacted in subtle ways with his biological explanation of reproduction. Aristotle held that any case of a biological generation presupposed the preexistence of the form of what came to be. This view is certainly verified by pre-paleontological experience. But it is also clear that this was a metaphysically fundamental principle for him. Matter could never organize itself into a functional organism of high complexity – that kind of organization could only be provided by a pre-existent instance of the kind reproduced.[36]

Similarly, there is ample evidence of an impulse to reproduce, often awe-inspiring in its persistence – consider the behavior of salmon, or of male black widow spiders. But Aristotle also argues eloquently that that impulse is an extension of the reproducer's basic goal of maintaining its being (*de An.* II.4 416b14–17). The biological evidence did not force this interpretation on him, but it did allow him to offer a naturalistic explanation of certain problems of predication.

Finally, the epistemological motive behind his insistence that the individual which comes to be and passes away can be eternal in form must be taken seriously. Without this, the world of generation would be incapable of demonstrative understanding (*Metaph.* Z.15 1039b20–1040a2). Aristotle's good sense told him that an explanatory science of living things was possible. His theory of formal reproduction lies at the very basis of his causal explanations of biological regularities.[37] And it is hard to imagine Aristotle referring to anything as 'being in the strongest sense' which was not a necessary and fundamental feature of the cosmos. The natural necessity of formal reproduction serves as the basis for Aristotle's biological science.

> For the non-random, the for-something's-sake is present in the works of nature most of all, and the end for which they have been composed or have come to be occupies the place of the beautiful. If anyone has thought the study of the other animals valueless, he should think the same about himself . . . (645a24–218; as Balme [1972] elegantly translates).

ACKNOWLEDGMENTS

In the summer of 1976 I was working on the last draft of my doctoral dissertation, which dealt with the relationship between Aristotle's biology and his *Metaphysics*. At that time I learned of a colloquium to be held that December on Aristotle's biology, at Princeton University. It was on that occasion that I met David Balme, whose work I had been grappling with for about three years. That meeting encouraged me to ask Professor Balme to be an external reader for my dissertation. I remember telling him some time later that the summary of the theme and argument of my thesis which he sent (along with pages of notes) to my committee, had captured its essence better than I could have. From the beginning of my work on the subject of Aristotle's philosophy of science, Balme's essays and commentaries have served as a constant source of stimulation and revelation. That this is so is nowhere clearer than in this chapter.

This chapter is a descendant of (though not one in form with) papers read to a symposium on Aristotle's biology at Marquette University on November 25, 1981, and to the Society for Ancient Greek Philosophy, at the APA Western Division meetings (Columbus), April 29, 1982. Many of the ideas in this final version are the result of the discussions on those occasions. I should like to single out John Cooper, Allan Gotthelf, Joan Kung, Julius Moravcsik, and Nicholas White for asking particularly vexing questions on these occasions.

NOTES

1. Among recent discussions, see Sorabji 1980, 145–146. In popular expositions, see Randall 1960, 226; Grene 1963, 136–137; G.E.R. Lloyd 1968, 88–90; Ackrill 1981a, 133–134.
2. Cf. *Metaph.* Z 1033b11–19, 1034b7–19, 1039b20–1040a8; H 1043b14–21.
3. *GA* II.1 731b21ff.; *de An.* II.4 415a22–b8; *GC* II 338b1–19.
4. Cf. Thales, DK 11 A12; Anaximander, DK 12 A9; Pythagoras, DK 28 B26; Democritus, DK 68 B9.
5. E.g., *Phaedo* 78d5, *Timaeus* 27d5ff., *Republic* 479a2, 484b2–3; *Symposium* 211a1–2. For an excellent discussion of the relationship between Parmenides and Plato on this question, cf. Owen 1966.
6. See *EN* VI.3 1139b21–24, *GC* II.11 338a1ff., *PA* I.1 639b24: cf. Bonitz 1870, 14b11–14.
7. *Metaph.* Z.15 1040a29; *Cael.* I.3 270a12ff., II.1 283b26–32.
8. At 731a25–27. This is a somewhat odd claim, for they also possess the nutritive capacity. In fact at *de An.* II.2 413b1 Aristotle claims they have no further capacity of soul than the nutritive. A verbal solution is provided by his claim that these are 'the same capacity' (416a19–20), but what this *means* is a complex story.
9. 'Reproduction' is an interpretative translation of *genesis* here, but is sanctioned by the companion passage at *de An.* II.4, where Aristotle explicitly discusses "making another like himself."
10. Again, for the explicit use of 'that for the sake of which' to describe being eternal in a way, cf. *de An.* II.4 415b1.
11. *Metaph.* Δ.28 only discusses three senses of *genos,* referring to races (e.g., the *genos* of the Heraclidae, *Metaph.* I 1058a29), to a continuous generation of the same form, and to the substratum or 'matter' of differentiation. Only the second can refer to items at the level of species. In addition, it is a peculiar feature of the word when used in this manner that it takes the genitive plural. That is, Aristotle will typically talk of the form of man, but the *genos of men.* This is inappropriate if one is thinking of the feature in virtue of which all men are men, but perfectly appropriate if one is thinking of the continuous sequence of individuals with that feature. Finally, three aspects of this current argument make taking *genos* as a continuous generation appropriate. (i) It renders innocuous the fact that this term ranges over 'men', 'animals', and 'plants' – one can switch easily from discussing the continuous generation of men to there being a continuous generation of animals having the same form. (ii) 732a1–2, while it adds content to 731b31–32, is intended to recall that prematurely stated conclusion, namely, that "there is a *genesis* of animals." (iii) Finally, the crucial point made in this passage is that what comes to be can be eternal in form. But that is so only if there is a continuous generation of organisms of the same kind. If *genos* is taken as I suggest, 732a1–2 says, in effect, 'because this good will result if there is a continuous generation of things with the same form, there is always a kind.' It provides a very natural teleological conclusion to the passage.

12. This passage is extremely tricky. Peck offers a reading in a note to his translation that requires the understood subjects of the protasis and apodosis to be different. While this gives the argument some force, it is grammatically unlikely. If one assumes the same subject for both clauses, but takes *toiouton* to mean 'eternal in number', the argument seems vacuous. I have suggested, therefore, taking *toiouton* as the abstract noun it often is in Aristotle. This seems plausible, given the extent to which Aristotle's technical metaphysical vocabulary plays a role in this discussion.

13. *De An.* II.4 415a28.

14. For the inspiration behind this passage's language, cf. Plato, *Symposium* 206e.

15. *Ph.* II.9 200a12–14; *PA* I.1 639b25–640a9. Cf. Cooper, 1987.

16. At any rate, a spatially continuous heap of this sort is not one *at all* in Aristotle's 'robust' sense of 'one'; see below, and *Top.* I 103a22.

17. Organisms are not actually mentioned, but it is hard to imagine what else the last line of *GC* II.1 1 could refer to, given that the elements are the contrast class and that the entire discussion as motivated by a difficulty about organisms. See Peck 1963, 573–574.

18. Thus a suggestion made some years ago by Rogers Albritton (1957, 706), that the universal form "is not one in number (but) only one in form," was rightly rejected by Aristotle from the outset. In fact Albritton then goes on to say that on his reading "men are one in form," which means both that the universal form of man, and the many men, are one in form. This is surely just the sort of problem Aristotle is trying to avoid.

19. This discussion makes clear reference to *Metaph.* Δ.6 at 1052a15–16.

20. Thus when in Z.16 Aristotle says earth, fire, and air are not one but mere heaps (1040b9), he is thinking of the robust sense of unity. See Balme 1972, 156.

21. His example in *Metaph.* Δ.6 is that a sandal cannot be put together in any which way and be one sandal (1016b14–16).

22. A passage which I must assume constitutes a slip on Aristotle's part is *de An.* I.5 411b19–20. He is discussing organisms which can be divided and remain alive. He claims the parts then have "the same soul in form, even if not in number." If soul is form, he is saying the division has taken us from one animal with numerically one form to many with many souls 'one in form'. He ought to have said many 'organisms one in form'.

23. *GA* V.1 778a29–35. On this teleological conception of essence, see Balme 1980 and chapter 7 below.

24. What I make of this distinction, however, is in agreement with neither Owen's analysis nor the general tenor of the *Zeta Notes*. Passages which I take to mark a clear distinction between the form as the organization (or organized capacities) of individual parcels of materials and the form as identical in many individuals, they are at great pains to explain away. See, for example, their note on 1039b25 (Burnyeat et al. 1981, 141–142), and generally on passages suggesting such a distinction. For an analysis more closely allied to my own, see Loux 1979.

25. E.g., 'the bronze sphere here' (1033b25), which he compares to Socrates and Callias: cf. 1036a1–4.

26. Compare *Metaph.* Λ.5 1071a26–29, which insists that within the same species there are different principles operating, not different in form, but different because the individuals differ from one another – *your* matter, form, and mover are other than

mine –, though in general they are the same in account. Deborah Modrak's suggestion (1979, 376) that the personal pronouns be taken only with 'matter' is possible out of context. However, Aristotle at 1071a20–24 distinguishes 'a man is the source of a man generally', from 'Peleus is the source of Achilles', and 'your father is the source of you', which suggests that Ross' reading is correct.

27. Not that these cannot be *known*. At 1036a5 Aristotle says sensibles cannot be defined but *are* known by perception and conception. On the importance of the distinction between 'knowing' and 'understanding' in Aristotle's philosophy of science, see Burnyeat 1981.

28. *Holōs:* not 'in the full sense' (Tredennick) but 'generally'. Cf. Alexander, *in Met.* 499.2–3.

29. For the force of the argument, see Burnyeat et al. 1981, 60–65.

30. Many passages in *GA* II suggest that even after it is in the embryo the heat transmitted by the semen is a potency of the male parent's form in other materials (cf. 735a12–14, 735a21–23, 737a18–22, 740b29–36).

31. I thus take serious exception to the argument found in Vlastos 1969, and supported in Gallop 1975, 183, that Aristotle takes Forms to be efficient causes and thus misses Plato's point. In fact, Aristotle treats them as Socrates does, i.e., as responsible for something's coming or ceasing to be F; and he claims that the form of the parent and what comes to be can do this, thereby making an independent and separate form unnecessary.

32. *Metaph.* Z 1040b5–16 rules out the elements because they lack the appropriate unity, and organic parts because they cannot exist separately.

33. Namely, at 1043b14, which reads, "In fact this being must either be eternal, or have passed away without passing away or have been generated without coming to be." He goes on to refer back to Z.8, and says that the form is neither produced nor generated, and that it is the composite which comes to be. Ross takes Aristotle to mean that substantial forms are eternal, whereas predicates in other categories are those which appear on the scene without a process of generation, a view which derives some support from 1034b17–20, where Aristotle argues that only substantial generation requires an actual, pre-existent form.

34. The relevant *PA* texts: Bats (697b1), Ostriches (697b14), Seals (697b1), Sea Cucumbers (681b1), Cetacea (697a17), Snakes (690b14), and Starfish (681b10–13).

35. Cf. *Metaph.* Z.8 1033b33–1034a2, on mules.

36. On the role of this principle in his argument for teleology in nature, see Gotthelf 1976.

37. I argue that this is basic to teleological explanations in chapter 9 below.

38. Cf. *Metaph.* Z 1028b8–10, 1034a4.

7

Kinds, Forms of Kinds, and the More and the Less in Aristotle's Biology

Aristotle is often characterized, by both philosophers and evolutionary biologists, as the fountainhead of a typological theory of species that is absolutely inconsistent with evolutionary thinking.[1] D'Arcy Thompson, on the other hand, in his remarkable *On Growth and Form,*[2] claimed that the idea of using quantitative methods to help understand morphological relationships among animals of different species took root in his mind during his work on Aristotle's biology:

> Our inquiry lies, in short, just within the limits which Aristotle laid down when, in defining a genus, he showed that (apart from those superficial characters, such as colour, which he called 'accidents') the essential differences between one 'species' and another are merely differences of proportion, or relative magnitude, or as he phrased it, of 'excess and defect'.[3]

A theory that asserts that species of a genus differ only in the relative magnitudes of their structures sounds very different, and might be thought to be incompatible with, a theory that claims that there are complete discon-

This chapter is an attempt to integrate ideas originally presented in "Aristotle on Genera, Species and 'the More and the Less'" (Lennox 1980) with ideas about the notion of being one in form which appear in "Are Aristotelian Species Eternal?" (chapter 6 above), a paper which was originally contributed to the Festschrift for David Balme (Gotthelf 1985a). A number of scholars have reached the conclusion recently that translating *genos* and *eidos* as 'genus' and 'species', unless these terms could be understood by all readers as merely conventional renderings of the Greek, is a dangerously misleading practice. Faced with this difficulty, I have opted in some contexts to transliterate and in others to render *genos* 'kind' and *eidos* 'form' (as in 'form of a kind'). This makes for some harshness, but the arguments presented by David Balme and Pierre Pellegrin have convinced me that some such move is required. The new title is not simply due to this change in translation, however. This chapter represents a reconsideration of the issues and texts of its ancestor, rather than a 'revision' of it.

tinuities between one form and all others. Can Aristotle consistently have held both these views? He can, and he did. To understand how he did so, one must understand the way in which he used the Academic technical notion of 'the more and the less' in his biology. It will be demonstrated that Aristotle treats variations between one form of a kind and another as differences of degree. Such a move conflicts with the sort of typological thinking traditionally ascribed to Aristotle by biologists and philosophers.[4] It will here be argued that Aristotle preserves the objective nature of biological kinds by stressing the teleological requirements of the lives of different organisms.

Being one in form and being one in kind are two ways in which numerically distinct entities can be the same, according to Aristotle. His many discussions of unity[5] lead us directly into the issue of how forms of a kind are differentiated from one another.

An individual is numerically one in virtue of being either naturally (e.g., a cat), artificially (e.g., table) or accidentally (e.g., a pool of water) physically continuous.[6] Two or more numerically distinct individuals may, nonetheless, be one in form (*hen eidei*). By this Aristotle seems to mean that they are structured or function in some sense identically, though the organization and functioning occur in discontinuous, self-subsistent bodies.[7] Organisms of different forms may yet be one in kind (*hen genei*).

In the next three sections of this paper (II–IV), I take up the issue of how Aristotle distinguishes forms of a kind; then I shall return to the question of the nature of the identity of animals that are one in form.

II

Aristotle offers two different, and not obviously related, accounts of what it is for different sorts of organisms to be one in kind. In the *Metaphysics* the kind is often described as the matter or substratum for differentiation into sub-kinds, forms. In the *PA* and *HA,* on the other hand, forms are said to be one in kind provided their parts for the most part only differ in degree, that is, by 'the more and the less'. If their differences are predominantly greater than this, they may be described as one only by analogy.[8]

It is only in the biological works that Aristotle uses the concepts of 'the more and the less' or 'excess and deficiency'[9] to express the nature of the relationship between biological (i.e., substantial) kinds and forms, though the basis for this application is established in the *Metaphysics.*[10] The logic of these concepts is clear in the *Categories* and plainly derives from the *Philebus*. I shall contend that these two apparently quite different accounts of the kind-form relationship are in fact closely related to one another. As a consequence of this argument, it should become clear that Aristotle's essentialism is *not* typological, nor is it in any obvious way 'anti-evolutionary'.[11] Whatever it was that Darwin was up against, it was not Aristotelian essentialism.

Here are two passages from the biology which make use of the notions of 'excess and defect' and 'more and less' in characterizing the relationship that holds among forms of a kind.

> *Parts of Animals* I.4 644a16–21: For all kinds that differ by degree and by the more and the less have been linked under one kind, while all those that are analogous have been separated. I mean that bird differs from bird by the more or by degree (for one is long-feathered, another is short-feathered) . . .

> *Parts of Animals* I.4 644b8–15: The kinds have been marked off mainly by the shapes (*ta schēmata*) of the parts and of the whole body, wherever they bear a similarity as the birds do when compared among themselves and the fishes, cephalopods and testaceans. For their parts differ not on the basis of analogous likeness, as bone in man is to spine in fish, but rather by bodily affections such as largeness and smallness, softness and hardness, smoothness and roughness, and such – in general by the more and the less.[12]

According to the account of kinds and forms of kinds in *PA* I and *HA* I, a kind is to be viewed as constituted of a set of general differentiae,[13] features common to every bird or fish, *qua* bird or fish. What distinguishes the members of one form of a kind from another is the way in which these general differentiae are further determined. The length, width, texture, density, arrangement, even the number[14] of this general feature may differ in measurable ways from one member of the kind to the next. Birds, *qua* birds, have beaks, for example. Different sorts of birds may have beaks of differing length, width, hue, hardness, curvature. It is these sorts of differences, throughout all the differentiae of the general kind, which differentiate one form of bird from another. Along any parameter one may choose, then, each organ will differ only by shades and degrees from one kind to the next. Aristotle's works don't come supplied with pictures, but Figures 7.1 and 7.2 beautifully illustrate Aristotle's account of the relationship among the features of different forms of a kind.

Aristotle's use of these concepts has its roots in Plato's *Philebus*. During the provocative discussion of the mixture of the limit and the unlimited, Socrates contrasts features of the world which are without qualification

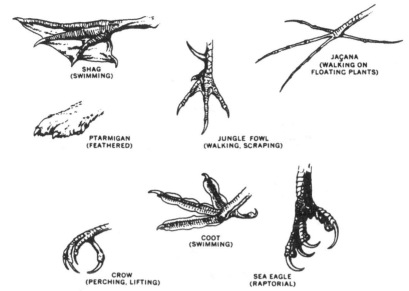

Figure 7.1. Types of birds' feet, indicating adaptations for locomotion and predation. (Source: J. A. Thomson, *The Biology of Birds,* as reprinted in *Encyclopedia Britannica,* 14th edn., Chicago 1970.)

unlimited from those which can be said to possess limit. The former group include

> drier and wetter, higher and lower, quicker and slower, greater and smaller, and everything that we brought together a while ago as belonging to that kind of being which admits of the more and the less. (25c5–8)

The latter

> don't admit of these terms, but admit of all the opposite terms like 'equal' and 'equality' in the first place, then 'double' and any term expressing a ratio of one number to another, or one unit of measurement to another . . . (*Philebus* 25a6–b1; cf. *Politicus* 283c3–284b2)

In the *Topics,* the phrase 'the more and less' has the flavor of a technical expression, and is one of the *topoi* under which various opinions can be challenged or supported.[15]

In the *Categories,* 'more and less' plays a central role in explicating and distinguishing the categories of substance, quantity and quality. For example,

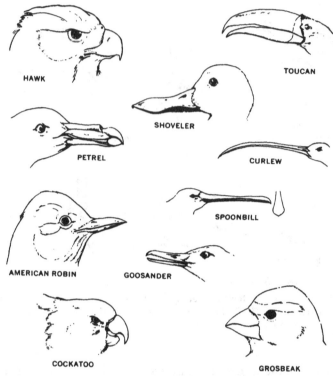

HAWK

TOUCAN

SHOVELER

PETREL

CURLEW

SPOONBILL

AMERICAN ROBIN   GOOSANDER

COCKATOO

GROSBEAK

Figure 7.2. Types of bills, indicating adaptations for feeding. (Source: J. A. Thomson, *The Biology of Birds,* as reprinted in *Encyclopaedia Britannica,* 14th edn., Chicago 1970.)

> Substance, it seems, does not admit of a more and a less . . . For example, if the substance is a man, it will not be more a man or less a man either than itself or than another man.[16]

Qualities, on the other hand, "admit of a more and less; for one thing is called more or less pale than another" (*Cat.* 10b26–8). This is true when "both [things] admit of the account of what is under discussion" (*Cat.* 11a11),[17] i.e. both are pale, though different shades of pale. Given that Aristotle treats the color spectrum as itself a continuum between dark and pale, one could presumably refer to an object, insofar as it was colored, as more or less dark or pale as well. Thus the claim is that, while Socrates cannot be more or less *human* than Callias, he may be more or less pale than Callias.

The *Categories* is innocent of the matter/form distinction, and free of the close association between form and differentia which is so central to the

Metaphysics. Indeed, the categorical status of *diaphora* in the *Categories* is left entirely unclear.[18] Like kinds, differentiae and their accounts may be said of forms and of individuals, but they are not granted even secondary substancehood. Nor are they ever mentioned as belonging to any other category.

It is probably with the *Categories'* discussion of substance and the more and less in mind that *Metaphysics* H.3 notes that,

> just as a number does not possess the more and the less, neither does the substance in virtue of the form (*kata to eidos*), but if it does possess the more and less, it is substance with the matter that does so. (1044a10–11)

The picture of natural substances as unities with material and formal aspects, the achievement of *Metaph.* H, suggests that the *Cat.* statement needs qualification: Socrates cannot be more or less a human than Callias in virtue of form. That is, the account which refers to them in abstraction from the different ways in which they actually embody human characteristics will not mention the more/less variations between them. But Socrates and Callias are "this matter and this form here, and humans are such taken generally" (*Metaph.* Z.8 1033b24–6, 10 1035b28–32, 11 1037a5–7); and as such – as substances with matter (*ousia meta tēs hulēs*) – they can differ by the more and the less.

Indeed, this conclusion is implicit in the chapter immediately preceding, *Metaph.* H.2. Assuming that being a substrate or matter is potential being, Aristotle turns to an exploration of the *actual being* of sensible things (1042b9–11). Taking a hint from Democritus, he notes that the same material substrate is able to *be* different things. The chapter is an exploration of the *differences* which make the same material *be* different sorts of things. 'Excess and defect'[19] is introduced as a general term for the ways in which sensible affections (*pathēmata*) can differ (1042b22–5), and in this context *animal parts* are discussed.

> In some cases their being will be defined in all these ways, by being mixed in some cases, blended, contained, and condensed in others, and by having the other differentiae as well, just as a hand or a foot is. So we must grasp the kinds of differentiae, for these will constitute the origins of being: differences in respect of the more and less, density and rarity, and the other such differences. For all these are excess and defect. (*Metaph.* H.2 1042b29–35)[20]

The parts of animals are ultimately constructed out of the four elements, constituted of the qualities hot, cold, wet and dry.[21] *PA* II.1 notes that the other affections of bodies – lightness, heaviness, density, rarity, roughness and smoothness and so on – follow from these (646a13–21). Uniform parts are constructed out of the four elements with these various second-order differentiae, and in turn such non-uniform parts as hand and foot are constructed out of these. Throughout, the relation of the less complex to more complex constituents in the process is teleological, the qualities acquired by the various parts at each stage being appropriate to and required for the functions to be performed by these parts in the organism's life (646b11–19). The uniform parts will differ along a limited number of qualitative parameters: harder/softer, more liquid/more solid, more flexible/more brittle – while the non-uniform parts constructed from them may have many such differentiae in various combinations for different ends.

> . . . for a different power (*dunamis*) will be useful to the hand relative to pressing and to grasping. (646b24–5)

The parts of animals, then, are materials differing with respect to the sorts of perceptible qualities acquired during development. And it is just these qualities which are said to differ by excess and defect, or the more and the less. But animal parts are also among the differentiae which constitute the nature or being of an animal. Thus, should one wish to distinguish one sort of bird from another, it will be in part by noting the differences in degree between the parts of one and the parts of another – thicker or thinner bone or blood, heavier or lighter body, thicker or thinner beak, and so on. The discussion of *Metaph.* H.2, then, suggests treating animal differentiae as constituted of just those qualitative features which the *Categories* had shown to differ by the more and the less.

Between *Categories*, *Metaphysics* H.3 and *Parts of Animals* I, we have four distinct statements about the relationship between substances and the more and the less.

1. Socrates is not more or less a human than Callias.
2. Socrates *qua* form does not have more and less, but Socrates *qua* form with matter does.
3. Socrates may be more or less pale than Callias, but only if both are, in general, pale.
4. One bird may differ from another bird by more and less, though both are birds.

There is a position Aristotle could have adopted which is able to incorporate all four of these claims, namely, the following.

For two individuals to differ in degree, they must both be the same general sort of thing. With respect to that *sort* they do not differ in degree. But the general sort is constituted of *features with range* – any sub-kind may have those features exemplified by different specifications of that range.

With this position in hand, our four claims can be dealt with in the following manner: there is a general *logos* of human being which applies equally to Socrates and Callias; but the precise way in which Socrates is the realization of one of the features specified in that *logos* may differ in degree from the way in which Callias is; for example, while the flesh of each is equally human, Socrates' may be more or less pale or soft than Callias';[22] and this is true at any level of generality – sparrows and eagles have the same *definition qua bird,* but the definition of bird specifies *features with range;* thus, while sparrow and eagle may not differ with respect to having wings, beaks, or feathers, they may differ in having the sensible properties of those organs specified on a different region of that range.

It will be my contention that the elements of such a position are to be found in the *Metaphysics* but only explicitly stated in the biological works.

There are, of course, differences in the three discussions of the more and less I've examined. *Metaph.* H indicates a willingness to speak of features which make something what it is – differentiae – as differing in degree from those which make something *else* what it is. The *Categories* is mute on whether this could be the case. Again, *Metaph.* H introduces the notion of an organ as an organized *complex* of all sorts of such differences. *PA* I takes the further step of referring to the substances themselves (e.g., birds, fish, crustacea) as differing in degree relative to some general kind to which they belong. Thus not only can the differentiating features be said to differ in degree from one form of a kind to another – the forms of the kind themselves can be said to differ by degree, or by the more and the less, from each other.

III

The suggestion I wish to put forward now is that the standard *Metaphysics* account of the relationship between a kind and its forms underlies Aristotle's willingness to see forms of a kind as differing only by the more and less from one another.

Aristotle's account of kinds in the *Metaphysics* is anything but extensional – that is, a kind is not primarily a class with members. Typically, a kind is represented as a substratum (*hupokeimenon*) or material (*hulē*) for differentiation.[23] Richard Rorty has suggested that the motivation for this account of kind is Aristotle's concern to stress the material/formal unity of his primary substances.[24] A second, related motivation, as *Metaph.* I strongly suggests, is the need to account for the identity of individuals which are different in form. What is the basis of our ability to see birds *as* birds, however different in form they may be?

Recent work on the concepts of form and kind can, I believe, illuminate, and in turn be illuminated by, the doctrine of kind as matter.

As Professor Balme has argued, the alleged equivocity of the term *eidos* is "both more extensive and more innocent"[25] than is commonly supposed. In fact, the research of Pierre Pellegrin suggests that 'equivocity' is an inappropriate description of the fact that *eidos* (or *genos*) can refer to organisms at different levels of generality. To be a form is to be a *determinate realization* of a kind. Socrates is a determinate realization of human nature, as water-fowl is a determinate realization of bird. The fact that things at various levels of generality can be, according to this account, forms does not render the concept equivocal.[26]

The claim that a kind is 'as matter' is the other side of this coin. Organisms which differ in form can be seen as different progressive determinations of a more general kind. To refer to a parrot and a blue jay as birds is to ignore (i.e., remove from my account) the way in which beak, feather, wing, legs, crop, etc., are differently realized in each. Similarly when I refer to two individual birds as parrots; in doing so I leave out of account the peculiar ways in which each realizes his parrot features. The kind is that which can be determined in various different ways – the generic potential, to use Montgomery Furth's expression. But this is just to say that the account of the kind will list what I previously referred to as *features with range* relative to the next stage of realization or actuality. Understood in this way, it is clear that the term has no fixed taxonomic reference.

The doctrine of kind as matter is introduced in three distinct contexts in the *Metaphysics:* in the discussions of levels of unity in *Metaph.* Δ.6 and 8, and I.3; in the discussions of the relationship between forms and their kinds in *Metaph.* Z.12 and I.7–8; and in the discussion of the meaning of *genos* in *Metaph.* Δ.28. The summary of this last passage will serve as a useful starting-point.

> Kind, then, is used in all these ways, in virtue of a continuous generation of the same form, in virtue of the first mover being like in form, and as matter (*hōs hulē*); for that thing of which the differentiae and the quality are such is the substrate, which we call matter. (1024b7–10)[27]

Similarly in *Metaph.* I.8 the kind is referred to as the matter of the things of that kind, and in a context which is of direct relevance to our discussion.[28] The subject of this chapter is the relationship between things which are different in form. It opens by declaring that

> That which is different in form from something is different in respect of something,[29] and this must be common to both; for example, if an animal is said to be different in form [from some other animal], both must be animals. Therefore the things which are different in form must be in the same kind. (1057b35–7)

The notion of kind as matter is introduced to clarify the role played by the kind which is common to its different forms, yet differentiated in each (1058a1–5). The animal itself, as he says, is both a horse and a man. Each is a different articulation of the features with range that constitute the nature of the kind animal.

The significance of this picture of the kind-form relation can be grasped by contrasting it with an alternative picture which commits the error Aristotle wishes to avoid. Horses and humans, according to this alternative model, would be identified as animals in virtue of one set of absolutely identical features in each, and distinguished from one another by an entirely different set of completely unrelated features. Certain aspects of modern technology exemplify this model, or could quite easily. Imagine, for example, that there were a 'basic' Porsche 911, and that the different 'models' of it consisted of this basic automobile onto which different features, *not present at all* on the basic model, are appended (FM radio, automatic transmission, air foil, racing stripes, mud flaps, etc.). That is, the kind does *not* consist of features with range, but rather completely determinate features; and the features of the forms of the kind are *not* determinate realizations of the generic features, but features 'added on' to the generic features. There is no unity to the kind and differentiae, of the sort Aristotle is seeking. The kind and differentiae appear in an account of a substance as two distinct elements, and a major concern of *Metaph.* Z–H is to explain how substances are unities in spite of this apparent duality in the account of their being. *Metaph.* Z.12 introduces its discussion of definition by division with the following puzzle: "That of which we say the account is a definition — why is it one; for example 'the animal biped' as definition of the human (for this is the account of man) — why is this human one and not many, animal and biped?" (1037b12– 14). And in *Metaph.* H.6 it is clear whom he takes to have held such a splintered view of substance.

> What is it, then, which makes the human one, i.e. why is it one but not many, for example both the animal and the biped, especially if there is, as some say, some animal itself and some biped itself. (1045a14–17)

On this view, it will be noted, the kind could exist without its forms. But this is just what Aristotle wishes to deny:

> If then the kind does not exist without qualification apart from the forms as forms of a kind, or if it exists it exists as matter (for the voice is a kind and matter, but the differentiae produce the forms, i.e., the syllables, from this), it is apparent that the definition is the account which derives from the differentiae. (*Metaph.* Z.12 1038a6–9)[30]

Aristotle insists on the parasitic nature of kinds because they constitute simply the potential for a number of distinct realizations. This is the conception of the kind-form relation which emerges from *Metaphysics* Z.12. Aristotle stresses that division ought to proceed by treating each previous stage of a division as the 'substrate' for the next stage – "divide by the differentia of the differentia," as he puts it (1038a10–11). If being footed is a differentia of animal, the next division should be into sorts of footed animals, not into footed animals that are winged or wingless – look for the differentia of footed animal *qua* footed (*Metaph.* Z.12 1038a10–16).[31]

The *Metaphysics* works throughout with simple, schematic examples which in certain ways suggest that Aristotle has not yet integrated the idea of a kind as the substratum of distinct realizations in its different forms, with the doctrine that the differentiae of animals within a kind vary only by the more and the less. The typical kind is 'animal', rather than the extensive kinds such as bird or fish of the biology – and the organs of forms of *this* kind will *not* typically vary only in degree. The kind itself is not said to be constituted of many general differentiae, and therefore the corollary of this conception, division proceeding simultaneously through many axes of differentiation, is never discussed.[32] Finally, in the biology the distinction between 'one in kind' and 'one by analogy' is typically drawn in terms of whether the majority of organs differ only in degree, or more extensively. The same distinction is drawn in the *Metaphysics,* but not on these grounds.[33]

The account of the relationship that holds among animals different in form but the same in kind which is so carefully worked out in *PA* 1.2–4 is based both on the concept of a kind as a substratum for differentiation into its forms, and on the concept of organic parts as materials differentiated by a complex of more/less variations in their perceptible affections. But while in the *Metaphysics* these ideas are introduced in different contexts and are not obviously related, in *PA* I.2–4 these two conceptions are integrated into a richer and more realistic understanding of the kind-form relationship.

IV

The general picture of division adopted by Aristotle in *PA* I.2–3, which is developed in the context of pointing out the errors involved in dividing dichotomously by one differentia at a time, is clear from the following.

> If man were merely a thing with toes, this method would have shown it to be his one differentia. But since in fact he is not, he must necessarily have many differentiae not under one division. Yet more than one cannot belong to the same object under one dichotomy, but one dichotomy must end with one at a time. So it is impossible to obtain any of the particular animals by dichotomous division. (644a6–10)

This indicates the artificiality of the *Metaph.* Z.12 discussion, and points the way to a more complex methodology.

> One should try to take the animals by kinds in the way already shown by the popular distinction between *bird* kind and *fish* kind. Each of these has been marked off by many differentiae, not dichotomously. (*PA* I.3 643b10–13)

These are general differentiae – the natures of the various forms of these kinds must be grasped by seeing how each of these general differentiae is realized distinctly in those forms.

> But the general differentiae must have forms, for otherwise what would make it a general differentia and not a particular one? Some differentiae certainly are general and have forms, for example *featheredness:* the one feather is *unsplit,* the other *split.* And footedness similarly is *many-toed, two-toed* (the cloven-hoofed), and *toeless* and undivided (the solid-hoofed). (*PA* I.3 642b25–31)

This picture of the differentiation of the general kind into its various forms is not viewed as a replacement of the idea that the kind is the substrate of differentiation, but rather an enrichment of it. In the midst of his discussion of the problems of 'dividing as the dichotomists do' he pauses to remind us that

> It is the differentia in the matter that is the form. For just as there is no part of animal that is without matter, so there is none that is only matter . . . (*PA* I.3 643a24–5)

Balme comments:

> Aristotle may be referring to the physical matter or to the logical genus. But it makes no difference here (does it anywhere?). Each is one way of considering *that which is potentially X*.[34]

Having introduced multiple divisions of general differentiae as the way to achieve an account of biological kinds, he next, in the passages I noted on pp. 341–2, notes that these differences will be in degree.

And thus we have returned to those passages from *PA* I.4 with which I began. To those, I now add a passage from *HA* I.1 which brings out clearly the relevance of *Metaphysics* H.2 to the biological discussion of forms of a kind.

> Some things are, on the one hand, the same, yet differ by excess and defect, namely those for which the kind is the same. And I call such things as bird and fish a kind; for each of these has a generic difference, and there are many forms of fish and birds. Now on the whole most of the parts in these cases differ besides by oppositions of their affections, e.g., color and shape, by the same feature being affected in some cases more and in some less, again by [the parts] being more or fewer, and larger and smaller, and generally speaking by excess and defect. For some are soft-fleshed and others hard-fleshed, some have long beaks and others short beaks, some have many feathers and some few feathers. What is more some even among these animals have different parts belonging to different animals; some have spurs while others don't, and some have crests while others don't. But speaking of the majority of cases, from which the entire body is composed, these parts are either the same or differ in the oppositions and according to excess and defect; for the more and less one might place under excess and defect. (486a23–b16)

A number of the details of this passage are worth noting, for they are crucial to the actual comparisons among animals which Aristotle makes in practice, yet are not stressed in the more theoretical passages in *PA* I.4. First, it is stressed that not *all* the parts of animals of different forms within a kind need differ by the more and less, or in degree. Some parts may be undifferentiated from one form to another, others may appear in some forms and not at all in others. Second, there is an attempt to distinguish between various sorts of difference in degree – between variations in affective quality, in magnitude, and in number, and examples are provided of each. Aristotle could have chosen the same part to exemplify all of these types of variations. Bird feathers will be many in some forms, few in others, longer in some, shorter in others, harder in some, softer in others, and so on. Once this sinks in, the complexity of the study of animal differentiae upon which he is here embarking reveals itself.

V

Based on ideas about the nature of kinds, forms of kinds, and differentiation explored in the *Metaphysics,* the zoology builds a rich and complex picture of how different forms of a kind may be related to one another – how, though animals may be different in form, they may be one in kind. But that picture may suggest that there is no clear, objective means of identifying organisms that are one in form. Two closely related forms *A* and *B* will differ only in degree from one another, and one might draw the conclusion from this fact that distinctions between members of *A* and members of *B* might be quite arbitrary. Groupings below the level of extensive kinds such as bird or cephalopod might plausibly be viewed as mere conventions.

Nonetheless, Aristotle does maintain on the one hand that the organisms of an extensive kind differ only in degree among themselves, *and* on the other that they fall into *indivisible forms,* groups of numerically distinct organisms which are one in form. Having dwelt on the non-typological aspects of Aristotle's thinking, I now wish to argue that he maintains the notion of sub-groups within kinds which are *formally identical,* which I take to be a sort of essentialism.[35]

Let us begin by asking Aristotle to be more precise about what it is for the members of a sub-kind to be one in form. Why should we not treat the organisms of an extensive kind as 'just one big happy family'?

It is certainly not that at a certain level of generality organisms cease differing in degree from one another. Aristotle does not restrict the notion of being one in form to perfectly identical twins. In fact, he is quite explicit that things which are one in form may nonetheless differ in degree, by the more and less, from one another.[36]

In the *Metaphysics,* in discussing the notion of being one in form, Aristotle makes this notion dependent on that of indivisibility in conception (*hē noēsis* – 1016b2, 1052a32, 35) or indivisibility in account (*ho logos* – 1016b33, 1052a33, 36), in contrast to things numerically one which are physically indivisible.

> Again, things are called one the account of the essence of which is indivisible
> relative to another which reveals the thing's essence. For just by itself every
> account is divisible. (1016a33–5)

This passage indicates that it is not having just any account in common that
makes many things one in form – the account must also reveal what it is to be
the thing in question. An account of the kind would do this, of course, but it
is *divisible* (into accounts of its various forms). Among accounts which make
clear what a thing is, there is one which relative to the kind being defined is
not further divisible and which is common to many things. Those things
which share such an account are one in form.[37]

This passage makes another important point. We are told that by itself
every account is divisible. 'By itself' in this passage is placed in opposition
to 'relative to another account which reveals the thing's essence'. That is,
this account may be further divisible, but not into further 'essence-revealing'
accounts. In any division, a point will be reached beyond which further
division will not further reveal features essential to the kind in question.

Can we provide any concrete, taxonomic specification of that point? I
think not. Two principles of Aristotle's approach to definition militate
against doing so. The first is his willingness to seek definitions for kinds at
various levels of generality.[38] What is incidental to being a bird may be
essential to being a wading bird, or a crane, or a sandhill crane. Thus a
feature is 'essential' or 'incidental' only relative to the kind being defined,
not in itself.

Take the example of a bird's legs as part of what it is to be a bird again.
Birds are by nature bipedal, but bipedal in a particular way: their legs bend at
the "knee" in the opposite direction to the other, true bipeds, human beings.
Birds may have long or short legs, legs with webbed feet or unwebbed feet,
and so on. Having *long* legs or *webbed* feet is not essential to being a bird –
and if we were giving an account of what it is to be a bird, dividing down the
leg axis further than 'backward-bending biped' would reveal nothing further
about what it is to be a bird. On the other hand, if we were seeking to give an
account of what it is to be a crane, reference to the length of leg would be
crucial and thus 'backward-bending biped' would be divisible into a 'more
revealing' account of the crane's nature.

In the above example division stopped short of quantitative (more/less)
differentiation. One might think this is precisely the point at which division
would cease to reveal anything but inessential variations. But we have seen
that Aristotle treats the forms of his extensive kinds as, for the most part,
differing only in this way. And this is not a merely theoretical claim which is

ignored in practice. Take the opening remarks of his chapter on the external part of birds in *PA* IV.12.

> Among the birds, the differentiation relative to each other is in the excess and deficiency of the parts, that is, according to the more and the less. For some are long-legged, some short-legged, some have a broad and some a narrow tongue; and similarly as well in the other parts. (692b3–6)

If such distinctions were, in some absolute sense, incidental, Aristotle would be unable to offer an account of anything below the level of the nine extensive kinds he recognized.

The second principle which rules out the possibility of specifying in advance a taxonomic level below which division will descend to the incidental also tells against more/less variations being necessarily incidental. It is that, at least in regard to living things, the 'essence/accident' distinction is a distinction between those features which are required by the kind of life an animal lives and those which aren't. If a crane is to survive and flourish, it *must* have, not simply 'long' legs, but legs of a certain length, defined relative to its body, neck length, environment, feeding habits, and so on.

> Whichever birds live in swamps and eat plants have a broad beak, this sort being useful for digging and for the uprooting and cropping of their food. (*PA* IV.12 693a15–17)

> Certain of the birds are long-legged. And a cause of this is that the life of such birds is marsh-bound; and nature makes organs relative to their function, not the function relative to the organ. (694b12–15)

At a slightly more specific level, similar explanations are offered in this chapter for a number of features of 'the crook-taloned birds' and the 'water fowl'. The stress is constantly on the way in which the differentiation of a feature is related to its possessor's life. For example, the discussion of beak variations begins, "Beaks differ according to the lives lived" (*PA* IV.12 693a11), following which are teleological explanations in terms of feeding habits, why some are sharp and curved, others straight, others broad and others long (693a12–22).

Aristotle usually does not consider lower levels of functional 'fine-tuning' to the environment than this, though he does occasionally. But these examples are sufficient to show that specific quantitative variations in the qualitative affections of their organic structures will often be part of an account of the *being* of an animal of a certain kind.

In *GA* V.1, Aristotle makes the essential-incidental distinction in just this manner. He is reviewing the affections by which the parts of animals are differentiated – that is, just those features which vary by the more and less. He insists that if the feature is neither a part of the kind's nature nor proper to each kind, then it neither came to be nor is for the sake of anything – nor should it be included in an account of the animal's being.[39] He makes it quite clear, however, that what sort of feature fits the above description cannot be specified out of context.

> For an eye is for something, while being blue is not, *unless (plēn)* this affection is a property of the kind. (*GA* V.1 778a33–4)

Aristotle apparently used the fact that a feature was a property (*idion*) of a biological kind as prima facie evidence that it exists for the sake of something. The above suggests that if a kind of animal were universally blue-eyed, being blue-eyed would in all likelihood have functional value. To take another example, while most forms of clawed crustacea have the right claw larger than the left, one group has the larger claw randomly distributed. Aristotle offers a teleological explanation for the difference in the former cases, but in the latter treats the size variation as a matter of chance, and not for the sake of anything (*PA* IV.8 684a25–32).

For two or more organisms to be one in form, then, they must all fall under one account, which is not further divisible into more specific accounts which specify features which are required by the organism's life. For any kind less general than Aristotle's nine extensive kinds, this will often include reference to increasingly narrow ranges on more-less continua. Whether an account is indivisible will depend on the generality of the kind in question relative to that account. And what ought to be included in such an account is to be determined by a study of teleological adaptation.

These last two points are tightly connected. For example, if the form under consideration is 'duck', an account common to all birds will not be indivisible, for it will not mention the peculiar adaptations to a duck's way of life – the bill, the webbed feet, the placement of the legs, the lack of tail feathers, the dense, more oily nature of the feathers, and so on. Suppose these were specified, however, yet we wanted an account of what it is to be a 'mallard duck'. The account indivisible relative to 'duck' would be further divisible relative to 'mallard duck', provided only that its life required peculiar features not possessed by other ducks leading different lives.

At some point in divisions within a kind, while we may still wish to distinguish groups of animals on various grounds, these will no longer be teleological. There is, for example, a reason why some humans are red-haired, some blond, and some brown-haired; but being one or the other is not a teleological requirement of blond, red-haired and brown-haired humans. Unlike other levels, this is not a division which is non-essential only relative to the form under investigation, but essential to the more specific kind. In this case, the division is non-essential *tout court*. Again, however, there is no *a priori* means of determining what this level might be – being a blond, might, under certain conditions, be a crucial adaptation to a specific lifestyle.

The same account of being one in form holds when one takes the activities of animals into consideration.

> Therefore we must first state the activities, both those common to all and those that are generic and those that are specific. (I call them common when they belong to all animals, generic when they belong to animals whose differences among each other are seen to be in degree. For example, I speak generically of 'bird' but specifically of 'man' and of every animal that has no differentia in respect of its general definition. What they have in common some have by analogy, some generically, some specifically.) (*PA* I.5 645b21–7)

The addition of the phrase "in respect of its general definition" to the notion of having no differentia is meant to indicate that there may be further subdivisions (say into those who farm and those who hunt), but these will not be part of the account of what it is to act in a specifically human manner.

Ultimately, then, the 'formal unity' of a number of animals turns on their having just the proper conformation of each part, and just the proper range of activities, for the life they have to live in a specific environment. This will involve their having a fully coordinated set of structures and activities suited to that life. Thus, while individuals of one form of a kind may differ with respect to one affection of one organ only by degree (or even not at all) from the individuals of another form of that kind, the overall unity of their differentiae will suit them to their life and no other, and distinguish them from the individuals of other forms suited to other lives in different environments.

VI

Let me provide a brief summary of the kind-form relationship sketched in this chapter. Below the level of analogical likeness every form of a kind may serve as a kind for further division. Second, at every level, the variations in the differentiae among forms of a kind will be predominantly the more/less variety. Third, at every level, the kind is 'matter' for its forms – and a form relative to the more general or common kind may be the matter/kind for further differentiation. This picture incorporates the research of David Balme and Pierre Pellegrin on the variable reference of kind and form in Aristotle's biology into the research that was reported in this chapter's ancestor. The account of kind as matter and form as differentia, and of the more and the less, there presented was still implicitly wedded to a taxonomic understanding of kind and form which I have now abandoned. The wedding was a mismatch, and I find the basic thrust of the earlier paper is much better suited to the 'level neutral' divisional account of these concepts provided by Pellegrin's *La Classification*.

What, finally, of Aristotle's essentialism? He appears to have seen every individual of a kind as having every feature of every structure quantified (in Professor Balme's sense) precisely, and to have viewed every kind from the most extensive to the most specific as a range of potential quantifications for the next more specific kind. To look at Socrates *qua* human is to ignore those (relative to being human) inessential details of his organic nature – snubness of nose, bulge of eye, bandiness of leg. Those all represent *possible* ways in which human noses, eyes and legs may be realized.

Yet at each level there will be an organization among the allowable ranges for each feature of each differentia which is essential to an animal's life, particularly its mode of feeding, cooling itself, and rearing its young.[40] This is not typological essentialism, if by that one means that members of a kind share in some bundle of qualitatively indistinguishable features and differ in certain other incidental features. Nor does it rule out the possibility of environmental changes requiring slight changes in the acceptable ranges of variations. Aristotle seems not to have considered this a serious possibility, and therefore did not discuss how it might possibly occur, and this is hardly surprising. The range and subtlety of the evidence needed to make natural evolutionary change plausible is extremely difficult to come by, and in fourth-century Athens could only have been a wild cosmological speculation.[41] Aristotle took the (then) much more reasonable view that organisms well-suited to stable environments breed true to form – that is, variation is maintained within the limits established by the requirements of that organism's life.[42]

NOTES

1. Sir Karl Popper's verdict is perhaps more strongly worded than most: ". . . every discipline, as long as it used the Aristotelian method of definition, has remained arrested in a state of empty verbiage and barren scholasticism" (Popper 1952, II 9). Among philosophers of biology, David Hull has argued for precisely the same

conclusion (Hull 1965–6). This position has been orthodoxy among neo-Darwinians, cf. Mayr 1963, 4; Dobzhansky 1970, 351; Simpson 1961, 46; Ghiselin 1969, 50–2. Recently, Professor Mayr has moderated his position considerably in the light of Professor Balme's research (cf. Mayr 1982, 11, 87–9, 149–54, 254–6, 636–8). Indeed, Mayr's is perhaps the most sympathetic treatment of Aristotle's biology by a historian of biology in this century. Mayr has separated Aristotle's non-evolutionary stance from his theory of natural kinds, whereas the texts noted above all assume that a typological essentialism on the latter topic was the primary reason that Aristotle adopted a non-evolutionary view of the zoological world.

2. Orig. ed. 1917; rev. ed. 1942; abrdgd. ed. 1971.
3. Thompson 1971, 274.
4. By 'typological essentialism' I intend a view that there is some one identical feature or set of features in which individuals of a kind share or 'participate', and in virtue of which the individuals are said to belong to the kind. Indeed, on this view the species name might be viewed as having primary application to this essence, and only secondary reference to individuals. Aristotelian forms are occasionally taken to be such shared essences. Thus one aim of this chapter is to discuss the evidence in Aristotle, and especially in his biology, for a non-typological essentialism which might be referred to as 'teleological essentialism'.
5. *Metaph.* Δ.6 1016b32ff.; Δ.9 1018a7ff.; *Metaph.* I.1 1052a15ff.; *PA* I.4 644a15–23, 644b1–16; *HA* I.1 486a15ff.; *Top.* I.7 103a6–16.
6. *Metaph.* I.1 1052a15–30. Natural things, which have an inherent source of their remaining one, are one in a more unqualified way than artificial unities or heaps.
7. *Metaph.* Δ.6 1016a32–1017a7; Z.8 1034a1–7; I.3 1054b16–17; 9 1058b6–12.
8. Lungs and gills, which are both used for cooling on Aristotle's understanding, are analogous structures (*PA* I.5 645b8); as are bone and cartilage (*PA* II.8 653b32ff.). In distinguishing extensive kinds from one another, this term usually refers to a relationship between structures which at a very abstract level perform a similar function for their possessors, but do so by different means, and are not structural variations on a common theme, i.e. are not open to more/less comparison.
9. *Huperochē kai elleipsis* is often rendered 'excess and defect', which has unfortunate connotations of a norm which properties 'fall away from' or 'overreach'. As will be clear, all Aristotle means by this phrase is variation *in degree* rather than by *discrete units.* Likewise with 'the more and the less'. The former phrase possesses greater generality and may have come into use later than the latter. Cf. *Metaph.* H.2 1042b32–5; *PA* IV.12 692b3–6; *HA* I.1 486a1–16.
10. Especially, as I shall argue, in *Metaph.* H.2.
11. Let me stress that what I mean by this is that his theory of natural kinds is not in any obvious way 'anti-evolutionary'. As I have argued elsewhere (chapter 6 above) there are many deep reasons for his not holding an evolutionary theory with respect to the origins of living things, but these do not include his theory of natural kinds, nor are they implications of it. I am not, as I have sometimes been taken to be, arguing that Aristotle could have easily adopted an evolutionary theory. He knew of a number of such theories, and (quite rightly, I believe, given the context) rejected them as implausible.
12. Cf. *HA* I 486a21–b16; II 497b1–15, 500b5; III 516b4–34, 517b22–3; IV 528b11–23; VIII 588a24–31; *PA* IV 692b3–6.

13. *PA* I.3 642b24–6. The contrast, well brought out in Balme's translation, is between *general* differentiae and their more determinate forms at a more particular level.
14. Cf. *HA* I.1 486b7–12. A clear example of Aristotle treating many/few variations as variations in degree of a common organ is *HA* II.13 505a8–20, which provides examples of fish that have one, two, four, five and eight gills per side, a discussion which begins: "And again, some fish have few gills, others a great number; but all have an equal number on each side."
15. *Top.* II.11 115b3–11; III.3 118a27–b10; IV.6 127b18–25; V.8 137b24.
16. *Cat.* 5 3b33. This seems to echo the suggestion of *Phaedo* 93d1–e2, 93a14–b6, that a soul taken by itself can't be more or less a soul than any other.
17. Cf. *Ph.* V.2 226b3; *Sens.* 7 448a13–16.
18. *Cat.* 5 3a21–b9; cf. Ackrill 1963, 85–7. The *Topics,* however, states quite clearly that the differentia signifies a certain qualification of the kind; cf. *Top.* IV.2 122b20–3; IV.6 128a25–9.
19. *Huperochē kai elleipsis.* From here on I will freely exchange this full but misleading translation for the less misleading paraphrase 'variation in degree'. Cf. note 8 above.
20. Cf. Furth 1988, 249–64.
21. The following is highly schematic. For a fuller discussion, cf. Furth 1988, 76–83.
22. Compare *Metaph.* I.9 1058b1–16.
23. Cf. *Metaph.* Δ.6 1016a27, 28 1024b8; Z.12 1038a6–7; H.6 1045a35; I.3 1054b30; 8 1058a23–4; *PA* I.3 643a24–7.
24. The role of the notion of *genos* as matter in the *Metaphysics* was the subject of an exchange between Richard Rorty and Marjorie Grene some years ago. Cf. Rorty 1973; Grene 1974; Rorty 1974. Grene finds a common thread in the work of Rorty, A. C. Lloyd and David Balme which sees in the notion of kind as matter no mere metaphor, but an integral part of Aristotle's account of natural substances. She argues that one possible "interpretation of *genos* [i.e. category] as 'material' . . . pertinent to the study of nature . . . is that 'matter' is simply the unity of categoreal context open for distinctions of more and less – a very metaphorical meaning, which cannot be identified with the concrete, worked up matter 'out of which' the sculptor makes his bronze statue or the father his son" (120). I can't see why the first meaning of 'matter' is very metaphorical or how 'bronze', as the name for what can become many different sorts of (bronze) thing and which is common to them all, is matter in any sense radically different from 'bird' as the name for what can become many different sorts of bird and which is common to them all. I think it is extremely important that the 'lexicon' entry under *genos* (*Metaph.* Δ.28) does *not* distinguish these two cases, nor do the discussions of being one in kind – nor do the passages in *Metaph.* I that Grene relies on for the first sense. Balme's reply to Grene, quoted in note 10 of her paper, seems to belie the necessity, at least for natural kinds, of the sort of distinction she wishes to find.
25. Balme 1987d.
26. Cf. Pellegrin 1982, ch. 2; Pellegrin 1987.
27. As Pellegrin has noted (Pellegrin 1985) the analysis in this chapter makes *genos* the most appropriate word for stressing the historical continuity of a kind: 'the continuous generation of things with the same form'. As Balme and Pellegrin docu-

ment fully, *genos* is by far the most common term to denote animal groups in the biological works.

28. Cf. *Metaph.* I.8 1058a23–5.
29. Following Ross 1924, 301, I take the *ti* here as an accusative of respect.
30. Compare *Philebus* 17a–c.
31. Cf. Balme 1987b, 73–80, on the methodological importance of this idea.
32. Cf. Balme 1987b, 73.
33. Cf. *Metaph.* Δ.6 1016a31–5; Δ.8 1018a12–15.
34. Balme 1972: 114 *ad* 643a24.
35. The following material in this section is excerpted from "Are Aristotelian Species Eternal?" (chapter 6 above). The argument developed there is that individuals secure eternality of a sort for themselves through reproduction, by creating something one in form with themselves. Crucial to this argument, and to the acceptability of the material reproduced here, is an understanding of what it means to be one in form. I argue that Aristotle is strongly opposed to a theory which explicates formal unity in terms of participation in numerically one form, and that he would find the suggestion of Albritton and others that forms of individuals may themselves be one in form unacceptable because open to 'Third Man' objections. I argue that it is matter/form unities that may be one in form, meaning that they share an account which specifies their essence, an account which, were it more precise, would not reveal anything further about their essence. Thus there is, to use Professor Balme's language, a level of formal description which is privileged from the *scientific* point of view. We may as well stop division at that point, for we will not learn anything by further divisions about that thing relevant to its being what it is.
36. Cf. *Metaph.* I.3 1054b6; *EN* VIII.1 1155b11–15; Balme 1987d, 296–7.
37. Compare *PA* I. 642b7–20, 643a8–12, 643a24–6; *Metaph.* Z.8 1034a5, 12 1038a16.
38. Cf. the evidence for this collected in Gotthelf 1985b.
39. *GA* V.1 778a29–35. On this teleological notion of essence, cf. Balme 1987d, 297.
40. Cf. the use of the concept of the more and the less with respect to psychological and physiological states at *HA* VIII.1 588a25–b2.
41. Such views were held by Anaximander (*DK* 12 A10, A30), Xenophanes (*DK* 11 B33), and Empedocles (*DK* 31 B9, B12, B17, B26).
42. Cf. Balme 1987d, 298–301. I wish to thank Allan Gotthelf for written comments and helpful discussions concerning the reworking of this material, and in particular concerning how to integrate the ideas from Lennox 1985a with those of this chapter's ancestor.

8

Material and Formal Natures in Aristotle's
De Partibus Animalium

I INTRODUCTION

To say that things such as animals and plants are natural is to say that they *have natures,* i.e. their very own sources or causes of change and rest (*Ph.* II.1 192b21–3, b32–34; *Metaph.* E.1 1025b19–22). The nature of a natural substance is, however, complex: both its constituent materials and its form are called its nature (193a9–11, a28–31, 193b7–8, 194a12–15).

What should the natural scientist's attitude be toward this complexity? Given Aristotle's insistence that ". . . form has a better claim than the matter to be called nature" (193b7–8; cf. *PA* I.1 640b28–29), one might expect him to argue that, as far as is possible, the natural scientist should study the form of his subjects independently of their matter. Yet he consistently stresses the importance of the scientist studying both together (*Ph.* II.2 194a12–b15, *PA* I.1 641a15–32, *de An.* I.1 403b1–19, *Metaph.* Δ.1 1025b31–1026a7, Z.11 1037a10–20).

These well-known claims and the tensions they embody provoke a number of questions. I want here to focus on just one. If we suppose that Aristotle is serious about this claim that the natural scientist should study *both* the material and the formal nature, what, in practice, does that mean? The programmatic remarks in the *Physics, De Anima,* and *Metaphysics* – just because they are programmatic – give us little guidance. Aristotle's zoological treatises, on the other hand, might. They constitute sustained scientific investigations of living things, composites of matter and form the substantial natures of which Aristotle never questions.

The focus in this chapter is on the *De Partibus Animalium.* The term 'nature' appears 265 times in that work.[1] To what does it refer, and what is it to know it?

The central theorem to be established here is that the *De Partibus Ani-*

malium is in large measure a study of the *interaction* of formal and material natures in living things. It will be argued that the formal nature is a goal-directed agent, the actions of which are active, selective and informative. By calling its actions *selective* I mean that given a certain material nature, each formal nature determines a *non-random* and *distinctive* realization for that material. Its actions are *informative* in the thermodynamic sense of that term. It determines complex patterns of distribution and organization of matter, of which matter is incapable on its own.[2]

On the material side, it will be argued that the actions of a substance's formal nature are severely constrained by its material nature. The material nature of an organism in fact plays a fundamental explanatory role in *De Partibus Animalium,* and in this chapter I will delineate the set of situations where a thing's material nature plays such a role.[3]

The argument of this chapter thus addresses two concerns of recent discussions of Aristotle's natural philosophy. First, it gives material natures both a more independent and a more central role in Aristotelian science than is typically suggested, by either functionalists or their critics. If functionalism is described simply as the view that biological structures are only contingently related to their functions, Aristotle cannot be a simple functionalist. To cite an example discussed at the end of this paper, it is essential to lunged animals that they have a structure for closing their windpipes when they swallow, though some do so by means of an epiglottis, others by means of collapsable opening. But to do so by means of an epiglottis, while in some sense 'contingent' for the general class of breathers, is essential to vivipara, and is so *because of their distinctive material natures.* Thus, whether an organ is essential for a given function will depend on the precise characterization of each.

Nor, on the other hand, can it be stated without qualification that "[f]or Aristotle, it is the existence of life which explains why animals have the physical constitutions they do, not the other way around."[4] In the explanations we will be examining, there are certain features of living things that *are* sufficiently explained by reference to their material natures; and there are certain material facts about certain kinds of animals that are as *explanatorily primitive* as are other facts about their living functions. Ultimately, the attitude toward material natures in Aristotle's science of living things is too complicated for either of these extremes.

Second, on the formal side, it will be argued that two problematic uses of the concept 'nature' are both references to formal natures. For convenience I will label the two types of problem cases 'Apparently Demiurgic Nature' and 'Apparently Cosmic Nature'. Let me briefly indicate the uses I have in mind.

1. There are seventy-plus cases in *PA* alone, where some specific outcome is attributed to a nature which distributes, treats, constructs, devises, etc., in order to achieve the outcome in question. It will be argued that, though the verbs of agency are often metaphoric extensions from the craft domain, the nature referred to as agent is simply an animal's formal nature, not an Aristotelian counterpart to Plato's Demiurge.

2. About a dozen times in *PA* alone, a quite general statement is made about this nature's activities, such as that nature does nothing in vain, or suits the organ to the function, not the function to the organ. I shall argue that these statements are 'first principles' in a rigorous sense: they must be assumed to insure the validity of explanations which appeal to the acts of formal natures, and within the science of biology they are primitive. At least twice in the corpus the most general such statement is referred to as an assumed starting point of natural science. It will be argued that these statements are abstract principles *about* the actions of formal natures that come at various levels of abstraction, or specification. They needn't be read as referring to a cosmic nature doing the best by arranging particular natures in various ways.

The discussion will center on a single chapter of *De Partibus Animalium,* chapter 2 of Book III. It is well-suited to my purpose, in that it uses the concept of nature often, and in all its problematic roles; and it includes a passage, with which we will begin, that sheds philosophical light on the use of the concept throughout Aristotle's natural science.

The logical form of my argument, in essence, is quite simple: Aristotle's philosophical analysis of the concept 'nature' in the *Physics,* and reiterated in *PA* I.1, leaves no room for a Demiurgic or Cosmic Nature over and above the formal and material natures of specific natural substances. It also insists that natural science seeks and acquires knowledge of both the formal and material natures of such things. Nevertheless, it is often assumed that the explanations in *De Partibus Animalium* appeal to Cosmic or Demiurgic Natures, and give little explanatory role to material natures.[5] A reading of the biological works that requires neither of these philosophically suspect assumptions is preferable on grounds of philosophical coherence: and such a reading is available, according to which Aristotle assumes nothing more than the formal and material natures of individual substances in his explanatory repertoire.[6]

II DISTINGUISHING FORMAL AND MATERIAL NATURES

One of the initial difficulties in such a study is that Aristotle rarely distinguishes nature as form and nature as matter in his biological *practice.*

For that reason, I will begin by focusing attention on the following passage, where something like such a distinction is made. It is a transitional passage during Aristotle's explanation of why animals with horns have them:

> So then, what the nature of the horns is for has been stated, and why some animals have such things while others do not; but we must say how, supposing the necessary nature is of such a character, the nature according to the account makes use of what is present of necessity for the sake of something. (663b22–24)[7]

A necessary step in understanding this passage is understanding the distinction at its heart between a necessary and a definitional nature. The first step, then, is to provide an argument that this is in fact a distinction between the animal's material and formal natures.[8]

Such an argument can be found in the philosophical groundwork of Aristotle's animal investigations, *PA* I. It is there insisted that animals have a material nature (640b28–9, 641a26), and a formal nature (640b27), and the latter is eventually identified with nature understood as an animal's substantial being (641a27). Understood in this way, the formal nature is claimed to be both the mover and goal of living things, and is eventually identified with their soul. As so often, having distinguished these two senses of 'nature', he treats one, 'formal nature', as primary, and having done so, occasionally uses the term 'nature' to refer only to formal nature. Nature, in this sense, is more than once *contrasted* with matter (e.g., 642a18: "the nature is more an origin than matter").

PA I.1, then, endorses the distinction, argued for in *Physics* II.2, between two natures, one material and the other formal, this latter variously identified with an animal's substantial being, the source of its movement, its goal, and its soul.[9] Aristotle concludes the passage in which the soul is identified as the primary object of biological study by saying that the natural scientist ". . . should speak about the soul more than about the matter, in as much as the matter *is* more because of this nature than the other way around – for in fact the wood *is* a bed and a tripod *because* it is these things potentially" (641a29–32).

PA I.1 provides reasons for understanding the '*necessary* nature' of our focal passage as the material nature. An argument for distinguishing two sorts of necessity is interwoven with the passages in *PA* I.1 which contrast formal and material natures. That turns out to be significant for the discussion of *PA* III.2, for 'the things that belong of necessity' in that discussion are not all of a kind.

Not long after developing the idea that an animal's formal nature is, in one of its guises, the goal of an animal's various structures and functions, Aristotle discusses a distinct type of necessity that materials possess relative to goals. He claims that it cannot be necessity of the two varieties he has discussed elsewhere – necessity associated with an object's constituent elements or with external force. He then goes on:

> But the third sort of necessity *is* present in those things which come to be; for we say that nutrients are necessary not in virtue of any of the other modes of necessity, but because it is not possible to be without nutrients. This necessity is, as it were, hypothetical. (642a7–9)

In this passage, an animal's parts and whole body are said 'to be for the sake of something', and its nourishment *necessary for* the continued existence of organisms. The material nature of an animal is, in this sense, a conditionally necessary nature – *if* there is to be an animal of a certain kind, it *must* (i.e. *cannot not*) have the proper nutrients. As we will see, the items in *PA* III.2 identified as present of necessity are best conceived of as fixed amounts of nutrients with specific elemental potentials.

Matter is, then, necessitated to be in a certain way, given a certain end; but it does not necessitate that end. If the formal nature of an ax is its functional ability to cut, hypothetical necessity takes the form of functional constraints on ax-matter: the function of the ax requires matter of an appropriate dispositional nature, shaped in an appropriate way. Matter, in this case, is the necessitat*ed,* form the necessitat*ing,* nature.[10]

In the world of nature, however, this cannot be the *entire* story. The craftsman is free to select the materials appropriate for the tool he wishes to fashion. The formal nature of an organism lacks this freedom (cf. *Ph.* II.2 194b7–9). The material nature of an organism is as much a 'given' as its formal nature. To provide a pair of examples from *PA* II–IV: even though, as will be discussed in detail later, blood and its analogue serve a nutritive function, it is not *hypothetically* necessary that some animals are blooded, others bloodless; nor that some animals have lard and others suet. There is no teleological argument grounding this differentiation – it is primitive. And it is primitive in two senses: [i] In blooded animals the maternal and paternal contributions to generation are already concocted blood of a certain sort, and [ii] blood (or its analogue) is the nutritive matter of the entire body. Similarly, one explains whether an animal has lard or suet not by reference to a distinctive function served, but by identifying the kind of blood it has – whether it is fibrous or serous in nature. What sort of fat develops is predetermined by the nutritive material of the organism in question. These facts are grounded simply in the given material nature of the animal in question. One explains them, at least, from the bottom up.[11]

That there is, in addition to conditional necessity, a *pre-conditional* necessity dependent on an animal's material nature is suggested in an enigmatic passage at the close of *PA* I.1, in which two sorts of necessity involved in respiration are discussed.

> . . . necessity sometimes signifies that if one thing is to be that for the sake of which, it is necessary that other things obtain, and sometimes that things are a certain way in accordance with their characters and natures. (642a33–35)

The first necessity mentioned here appears to be that required by a goal. The second resides in the character and nature of things, and is not merely conditional on a goal. *PA* III.2 provides ample evidence that this is, at least in part, a necessity rooted in the material nature of an animal, which constrains, and perhaps acts independently of, the actions of its formal nature.[12] *PA* I.1 thus provides grounds for supposing that the necessary nature of our passage is the material nature of the animal. But that discussion also raises the question of what *sort* of necessity it is that the necessary nature of our focal passage possesses? This question will be taken up after the actions of the formal nature are examined.

III FORMAL NATURES

As mentioned earlier, there are two particularly worrisome uses of the concept 'nature' in *De Partibus,* referred to earlier as Apparently Demiurgic and Apparently Cosmic. The question these uses of the concept raise is whether they cohere with Aristotle's philosophical analysis of the concept. Aristotle's *De Partibus Animalium* is 58 Bekker pages long. In the course of those 58 pages, the subject 'nature' is used with verbs of agency – using, distributing, giving, treating, sketching – 79 times, not counting the times when it is the *implied* subject of such verbs. Are these 'natures' material natures, formal natures, or something else entirely? Answering that question is one of the purposes of this paper.[13]

PA III.2 begins by explaining what horns in general are, and why certain animals have them and others do not, after which it moves to account for variations in horns among the animals that have them. To be a horn is to have a proper function (662b27), attack and/or defense, and only in vivipara do hard head appendages perform such functions. Things resembling horns but lacking this function, Aristotle argues, are called horns only metaphorically.

But not even *all* vivipara have horns (662a31–35) – those with toes do not, "for nature has provided to some of these claws, to others fighting teeth, and to yet others some other part sufficient for defense."

In fact, only the *hoofed* vivipara have horns. Aristotle thus identifies his first universal predication: *All horned animals are hoofed vivipara.* This is not, however, a reciprocal universal predication, since some hoofed vivipara have other means of defense – speed, bulk – and therefore do not need horns. But to say that these vivipara lack horns because they don't need them is no explanation without a premise that animals lack what they do not need. Aristotle provides this premise explicitly at 663a17–18: "Nature has not provided to these more than sufficient defenses." This is the first instance of an 'apparently cosmic' reference to nature's actions, though its action is restricted to the hoofed vivipara. Elsewhere, however, it is made clear that natures *never* provide more than sufficient defenses.

Thus in the opening arguments of *PA* III.2, nature, conceived of as an agent, has already played a crucial explanatory role in the discussion.[14] Indeed, by the time it is characterized as "making use of the necessary nature" in our focal passage, this nature/agent has already done the following:

(i) Provided different means of defense to different vivipara (662b33–663a4)
(ii) Added a different means of defense when the horns are useless as such (663a8–11)
(iii) Provided no more than sufficient means of defense (663a17–18)
(iv) Removed an excess of earthen material from above, making the animal single horned, and given the excess to the hoofs in certain solid-hoofed animals (663a28–33)
(v) Acted rightly in producing the nature of the horns on the head (663a34–35)

These claims, in turn, presuppose certain more abstract principles about the actions of natures, claims which appear to have the status of first principles in *De Partibus:* that is, they are fundamental, as here, to the cogency of particular explanations, and are not themselves explained. As we are told at *De Incessu Animalium* 2:

> The starting point of investigation is hypothesizing things, which we are in the habit of using often in the study of nature, assuming things of this character are present in all the works of nature. One of these ⟨hypotheses⟩ is that nature does nothing in vain, but always ⟨does⟩ what is best, from among the possibilities,

for the substantial being of each kind of animal; wherefore if it is better a certain way, that is also the way it is according to nature. (704b11–17)[15]

I believe such abstract hypotheses about nature's agency identify essential features common to the actions of formal natures. For example, to say that nature doesn't provide more than sufficient means of defense is a specific application of the more abstract claim, repeated a number of times (always syntactically linked to a specific explanation) in *PA* II–IV, that nature does nothing 'superfluous' or 'without a point'.[16] Both of these claims can be read as abstract principles about the actions of the natures of members of particular kinds of animals. And since there is no provision in Aristotle's philosophical analysis of the concept of nature for reading them as references to a cosmic nature, the onus of proof must fall on those who would read them thus.

Nevertheless, the concept of nature discussed in *Physics* II.2, and endorsed in *PA* I, leaves us with a number of options. Which nature is it that is moving the material shared by various animals around so that it is deployed 'reasonably' and 'correctly'? This cannot be the material nature shared by hoof and horn, since the nature in question acts to deploy the material nature in different ways. Nor can it be the formal/functional nature of *horns,* since the nature in question determines whether a certain animal has horns at all, and whether it has one or two. The view that makes best sense of this and many other similar passages is that the agent is the formal nature of the animal. Its actions distribute the materials of each animal (which *qua* materials are similar in animals within a kind) in just those ways appropriate for the animal's well-being as the form of animal it is. Both in *PA* I.1 and *de An.* II.4, however, the formal nature can be taken to refer either to the end for the sake of which changes are taking place, or the motive source of those changes. The argument for taking it in the latter way here (and in the many similar passages throughout *PA* II–IV) is straightforward: in our central passage, we are told that this 'definitional nature' is making use of the things present of necessity *for the sake of something.* Thus in this formulation (again, typical of these passages), the nature that is an agent is distinguished from the goal for which it acts.[17]

There is another problem with this answer. The problem is well-illustrated by the discussion in *PA* III.2. In that discussion, the nature (in the singular) provides different organs of defense to different kinds of animals; different kinds of horns to different kinds of animals; and solid hoofs to some animals, split hoofs to others. Furthermore, it never provides more than sufficient organs of defense to *any* animal. That is, the *extensional scope* of 'the nature' in this, and other similar, discussions, seems to be 'trans-specific'; it appears to *coordinate* the features of the organic world in various ways, and the natures of individual composite substances could hardly do that.

This is, I believe, the chief consideration inclining some commentators to picture a single nature, an Aristotelian demiurge, as it were, insuring the appropriate outcomes in specific cases. (They might be encouraged to do so by passages such as 654b32, where Aristotle says that nature has crafted animals from flesh like a sculptor!) Anyone taking this route must admit that there is nothing in Aristotle's *philosophical* reflections on the concept of nature (e.g., in *Ph.* II.1, 2) that answers to such a nature as this.

But the account of 'formal nature' just sketched accommodates these worries about scope without leading to the 'demiurgic' interpretation. When, for example, it is claimed that 'the nature' provides claws to some four-legged vivipara, fighting teeth to others, and horns to yet others, the subject may be taken distributively: the nature of certain such animals produces claws, of others horns, and of some of these a single horn. But in each case, an animal's formal nature produces only one means of defense, and indeed the best possible means for each animal. These axiomatic statements are thus conceived as general principles characterizing essential features of the actions of formal natures.

A case has been made, then, that it is possible, even plausible, to read these passages thus; and discussions like *Ph.* II.1, 2 and *PA* I.1 don't provide any obvious grounding for the 'demiurgic' reading. Yet one would still like Aristotle to address the question directly. Since the overwhelming number of passages where this 'agent nature' plays an explanatorily central role are in *PA* II–IV, this is surely the place to look to test such a proposal.

It is therefore frustrating that, while the concept of nature is perhaps the most pervasive explanatory concept in *PA* II–IV, Aristotle rarely bothers to signal its referent, or to connect its use to the general philosophical analysis in *PA* I or elsewhere. The importance of the transitional passage in *PA* III.2 is that he does so in a way that provides a map of the concept's use throughout his biology.

That passage tells us we are about to discuss how the formal nature makes use of the necessary nature for the sake of something. If we now look to this discussion, I believe the reading I have been arguing for is confirmed.

First of all, the bodily and earthen belongs more to the larger animals, while we know of no completely small horn-bearing animal – the smallest of those known is a gazelle. . . . That which is boney in the bodies of animals is in origins earthen; which is also why it is most abundant in the largest animals, to speak with a view to what occurs for the most part. For the residual surplus of such body present in the larger of the animals is made use of by nature for protection and advantage, and this surplus, flowing of necessity to the upper region, is distributed in some cases to teeth and tusks, in other cases to horns. (663b25–36)

Earlier in this chapter, Aristotle noted certain interesting correlations between split hoofs and double horns, and between solid hoofs and single horns, and explained why nature *provides* an excess of something to the hoofs while *removing* it from 'above', i.e. from the head. All we are there told about what is being distributed is that hoof and horn share 'the same nature'. This discussion relies on the account of bone and kindred materials in *PA* II.9, where we are informed that this shared nature is 'the bodily and earthen'.

In many larger animals, it is claimed, there is a 'residual surplus' of this earthen, bodily material, and that "nature *makes use of* it for protection and that which is advantageous." The repetition here of 'makes use of' with 'nature' as its subject argues for this being the formal nature of the animal introduced earlier – and thus that this earthen body, out of which horn or hoof is constituted, is the necessary nature characterized earlier as made use of for the sake of something.

The claim that this bodily and earthen material nature is *'flowing upward of necessity'* makes sense, since it is the animal's formal nature that is the agent, not the elemental nature of the material. That the nutritive actions of an animal's soul can cause its earthen nutrients to move upward is not problematic – indeed, in every animal with teeth, this *must* continuously take place. Further, it is virtually certain that the surplus 'earthen and bodily' residue is in a liquid form, since it is Aristotle's view that "all of these ⟨uniform parts⟩ take their growth from liquid" (*PA* II.2 647b26–28). In fact, prior to being converted to teeth or horn, this liquid *must* be blood, since we are told in *PA* II, "that the nature of the blood is . . . the matter of the entire body; for the nourishment is matter, and the blood is the final nourishment" (651a12–15).[18] Thus in this passage the earthen and bodily surplus flowing upward is almost certainly blood of an earthen character, obeying the dictates of nutritional dynamics, not terrestrial mechanics.

This passage provides a relatively rare glimpse into Aristotle's conception of the dynamics of form and matter in a composite substance. The nature making use of this material either for horns, larger teeth or tusks is the formal nature of the animal in question – the causal agent dynamically producing and maintaining the composite substance.[19]

An earlier discussion of eyelids and eyelashes confirms and deepens this understanding of the concept of formal nature as agent. At the close of that discussion, Aristotle makes the following puzzling claim:

> So it is necessary, because the evaporating fluid is bodily – unless some function of the nature redirects it to another use – that hair comes to be in such places from necessity and on account of such a cause. (*PA* II.15 658b23–26)

Apparently, a bodily material in fluid form flows to the ends of the blood vessels of the eyelids, and hardens into hair when the moisture in the fluid evaporates. From a philosophical perspective, there are a number of problematic formulations in this sentence that catch the eye. First there is the curious juxtaposition of 'it is necessary' with 'unless'. If one omits the material I have set off with dashes, there appears to be the assertion that a material level process taking place on the surface of eyelids necessitates the formation of hair. But this happens only on the assumption that this bodily fluid is not redirected elsewhere. The necessity involved here, then, is conditional or 'hypothetical' necessity.

This leads to the second problematic formula, for this 'redirection' of material takes place *for another use* and by means of 'some function of the nature'. This should be read in the context of Aristotle's claim that eyelashes are present, in animals that have them, for the sake of keeping the eyes free of debris (658b17–18). Not all animals are exposed to such debris, however; and animals with different material constitutions solve the 'debris' problem differently. Nevertheless, eyelids have a use, and redirection of the material for *another* use presumably occurs in cases where the formal nature of the animal has a more compelling use for a limited amount of such material. This suggests, and we will see further evidence for this suggestion momentarily, that the quantity of material of various kinds available for the construction and maintenance of parts constrains the actions of formal natures. Presuming, for example, there are animals where both eyelashes and some other structures made from the same material might be beneficial (if not absolutely necessary), this passage suggests that the distribution of limited material may take place according to a set of functional priorities. Thus a picture of a rather complex biophysics begins to emerge, one law of which we may call the *Principle of Functional Priority*.

> Given the presence of a certain material, it will be directed to location L for the sake of natural function F, unless function F is subordinate to function F*, in which case it will be re-directed to L*, the location of F*.

Formulating this principle explicitly highlights the fact that the necessity of 'the necessary nature' is subservient to a hierarchy of ends. It is an animal's *formal* nature that determines the placement and structural design of the material. In passages such as the above, essentially the same material nature is placed and structured in entirely different ways by distinct formal natures. The material's *own* nature doesn't dictate its location, structure, or function.

The Principle of Functional Priority is assumed in Aristotle's explanation of yet another (non-reciprocal) universal regarding animals that have horns – all of them lack upper front teeth.[20]

> That is why none of the horn-bearing animals is possessed of both sets of teeth (for they do not have upper front teeth); for taking from there, nature adds to the horns, and the nourishment given to the upper front teeth is expended in the growth of the horns. (663b36–664a2)

Once again this is easily read as a reference to the formal natures of horn-bearing animals, which in some animals provide the appropriate nourishment to the teeth, but in *these* animals direct it toward "the growth of the horns." Again the Principle of Functional Priority is assumed: the function of horns overrides that of certain teeth in these cases, and thus the material is redirected for the sake of the overriding function. Aristotle never suggests that a full complement of teeth would not be beneficial to such animals – the force of his argument is that there is insufficient nutrients of the appropriate kind for both horns *and* a full set of teeth.

At least four unargued premises are involved in this explanation for the lack of certain teeth in horned animals – and at least two are assumptions about an animal's material nature.

1. Large viviparous quadrupeds have surplus earthen material.
2. There is a portion of such material that 'flows upward necessarily'.
3. Nature doesn't provide more than sufficient defenses to any organism (cf. 663a16–17).
4. Nature makes use of the excess residue for the most important end, i.e. defense and advantage.

Only the fourth premise refers explicitly to the goal-directed agency of the formal nature. The first two identify fundamental material facts about the kind being investigated, facts which impose *constraints* on what the formal nature can achieve. Further, even the third may be a consequence of these constraints. In this explanation it doesn't appear that nature has the option of providing tusks *and* horns, or a full set of teeth *and* horns, to the same animal – there isn't sufficient material to do so. That is, Aristotle's assumption that nature is non-redundant might express, *not* a prior assumption regarding the inappropriateness of redundant defenses, but rather the prior assumption that insufficient materials are provided for redundant defenses.[21]

IV MATERIAL NATURES

We have discussed at length the role of the formal nature in the production and maintenance of horns. But *PA* III.2 also reveals much about the study of the material nature of living things. Prior to our focal passage, it is claimed that hoof and horn "have the same nature" (663a29), and argued that the splitting involved in the construction of the split hoof is

> in virtue of a deficiency of this nature, so it is reasonable that, in the solid-hoofed animals, nature has taken the excess from above and made a single horn, and given it [the excess] to the hoofs. And it also did the right thing in producing the nature of the horns on the head. (663a31–35)

There are three distinct uses of 'nature' here. The nature moving the excess material from one locale to another is, on the account of this essay's previous section, the formal nature of the animal being investigated. The 'nature of the horns' may refer either to the 'form' of the horns or to horns as a material/formal complex, an issue I put aside for the moment. This leaves the claim that hoof and horn have *the same nature,* and that there is a *deficiency of this nature,* to be considered.

It will help to begin with a discussion in *PA* II.9 on bone. At the close of that discussion, Aristotle takes up parts which are *like* bone 'to the touch', among which are included hoof and horn.

> All these ⟨uniform parts⟩ the animals have for the sake of protection; for the whole ⟨organs⟩ constituted from these ⟨uniform⟩ parts, and synonymous with them, e.g., the whole hoof and whole horn, have been constructed for the safety of each of these animals. . . . All these parts have an earthen and hard nature of necessity; for this potential is of a defensive sort. For this reason too, all such parts are present more in the four-footed and viviparous animals than in humans, on account of their entire system being more earthen. (655b4–16)

The uniform parts here discussed are said to have *an earthen and hard nature of necessity.* But the justification for this claim shows that *this* necessity is hypothetical, deriving from the fact that an earthen and hard nature has the appropriate potency for protective purposes.

But this passage also indicates that the 'same nature' of horn and hoof referred to in our focal passage is their shared *material* nature. The role of that nature in the explanation is, however, complex. Horn and hoof have the same material nature, i.e. they are 'hard and earthen'. But such material is *present for the sake of* constituting parts which *must have* a material propensity suitable for defense. Given parts of such a functional nature, such material is present 'of necessity'. That 'material, necessary nature' can be *described* without reference to such functional requirements, but its presence in an animal cannot be *explained* without such a reference.

But it is also the case that each animal has a fixed quantity of different sorts of basic material ingredients out of which the formal nature constitutes its parts. These 'givens' of an animal's material nature constrain the actions of its formal nature. Large viviparous quadrupeds have only a certain amount of 'earthen and bodily' material for hard head appendages, and that means that, should the formal nature be such as to constitute horns, certain teeth will be missing. *Small* vivipara, on the other hand, begin with no such surplus material. When Aristotle states that nature does the best *from among the possibilities* for the substantial being of each kind of animal (*IA* 2 704b15–16), he acknowledges a *pre-conditional* necessity as such a constraint. Aristotelian formal natures work (and in *this* respect *are* akin to Timaeus' Demiurge) within the bounds of a comprescent material necessity.

V PRELIMINARY RESULTS

What, then, does Aristotle's practice suggest is involved in knowing an organism's material nature?

First, Aristotle uses the term to refer to the material common to different organic structures – often the elemental potentialities of nutritive blood rather than to the more 'proximate' uniform part or parts out of which they are immediately constructed: that is, to a fluid which is potentially earthen and hard, and which is capable of becoming hoof, nail, tusk, tooth, palate or horn (but not, apparently, brain, or flesh). Each of these non-uniform parts must be constituted of an earthen uniform part, because of the dispositional properties required for the varied tasks they perform.

Second, to know the material nature of a part, in *that* sense, is to know *that* such material is present, and to *explain* its presence by appeal to conditional necessity: *if* there is to be an animal which defends itself with horns, there must be horns; and if there are to be horns, there must be a certain amount of earthen nutrient out of which to compose them.

Third, in a significant number of explanations in *PA* II–IV, of which the explanation of horns is one, the amount and kind of elementally distinct materials available in the nutritional make-up of animals is taken as a given. I have argued that this evidences a *preconditional* necessity that places *fundamental material constraints* on the actions of the formal nature. For example, it is taken for granted, without further explanation, that in animals with horns, their production is achieved at a cost to dentition. This loss of grinding teeth has further explanatory implications – these animals have multiple stomachs due to the absence of teeth suitable for grinding up nutrients.[22]

The next chapter of *PA* III, chapter 3, provides three fine examples of the role of matter in biological explanation. These can be used to extend the results of our discussion.

VI MATERIALS, DISPOSITIONS, AND FUNCTIONS IN *PA* III.3

The following three passages explain why an organ is constituted of the material, or materials, that it is.

A. The esophagus:

Since the organ connected with breathing from necessity has length, it is necessary that the esophagus be between the mouth and the stomach. *And the esophagus is fleshy, with a sinuous elasticity* – sinuous *so that it may dilate when food is ingested, yet* fleshy *so that it is soft and yielding and is not damaged by being scraped by the food going down.* (664a29–36; emphasis added)

B. The larynx and windpipe:

That which is called the larynx, and the windpipe, *are constituted from a cartilaginous body;* for they are not only for the sake of breathing, but also for the sake of vocalizing, *and that which is to produce sound must be smooth and hard.* (664a36–664b3; emphasis added)

C. The epiglottis and its analogue:

The windpipe, by being positioned, as we said, in the front, is interfered with by the food; but for this *nature has constructed the epiglottis.* Not all the live-bearing animals have this, though *all those with a lung and skin with hair, and that are naturally neither covered with scales nor feathers, do.* (664b20–25; emphasis added)

The animals being spoken of do not have the epiglottis on account of their flesh being dry and their skin hard, so that in them such a part, constituted from such

196

flesh and such skin, would not move easily; rather, the closure would occur as quickly from the sides of the windpipe itself as from an epiglottis made of their own flesh, such as the animals with hair have. (664b36–665a6)

There are a number of features common to all of these explanations.

[i] In each of them, the material nature of the organ – that it is cartilaginous, fleshy and sinuous, or just fleshy – is one of *the facts to be explained* about it.

[ii] Each of these explanations involves *two* inferences. First, from premises which identify [a] the (or a) function of the organ, and [b] the dispositional properties required to perform that function, it is established that the organ must be made of material with that dispositional property (or with those dispositional properties). Second, from *that* conclusion, and a premise that notes that a certain uniform part (or uniform parts) has the relevant dispositional property (or properties), it is concluded that the organ must be made of that (or those) uniform parts. Schematically:

> The function of part P is to f.
> Anything which fs must have disposition d.
> Hence P must have d.
> Uniform part UP has d.
> P is UP.

[iii] Represented in the terms of the *Posterior Analytics*, the middle term of the first inference refers to (one of) the functions for the sake of which the organ is present, while the middle term of the second refers to the dispositional properties of the material needed for that function. That is, the dispositional properties of the uniform part, rooted in its elemental make-up, constitute one of the premises in the explanation.

[iv] Finally, the preliminary conclusion – that the organ must have certain dispositional properties – becomes the major premise in demonstrating the material conclusion. It 'feeds' information about the nature of the material required into the explanation of why the organ is constituted of the tissues it is. The two inferences are thus tightly interwoven into a single explanation.

The last of the three explanations, while sharing certain of these features, has a somewhat different character. At 665a7 Aristotle says he has given the cause which explains why some animals have an epiglottis and some do not. The [formal] nature has remedied the unfortunate positioning of the windpipe by devising (664b22, 665a9) the epiglottis. The function of the epiglottis explains its presence. Yet many animals (all the ovipara, in fact) with the same functional requirement have no epiglottis – how can *this* be explained? Aristotle indicates that what needs explaining is the presence of a different mechanism for closing the windpipe in those animals covered with feathers and scales – or, alternatively, why these animals lack an epiglottis.

The explanation appeals to the material nature of the flesh of the blooded ovipara – it is dry and hard. An epiglottis made of such flesh would not open and close quickly. And if it could not do that, it would not be able to perform the requisite function well. Here *absence* of a structure is explained by the *absence* of the uniform material out of which the structure must be made, if it is to be able to perform its function. (In fact there is also an explanation for the alternative mechanism.)

In other respects, however, this explanation parallels those of the material natures of the esophagus and windpipe. The goal of closing the windpipe requires an organ to perform a certain action; that action's performance depends upon the organ having the dispositional properties of soft flesh; but the uniform parts of the ovipara lack those dispositional properties; hence the ovipara lack an epiglottis. Again, schematically:

> The function of the epiglottis (E) is to g.
> Whatever is to g must have soft flesh (s).
> E must have s.
> s belongs to no Ovipara (O).
> E belongs to no O.

Such a demonstrative presentation again reveals two levels to the explanation. The placement of the windpipe behind the esophagus explains the necessity for all breathers to be able to close their windpipe when eating. But that functional requirement is fulfilled by two different mechanisms in the two distinct classes of breathers. The basic features that explain the presence or absence of an epiglottis in particular are the different material natures of the flesh and skin in these two classes.

Such passages as these provide further insight into the claim that the natural philosopher should study both the matter and the form, and the form insofar as it is that for the sake of which the item to be understood exists (cf. *Ph.* II.2 194b7–15; *de An.* I.1 403b1–12). He must study the matter if he wishes to understand why a non-uniform part is constituted of a certain sort of material, for its being constituted of a certain material limits its changes and actions, permitting it to act in certain ways and not others. Further, even in the premises of such explanations there are appeals to the dispositional properties of the uniform parts in question, properties rooted in their elemental nature. These are the properties in virtue of which a uniform material is the only, or best, material for a given function. Those properties will play the role of the basic middle explaining why a particular uniform material, rather than some other, is required in the construction of a particular non-uniform part.

VII A PROBLEMATIC PASSAGE

For the present, I will restrict my consideration of problematic passages to one well-known one within the *PA* itself. In discussing the placement of the mouths of dolphins and selachians, beneath the 'snout', Aristotle remarks:

> Nature appears to have done this [place the mouth underneath] for the sake of preserving the other animals. . . . But it also ⟨appears to have done this⟩ in order that they not follow their gluttonous desire for food. . . . And in addition to these reasons, it is impossible for the nature of their snout, being rounded and thin, to be divided. (696b28–34)

The first thing to notice about this passage is that the first two explanations are both qualified as statements of appearances. The second thing to note is that the one part of the explanation that is not so qualified states that it was impossible for the mouth to be at the end of the snout anyway. The third thing to note is that one of the two 'apparent' explanations refers to the good of the animals in question – the design serves to prevent their succumbing to gluttony. Finally, though I wouldn't let anything hang on this reading alone, it is just possible to read the first sentence to say "Nature appears to have done this for X, but in fact it has done it for Y." At any rate, given that the overwhelming preponderance of passages in *PA* II–IV tell against supposing that 'nature' here refers to a cosmic nature designing one animal one way in order to help other animals escape, and given the many reasons to doubt that Aristotle takes the first proffered explanation seriously, this passage cannot be used as a serious piece of evidence against taking nature in *PA* II–IV as I have been suggesting.[23]

VIII CONCLUSION

The concept of nature is all pervasive in *De Partibus Animalium,* but it refers to very different things. The lack of referential unity *could* arise from an unreflective use of a term in common use. Yet we know that Aristotle reinvents this concept once he begins to view natural substances as matter/ form composites. In the *Physics* and *Metaphysics* nature is characterized as *systematically* differentiated, referring to either the material or the formal sources of change of a composite substance – and once 'the nature of a thing' refers to its form, it also may refer to the inherent sources of a thing's functions and to the end for which those functions are performed.

De Partibus Animalium I.1 and 5 *suggest* that just this referential pluralism will be a feature of the use of the concept of nature in biological science. But *PA* I is *philosophy,* not the biological nitty gritty, and that leaves us wondering, first, whether the nitty gritty is in line with the philosophy and second, if it is, what the nitty gritty will actually look like. I have argued that the explanatory use of the concept of nature in *PA* II–IV is in line with *PA* I; and yet its protean richness would have been hard to predict based on *PA* I's sketchy remarks.

What it actually looks like may well have implications for interpretations of Aristotle's metaphysics of natural substance. In particular, attempting to preserve the unity of the composite by strongly identifying matter and form in the actual composite may have trouble accommodating the dynamic interaction of material and formal natures in the explanations we have looked at, as well as the extensive role played by material natures in those explanations. Conversely, downplaying the importance of material natures in Aristotle's science can be accomplished only by ignoring a sizable portion of the biological corpus.

While all of the references to nature as an agent of construction and maintenance in *PA* II–IV can be accommodated by Aristotle's notion of a *formal* nature, these references take two quite different forms. Those statements I have described as 'first principles' are statements about an essential feature common to the actions of formal natures. Others describe the actions of a specific formal nature in structuring and maintaining a particular part. This latter use is best understood as a reference to the formal nature as a goal-oriented efficient cause, an efficient cause which acts *selectively* and *in accordance with instructions* inherited from its parents. It is part of the task of *De Generatione Animalium,* of course, to explain this process of instructional inheritance.[24]

NOTES

1. Cf. Bodson 1990, 251–253, 348. Only two other nouns are used over two hundred times – 'animal' and 'part'.
2. It should be noted that this discussion focuses on the interactions of formal and material natures in fully developed living things, rather than on the process of

development. *PA* II–IV is interested, in my understanding, in how the formal nature of a living thing actively maintains, especially via its control of nutritional material, the living organism. It *presupposes* that the parts have come to be for the sake of performing these activities. On the importance of this presupposition for Aristotle's justification of teleology, see Gotthelf 1987b, 213–214n1, and 238–242.

3. While I agree with Cooper 1987 that the natures of the elemental materials are presupposed in explanations that appeal to hypothetical necessity, my thesis should be distinguished in two respects from his. First, Cooper asserts that the "Democritean necessity 'seated in' the material nature is presupposed by, *and does nothing to explain,* the operation of hypothetical necessity" (261–263), while I argue that an organism's material nature places constraints on the operations of its formal nature in achieving its ends, and thus *does* play an explanatory role in the operations of hypothetical necessity. Second, I will argue that in a significant number of cases, biological explanation appeals to unqualified material necessity, outside the context of appeals to hypothetical necessity.

4. Burnyeat 1992, 22. And note his description of Aristotle's 'physical thesis': "It is a physical thesis to the effect that the flesh, bones, organs etc. of which we are composed are essentially alive" (26). On this point, see the response to Burnyeat found in Whiting 1992, 76–91, esp. 82–88. In general terms the results of this paper would seem to confirm the approach to animal bodies she defends.

5. One needs to distinguish the idea of a Demiurgic Nature, not present in particular organisms, that coordinates the actions and interactions of the universe, from that of a Cosmic Teleology, i.e. a single overarching goal toward which all natural processes aim – though there is no doubt a tendency for believers in the latter to imagine the presence of the former. The most recent defense of both is Sedley 1991, 179–196. Sedley argues that "the nature which is exhibited by the anthropocentric natural hierarchy must not be so much individual nature as global nature – the nature of the entire ecosystem, so to speak" (192). His primary support is drawn from *Metaphysics* Λ.10, 1075a11–25. Because he thinks this global teleology is ultimately man-centered, however, he doesn't cite, as he otherwise might, the dozens of passages in *PA* II–IV where nature is described as an agent. Indeed, he claims the kind of natural hierarchy he defends "never surfaces . . . in the zoological works." He is, of course, right that virtually no part of any other animal is argued to be present because of its benefits to humans – in *PA,* the appeal is virtually always to the needs of the organism with the part. This is presumptive evidence against Sedley's thesis that Aristotle's teleology is anthropocentric. For many other previous defenders of "a broader, interactive teleology" see the references in Sedley 1991, 179n2 to the papers of Owens, Lloyd, Furley, Kahn, and Rist.

6. A useful review of many of these passages both in *PA* and the other biological works can be found in Preus 1975, 225–248. Preus also concludes, though on quite different grounds, that the biology doesn't support any notion of a 'transcendent, cosmic nature'.

7. On Aristotle's use of the language of nature 'using' and 'making use of' various materials, see Preus 1975, 225–235.

8. Wolfgang Detel points out in discussion that, superficially at least, this passage is paradoxical. It claims we have already stated what the nature of horns is for; why, then, must we go on to discuss how nature makes use of the necessary nature for the

sake of something? Is that not the same inquiry? As I read the chapter, it is not. Up to this transitional passage we have been told the final cause, i.e. the function, of horns. But we have been told nothing about *how* horns are produced or maintained by the interactions of formal and material natures. It is this that Aristotle is about to explain. Detel also pointed out the difficulty of understanding the genitive absolute at the heart of this complex sentence. If one reads it as I have, conditionally, and takes 'the necessary nature' as subject, it establishes an assumption which governs the inquiry. If one were to read it temporally or causally, the point would be subtly different, though not in ways that would affect my argument substantially.

9. Aristotle's point is, of course, that the soul of the entire organism, as its form, is that for the sake of which the body, its parts, and their particular functions, are present. In *PA* II–IV, the formal nature functions, it seems, principally as a goal-directed *efficient* cause, rather than as goal. This follows from the fact that the nature under consideration always acts for the sake of something. This doesn't rule out, of course, its also being a goal (see notes 14 and 17, below). The formal nature of an organism in this sense refers to those dispositions to organize the distribution and assimilation of nutrients in the way that maintains the organism (and so itself; and so the formal nature, in a different sense, is the goal for the sake of which the formal nature acts in the ways it does).

10. Cf. Cooper 1987, 243–274; Freeland 1987, 392–407.

11. This *may* be relevant to the fact that in three texts in *PA* IV (678a26, 693b2–13, 695b17–25) an animal's being blooded or bloodless is said to be "in the account of its substantial being," on which see Gotthelf 1985a, 43–50. If Aristotle is inclined to decide what is in the account of a thing's being on the basis of explanatory primitiveness, and if he is willing, in natural science, to include matter in definitions, then we could expect that being blooded or being bloodless would indeed be in the substantial being of animals identified at a sufficiently general level.

12. This is what Cooper refers to as 'Democritean' necessity, the necessity associated with material (especially, I take it, elemental) natures acting according to their natures. This he argues to be presupposed by the conditional necessity associated with goals. I'll be pressing the case for two additional claims: [i] that certain material givens aren't merely *presupposed* by a formal nature's goal-directed actions, but *restrict* those goal-directed actions of formal natures in specific ways; and [ii] that certain biological explananda are explained solely by the elemental material potentials of the basic nutrients of the animal in question.

13. David Charles calls attention to the importance of such an investigation for interpreting Aristotle's teleology in Charles 1991, 127n29.

14. Robert Bolton suggests in discussion that all such references must be taken metaphorically, since formal natures are never efficient causes. This runs counter to a number of claims to the contrary, however. I have already discussed *PA* I.1 641a27–28, where nature spoken of as substantial being is said to be both nature *as mover* and nature as goal. In very similar words, *de An.* II.4 415b9–12, 415b22–416a9 argues that soul is source and cause in the sense of substantial being, that for the sake of which, *and origin of movement,* and then goes on to make it clear that insofar as the capacity to nourish is an aspect of soul, an organism's growth and nourishment is due to the activation of these capacities. Thus while it is clear that

the language to characterize nature's actions is borrowed metaphorically from human craftsmanship, the idea that formal natures (souls) are agents is not to be taken metaphorically.

15. Compare *GA* V.8 788b20–25, where a shorter version of the same axiom is again said to be something we 'hypothesize', but in this case it is stressed that we hypothesize it "based on what we see."

16. Cf. 658a10, 661b4, 691b4–5, 694a15, 695b19–20. This principle in turn may be viewed as a specific way in which nature does the best possible (as suggested in the *De Incessu* passage quoted in the text, and repeated at 687a16); and there are a number of passages in *PA* II–IV which seem to be less abstract than a statement of the principle but more abstract than a particular application of it, e.g., that nature gives parts most to those most able to use them (661b30, 684a20), puts the most honorable parts in the most honorable place (665b20), uses distinct organs for distinct functions where possible (683a20–26, 688a24), or makes organs for functions, not vice versa (694b14–15).

17. It is plausible, however, that, since the actions of the nutritional soul not only support all other parts and functions of the organism, but are self-supporting as well, they are at least partly constitutive of the end for the sake of which they are directed.

18. Cf. Freeland 1987, 398–404. The context makes it clear, I believe, that Aristotle is discussing the ongoing supply of the appropriate materials to the appropriate organs in different kinds of developed organisms, *not* the embryological process. He surely knew that horns and teeth don't develop until after birth – depending on the species, often quite a long time after.

19. There is a very important, but wider, question about whether the teleology involved in the animal's formal nature *using a residue* (as here) is the same sort of teleology as that associated with the development of organs such as lungs, hearts or eyes out of *non-residual* materials. I won't, however, be able to discuss this issue here. See, however, Kullmann 1974, 196–200; Kullmann 1985, 174.

20. It is a non-reciprocal universal because of the camel, which has the same dentition and digestive system as the animals with horns, but not the horns (cf. *PA* III.14, 674a22–674b7, and the illuminating discussion in Gotthelf 1987a, 178–185).

21. Further evidence for this reading comes from 655a28ff., in which it is claimed that "nature cannot simultaneously distribute the same excess to many locations."

22. There are many other examples: absence of outer ears is accounted for by reference to the hard earthen material of animals that lack them (657a20–21); similarly with lack of eyelids (657b7–11). Insects see poorly because of their hardness (657b36); the presence or absence of the epiglottis is explained by differences in the material of the flesh in different animals (665a2); the presence of viscera is explained as due to the presence of blood (657b7ff); rennet is explained by reference to the thickness of the milk in ruminants (676a12); bloodless animals lack viscera because they lack blood (678a32–35); the sepiae give off ink and have a bone because they are earthen in nature (679a20–24); insect eating habits reflect their cold nature (682b1).

23. Compare the discussion of this passage in Balme 1987c, 278–279. I don't detect the 'sarcasm' here that Balme claims to, nor does Aristotle explicitly say he is replacing

the 'apparent' explanations with a 'proper' one – indeed it is perfectly natural to take the Greek to reflect a lack of commitment to either explanation on Aristotle's part.

24. This chapter has benefited in many ways from discussions after its earlier presentation to the Center for Philosophy of Science, University of Pittsburgh, the Boston Area Ancient Philosophy Colloquium, and a Conference on Aristotle's Biology organized in Bad Homburg by Professor Wolfgang Kullmann. Especially helpful were the comments of Nicholas Rescher, Bob Olby and Mary Louise Gill on the first occasion; of my commentator Bill Wians, and of Charles Kahn and Arthur Madigan on the second; and of Robert Bolton, David Charles, Alan Code, Andrew Coles, David Depew, Wolfgang Detel, Mary Louise Gill (again), Allan Gotthelf (who also provided valuable written comments) and Heinrich von Staden on the third.

9

Nature Does Nothing in Vain . . .

I   INTRODUCTION

Early in chapter 8 of *Exercitatio Anatomica de Motu Cordis et Sanguinis in Animalibus,* William Harvey recounts the consequences of his theory of the heart's motion that led him to conjecture that the blood circulates back and forth between the venous system and the arterial system through the heart. Among these was

> . . . the symmetry and size of the ventricles of the heart and of the vessels which enter and leave them (since Nature, who does nothing in vain, would not purposelessly have given these vessels such relatively large size) . . . (Harvey [1628] 1963, 57)

And in the final chapter, considering both animals without hearts and the timing of the heart's development in those with them, Harvey concludes that

> Nature, perfect and divine, making nothing in vain, has neither added a heart necessarily to any animal nor created a heart before it had a function to fulfill . . . (Harvey [1628] 1963, 83)

Harvey learned his Aristotelian naturalism while studying at its Renaissance epicenter, Padua's School of Medicine and Philosophy. His teacher, Fabricius d'Aquapendente, was pushing forward with what Andrew Cunningham has termed the "Aristotle Project," the conversion of surgical dissection, traditionally subservient to medicine, into a universal, and comparative, functional anatomy.[1]

Is the repetition of Aristotle's phrase "nature does nothing in vain" simply a symbolic mantra, chanted by Harvey and others to indicate their philosophical allegiances? Or does it play some important part in his reasoning? That is

205

actually a different question than the one I am asking, but I will sketch an answer in this chapter's closing section. Here I want to ask, of Aristotle himself, the question of what role this principle plays in his philosophy of biology and biology.[2]

The term 'nature' is used over 250 times in *PA* II–IV. Approximately one-fourth of the time, 'nature' is the subject of a verb of agency, the verb often borrowed from human, artistic contexts. In the previous chapter, I argued that this is in fact Aristotle's formal nature, rather than what it appears to be at first glance – some Aristotelian equivalent of Plato's Divine Craftsman, organizing the world for the best.[3]

Among this problematic subset of uses of the concept are seven cases of the general claim that nature does nothing in vain. In this paper I will examine the variations in these phrases, and the actual role they play in the contexts in which they appear. But before doing so, it is critical that we look at two passages in the biological corpus which make an important meta-theoretic claim about the role of this very proposition in natural science.

II HYPOTHETICAL STARTING POINTS

These two passages explicitly refer to the assertion that "Nature does nothing in vain" (**NP** from here on) as a starting point of natural inquiry.[4] It will be noted that each of them asserts more than that – indeed, as we review the various statements of this principle, the variations will be helpful in providing information to guide interpretation.

On Animal Locomotion 2 704b12–18.

The starting point of our investigation is achieved by supposing ⟨principles⟩ which we are accustomed to use often in natural inquiry – assuming this is the way things stand in all the works of nature. One of these ⟨principles⟩ is that nature does nothing in vain, but always, given the possibilities, does what is best for the substantial being of each kind of animal; accordingly, if it is better in a certain way, that is also how it is by nature.

According to this passage, not only does nature do nothing in vain – it does the best, where 'best' implies a selection from a range of possibilities, and is relative to each kind of animal's being. This principle also apparently licenses an inference: if, among a set of possible results, one is better than the others for the animal's being, then that is the result that will be generated by nature. Thus 'best', in this context, means 'best among the possibilities'.

This additional content will require the exploration of two difficult questions.

(i) Since what is best is relative to the being of each kind, explanatory appeals to the principle that a feature is as it is because it is better would seem to depend on prior knowledge of the *substantial being* of each kind of animal, which raises questions about what form that knowledge takes and about how it is to be acquired. Using **NP*** as an assumed starting point of inquiry doesn't presuppose such knowledge, but using it in a scientific explanation, given Aristotle's requirements on scientific explanation, does.

(ii) Since the best that nature produces is one of a range *possibilities,* it is also natural to wonder what *sense* of possibility Aristotle has in mind, and what restrictions there are on the *range* of possibilities.

The inquiry for which **NP*** is claimed as a starting point is *natural* inquiry, and Aristotle believes that what happens by nature happens *either always or for the most part.*[5] This is not a belief based simply on the observation that similar things do not always behave in similar ways. It rests on Aristotle's metaphysical account of natural objects, according to which they have propensities to behave in kind-specific ways, but can be prevented by abnormal external circumstances from doing so. Thus, on a variety of occasions, 'for the most part' is explained by noting that things happen in accordance with their natures *unless something prevents them from doing so.* In the absence of intervening processes, that is, Aristotelian substances act in accordance with their natures (*Ph.* II.7 198b4–6, II.8 199a8–11, 199b16–19, 23–26). Thus, in saying that nature produces what is best, given the possibilities, Aristotle *might* have in mind the idea that nature is restricted to doing what is best given the circumstances created by various intervening variables.

A close study of these passages, however, favors a different form of constraint on the possible actions of formal natures. In the passages we will look at, the range of possibilities is represented by the generic features of the more extensive kind to which an animal belongs. Each formal nature does what is best, within that range of possibilities represented by its wider kind(s).[6] Among these 'generic' restrictions are those related to the kinds and quantities of materials the formal nature is provided to work with in achieving the good.[7]

The wording of **NP** in the second passage raises two further questions.

Generation of Animals V.8 788b20–25

But since we are supposing, doing so based on what we observe, that nature
neither falls short nor produces anything pointless among the possibilities in
each case, it is necessary for animals that are to take in nourishment after
suckling to have instruments for the chewing of food.

Here the idea of a range of possibilities continues to be part of the
principle, but an alternative formulation is added to its core – not by 'pointless', which simply renders into English the Greek adjective corresponding
to the adverb translated as 'in vain' in our first passage; but by the idea that
nature does not 'fall short'. In the teleological conception of nature Aristotle
defends, biological processes have proper completions or ends, and one can
evaluate such processes in accordance with that standard. In the case Aristotle is here investigating, for example, nature would have 'fallen short' of
achieving its goal if immature animals that were beyond suckling, but not yet
adults, were not provided with teeth needed for chewing, since such teeth are
part of the proper development of such animals. Conversely, had such teeth
been provided when they were not needed, nature would have acted in vain.
The notion that nature doesn't fall short, then, may be viewed as a quick nod
in the direction of the principle that nature does what is best for the being of
each animal. This example also illustrates the fact that this notion of an end
doesn't simply refer to the 'mature' state of an organism of a certain kind. At
each stage of development, features appropriate to that stage should emerge
if the animal's formal nature is operating properly.

In the *IA* passage, Aristotle identifies **NP*** as a *starting point* (sometimes
rendered as 'first principle'), and in these passages **NP** and **NP*** are said to
be *posited* and *assumed* in natural inquiry. In the *GA* passage, he adds that
this positing rests on observation – that is, it is not an *a priori* postulate.[8] The
first passage is clearly a meta-theoretic claim about the nature of starting
points in natural science (indeed, two other such starting points are mentioned in this passage). But the second makes its philosophical points in the
midst of *using* **NP** in an inference, one establishing the necessity that young
animals have teeth even if, for other reasons, they must be replaced by
mature teeth later on. The role of this principle in such explanations will be
investigated by examining a representative sample in *PA* II–IV, in due
course. But first, I want to ask whether there are reasons to believe that the
claim that these principles are suppositional starting points is related to
Aristotle's views about the starting points of a science in the *Posterior
Analytics*. The repetitive reference to them as 'suppositions' and 'assumptions' is sufficient grounds to consider this a potentially fruitful endeavor.

III DEMONSTRATIVE STARTING POINTS IN THE *POSTERIOR ANALYTICS*

Famously, Aristotle opens his account of scientific knowledge by claiming that to know a fact scientifically, one must be able to demonstrate it, and that to demonstrate one must proceed from true, primary and immediate (or 'unmediated') premises which are better known than, prior to and causes of the things demonstrated (71b26–33). In explicating what it is to be primary, he distinguishes (at 72a15–25) a number of different sorts of starting points: (i) an axiom is a presupposition of any proof; (ii) a posit is also an immediate starting point of a proof, but may be restricted to proofs in one or a limited number of sciences (e.g., in geometric sciences, but not in arithmetic). Among posits, he further distinguishes a *supposition* from a *definition*. The former assumes that its referent 'is or is not'; the latter makes an assumption about what its referent is, but not about whether it is or is not.

One of the interpretive difficulties regarding these claims arises from apparent conflicts between this discussion of first principles and a later one in chapters 9 and 10. Since Aristotle's meta-theoretic claims about **NP*** suggest that it is a supposition, the crucial issue for us is whether *APo.* has a consistent story to tell about suppositions, and if so, what that story is.[9]

The claim that **NP** is a supposition of the sort being discussed in *APo.* I conflicts with a view of suppositions defended recently by Richard McKirihan Jr.[10] He argues that suppositions are *restricted* to positing the existence of the kind to be investigated by a science (e.g., figure or animal). Jonathan Barnes, on the other hand, defends a view of suppositions more congenial to viewing **NP** and **NP*** as such. According to Barnes, suppositions are claims about what is or is not the case; claims about the existence of the genera to be studied may be among them, but suppositions are not limited to such claims.[11]

In favor of Barnes' position is that Aristotle explicitly distinguishes hypotheses from both definitions and axioms in that the former are "among the premises" of demonstrations (*APo.* I.10 76b35–39, 77a10–12). This indicates they are to have propositional form and content.

An example from *Physics* VIII.3 further supports Barnes' reading. Aristotle begins his discussion of whether there needs to be an unmoved mover by noting that one of three things must be true — either all things are always at rest, or all things are always in motion; or some things are in motion and some at rest. He dismisses the first option on methodological grounds:

> Further, just as in arguments about mathematics objections that involve first
> principles do not affect the mathematician — and similarly with the other

sciences – so, too, objections involving the point just raised do not affect the natural scientist; for it is a supposition ⟨of his⟩ that nature is a source of motion. (253b2–6)

Since the claim 'nature is a source of motion' is said to be a supposition of natural inquiry, the notion that everything is always at rest cannot be taken up *within* natural inquiry (compare *Physics* I.2 185a12). To entertain such a notion, one would have to give up – at least temporarily – natural inquiry, which starts with the assumption that nature is a source of motion.

Now the clearly meta-level, methodological nature of this remark speaks strongly in favor of Aristotle having written it with the strictures of the *Posterior Analytics* in mind. And indeed, it is the position of the *Analytics* that the starting points of a science are incapable of demonstration within that science (cf. *APo.* I.9 76a17–25, 12, 77b3–7). But the claim that "nature is a source of motion" while quite clearly presupposing the subject matter of natural science (i.e. nature), also presupposes something to be the case about nature – it is not merely a positing of the 'kind'.

At this point one might respond that, since Aristotle *defines* nature as an inherent source of motion and rest, the above may be viewed as a *definitional* posit, rather than a supposition. Yet, though this is a reasonable suggestion, there are two points against it. First, the principle is explicitly referred to as a supposition, not a definition. Second, this chapter is considering which of three options – everything is necessarily always at rest, everything is necessarily always in motion, some things are in motion and some at rest – is the case. That the natural scientist, *qua* natural scientist, must presuppose that nature is a source of motion, is taken as a sufficient reason for him to reject the first option. This requires that he treat this posit as a basic truth, not merely a meaning postulate, of his science. And this is not the form in which definitions are presupposed.

In a completed science, of course, these two sorts of posit may not be all that different. Aristotle allows that investigation may begin with a postulated definition which only reflects awareness of the significance of the definiens; but final definitions are statements about the actual natures of things, and hence about what is or is not the case.

Nevertheless, the evidence favors the view that Aristotle's philosophy of science holds that among a science's starting points are propositions regarding what is or is not the case about the subject of that science, propositions required by many of the proofs of that science which are not themselves the subject of proof. These are the starting points referred to as the suppositions of that science.[12] The two passages we began with would seem to be claiming that **NP*** is such a supposition. If, in addition, it plays the role of a supposition of this sort in zoological practice, this strengthens this reading.

IV NP AS A PRESUPPOSITION OF ZOOLOGICAL SCIENCE

A question that has plagued discussions of scientific starting points or 'first principles' in the *Analytics* is whether they are to be viewed as actual premises in demonstrations, or whether they are more like 'inference tickets', rules of inference or rules for the proper formulation of premises. The answer to this question would seem to be obvious when it is axioms that are under consideration, since it would be odd to hold that every demonstration used the laws of non-contradiction and excluded middle as premises.[13] But it is, as we will see, a more interesting question when it is the *posits* that are under consideration, i.e. definitions and suppositions. Rather than explore this question in logical space, I shall do so in zoological space, looking at the actual use of **NP** and **NP*** in a sample of zoological contexts.

Aristotle's critical remarks about Empedocles and Democritus attempting to do biology without teleology both use their accounts of teeth to show the error of their ways.[14] I will thus begin by looking at the explanatory role of **NP** in a number of passages in which Aristotle is explaining something about teeth.

PA III.1 661b18–25

Of those animals that have teeth for both defense and attack, some, like boar, have tusks, while others have sharp and interlocking teeth, for which reason they are called 'sawtoothed'. For, since the strength of these animals is in their teeth, and this strength can be secured by sharpness, the teeth that are useful as weapons are arranged to fit side by side, so as not to be blunted by rubbing against each other. Now none of these animals has both sawteeth and tusks, because nature does nothing in vain or superfluous . . .

Among animals with teeth, Aristotle distinguishes a subset that uses them as tools of defense and attack. He sees tusks and serrated teeth as different tools for performing this function, and apparently thinks that having both would be redundant. The explanation seems to have the following form:

Being both tusked and saw-toothed is superfluous or 'in vain'.
Nothing superfluous or 'in vain' is produced by nature.

Nothing both tusked and saw-toothed is produced by nature.

Two features of this explanation are important for my claim that we should see **NP** as an *Analytics*-style suppositional starting point: **NP** is explicitly a premise in the explanation, indeed the only premise explicitly stated, and one introduced with an explanatory particle; and it is presupposed without argument. The first premise in my reconstruction is not actually stated in so many words by Aristotle. What *is* argued is that serrated teeth and tusks play the same functional role in different kinds of animals; and what Aristotle thinks he has inductively established is that no animal has both. The assumed premise that, with **NP**, *explains* this is one which claims that having both these, functionally equivalent, parts is superfluous. The following passage, from *On Respiration,* in which **NP** is used, makes this point explicitly. Aristotle is explaining the fact that no organism has been found with both gills and a lung. Having indicated that both are for cooling, he goes on:

> One organ is useful for one thing, and in every case one mode of cooling is sufficient. So, since we see nature doing nothing in vain, while were there two ⟨organs for cooling⟩ one would be in vain, for this reason some have gills, some have a lung, but none has both. (*Resp.* 10 476a11–15)

Though in our passage the second premise here is left implicit, the line of argument would seem to be identical.

A few lines later after this use of **NP**, Aristotle pauses to make the following remark:

> A general principle needs to be assumed, which will be useful both in these cases and in many others to be discussed later: Nature provides each of the organic parts for attack and defense – stingers, spurs, horns, tusks, and any other such part – either only to those able to use them or more to them, and provides them most to those able to use them most. (661b28–32)

He then immediately *uses* this principle to explain why males either alone have such 'weapons' or at least have them more than females. This is *not* the premise that having more than one such weapon is superfluous, but rather that an animal having one it couldn't use, or having one to a greater extent than could be used, is superfluous. But here again we find a teleological principle about the actions of nature which is (i) universal, (ii) assumed rather than demonstrated and (iii) a *specification* of **NP.** That is, Aristotle appears to be assuming that it would be *superfluous* for animals which are unable to use a part to have it, and superfluous to have a part out of proportion with the ability to use it. One reason to think this is what he has in mind is that he pauses to state this general assumption immediately after stating **NP** in the previous explanation. Nevertheless, while not proven here, this principle could be – simply by showing it to follow from **NP** in its unrestricted form. Thus, its place in the argument makes it clear why Aristotle would not identity it as a first principle.

Two passages late in *PA* IV are closely related to these two in both content and logic.

PA IV.11 691a28–b5

In humans and the fourfooted vivipara, the jaws move both up and down and to the side, while in the fish, birds and fourfooted ovipara they move up and down only. This is because the vertical motion is useful for biting and cutting, while the motion from side to side is useful for grinding. Therefore the motion from side to side is useful for those animals that have molars, but is of no use to those that lack them, for which reason they are absent in all such animals; for nature does nothing superfluous.

In this explanation, the explanandum is the difference in jaw movements between the vivipara and the ovipara, on which Aristotle, to a first approximation, is correct. The mammalian radiation is in part distinguished by newly evolved characteristics in jaw and dental anatomy. Aristotle doesn't discuss the anatomical differences in the jaw bones that makes these behavioral differences possible, but, again, he correctly sees differentiation of teeth into molars, incisors and cuspids as connected, biomechanically, to these differences in motion.

Explanations are provided for why animals with grinding teeth have horizontal jaw motions, and for why animals without horizontal jaw motions don't have teeth suitable for grinding food. The explanations sketched can be reasonably reduced to the following:

A. Horizontal jaw motion is useful for animals with grinding teeth.
What is useful for animals with grinding teeth is produced by nature.

_____

Horizontal jaw motion is present in animals with grinding teeth.

B. Horizontal jaw motion is useless for animals with no grinding teeth.
What is useless for animals with no grinding teeth is not produced by nature.

_____

Horizontal jaw motion for animals with no grinding teeth is not produced by nature.

As with every case I have examined, **NP** is used as a premise in an explanation for the *absence* of a feature which, if present, would be superfluous.[15] But what would make a researcher feel the need to proffer such an explanation? After all, there is a virtual infinity of properties that any organism does not have – trying to explain such things would appear to be the mission of a fool.

To deal with this question will provide an answer to the second question raised earlier, regarding how Aristotle conceives of the space of possibilities within which nature operates. Suppose that space to be limited by a *commonly possessed property,* in this case, teeth. The general problem may be thought of as explaining why the teeth of all animals with teeth are not the same. A differentiation of teeth into molars and incisors, present in one class of toothed animals, is absent in all the others. If this differentiation is explained by establishing that it serves a biomechanical specialization in the one group, its absence in the rest, given their background commonalities, may be explained by its absence in the others.

Thus what is possible within a kind is established inductively, through a study of the way, to use our example, teeth are arranged in the various kinds of toothed animals. With the possibilities thus delimited, there are then a limited set of questions having to do with what is absent in animals. Some animals have teeth and tusks (thus, it is possible); why don't those with serrated teeth have them? Some animals with two legs have wings (thus it is possible); why don't humans (cf. *IA* 11 711a1–6)?[16]

Even so delimited, however, such explanations are only reasonable if it is reasonable to think animals will, other things being equal, have parts that are useful or suitable to them, and lack parts that are useless or detrimental to them. Aristotle adopts as a starting point a slightly more robust premise – that nature never produces what is superfluous or 'in vain', but rather, given the possibilities, what is most suitable for the being of each kind of animal. *Physics* II.8 and *Parts of Animals* I.1 provide *philosophical* justification for this principle, but actual natural inquiry must adopt it as a given: it has no deeper, biological, explanation. If it appears to be violated, that is because, as *GA* IV.4 770b17–18 nicely puts it, "the formal nature doesn't master the material nature." Such productions of monstrosities are rare and, by the same token, unnatural. If they happened regularly, as he says in criticism of Democritus' account of the production of teeth, " . . . nature would be failing to do one of the things possible for it, and nature's production would develop contrary to nature" (*GA* V.8 788b26–28). But, as he says in explaining why humans don't have wings, "nature makes nothing contrary to nature" (*IA* 11 711a5–6). This is not an *argument* against Democritus' account – it is merely a reminder that nature does not fail in doing the things possible to it.[17]

It will be noticed that the two arms of this explanation are distinguished by the fact that the second, upon which we've been focused, uses only **NP**, while, as I read the 'because' in the first, its force derives from the additional, positive idea that nature does what is best among the possibilities for the animal in question. Thus, if it is better (more suitable) that the animal with grinding teeth as well as incisors have a jaw that moves horizontally, so it will be, if nothing prevents it. We will look more closely at the use of this supposition in the next section. In general, while **NP** is used to explain the absence of a possible feature, the positive content of **NP*** is used to explain a trait's presence. Again, observation establishes a range of possible ways in which a generic feature can be present; **NP*** underwrites an explanation of the presence of the actual feature on grounds that it is the best among that range.

The importance of **NP*** to an etiological concept of teleology like Aristotle's is hard to overstate.[18] One can perhaps demonstrate that a feature of some kind of animal or other is either of no use, or that, given others of its features, it is redundant. Furthermore, questions about correlations between parts, such as whether good fliers lack spurs, or whether talons and spurs are ever found together, or whether animals incapable of using defensive parts have them anyway, are established inductively. But if one wants to *explain* such correlations by appeal to the animal's nature, a premise must be in place to the effect that parts which are useless or superfluous are not produced by that nature.

Aristotle explicitly identifies **NP*** as a suppositional first principle of the study of nature. According to the *Posterior Analytics,* this ought to be a universal statement about what is or is not the case that cannot be derived deductively from any more basic or certain truth within that science, and the truth of which must be presupposed in many of the explanations in that science. A *philosopher* can argue with those who deny it (see *Physics* II.8, which argues for this one), but as a *zoologist* you must presuppose it as a starting point, because there is no zoological explanation for it – it is basic. **NP** and **NP*** satisfy these criteria. To this point, we've seen **NP** in action. Doing so has provided evidence that it is a demonstrative supposition, and that as such it plays the role of an explicit premise in the demonstrations of natural science. But what of **NP*?**

V NATURE DOES BEST

We have so far principally looked at examples of **NP,** the weakest, negative version of the principle in question. But the principle laid down in *IA.*2 and *GA* V.8 (**NP***) also has a *positive* component, asserting that *nature also does what, among the possibilities, is best for the being of each kind of animal.* If it can be established that, for some animal A, having *p* is possible, and that among the alternatives *p* is best for A, then it follows that A must have *p.* Note again that it is the need for explanation that makes this supposition necessary. If A in fact has *p,* which can be established by observation, nothing more is needed to establish that having *p* is possible. Furthermore, it might be argued that having *p* is better than various other possibilities simply on engineering grounds. Nevertheless, *to use* the claim that having *p* is better than the other possibilities to explain why A has *p,* one will need an additional principle like **NP*.** Put simply, one needs such a principle to go from the conjunction,

'A has p' & 'p is the best possible thing for A'

to the explanation

A has p because p is the best possible for the being of A.

One needs a background principle to the effect that in nature animals will be provided with features that are the best ones *for them,* given the possibilities. The background principle that plays the analogous role in modern biology is the principle of natural selection – among the available genotypes (i.e. among the possibilities), the one producing the best adapted phenotype for a common environment has a better chance of surviving and of increasing its representation in future generations. In Aristotle, it is a principle about the agency of formal natures. The processes of development and self-maintenance, upon which an animal's viability and flourishing depend, are directed by the formal nature of that animal. Explanatory bedrock, for Aristotle, is that such formal natures produce nothing in vain, but the best, given the possibilities for *beings of their kind.* **NP*** is simply a universal statement of this idea, one which generalizes over all organic natures.

In the previous section, we dealt with the issue of establishing the range of possibilities. Here we need to come to grips with the second issue raised earlier, how **NP*** interacts in explanation with knowledge of the substantial being of the kind in question. To deal with this issue, I want to close by looking at a passage, again from *PA* IV, in which **NP*** is used in a manner that ties it tightly to claims about the being of the kind of animal in question.

216

PA IV.13 695b17-27

Fish do not have distinct limbs, due to the fact that the nature of fish, according to the account of their substantial being, is to be able to swim, and since nature makes nothing either superfluous or pointless. And since they are blooded in virtue of their substantial being, it is on account of being swimmers that they have fins, and on account of not being land-dwellers that they do not have feet; for the addition of feet is useful in relation to movement on land. And they are not able to have four fins and at the same time feet or any other such limb; for they are blooded. The water newts, however, though they have gills, have feet, for they do not have fins, but a flaccid, flattened tail.

Fish are *essentially* blooded and *essentially* swimming animals. As blooded animals, they are unable to have more than four limbs,[19] while as swimmers it is valuable that their limbs be suitable for swimming. At this point Aristotle seems to have all the explanation he needs – if blooded animals are limited to four appendages, and it is best that their appendages be fins, then it would seem to be ruled out that they have feet. Yet Aristotle goes on to argue that it would be *superfluous* for fish to have walking limbs. If the possibility has been excluded, why go on to explain that they lack feet because having them would be superfluous?

One possible answer lies in the fact that, though the account of their substantial being establishes that they be limited to four limbs, and that they need fins, that account nevertheless isn't sufficient to exclude their having feet, as is established by the existence of birds – the 'four appendage' requirement is met in birds by two limbs they need because they are essentially flyers and two legs which allow them to walk.[20] By arguing that legs are useless to fish, he explains why the combination of their blooded nature and their swimming nature implies four *fins*. This is the point of his comment that fish don't walk, and feet are for walking.

In this explanation, **NP*** is combined with propositions identifying features which Aristotle claims are in the account of the substantial being of fish. The 'nature' that does nothing superfluous, but what is best, is the *formal* nature of this or that kind of animal. It cannot be established that fish must have four fins simply by an appeal to their substantial being, that they are blooded creatures that live their lives in water. Their generic being permits various possibilities, and among those possibilities, their formal being realizes what is best. Given that fish are blooded, it is not possible that they have more than four 'limbs'; but differentiation within the blooded animals shows that it *is* possible to have those limbs designed in ways appropriate to different environments, e.g., to have them organized as fins rather than wings or legs, or to have a combination of fins and legs (or who-knows-what). Yet it is apparent that the nature of each kind of blooded animal produces limbs in the best way possible, within the limits set by being blooded.

If, in this passage, we need to read between a few lines for this interpretation, this is not so in *IA* 8, where Aristotle explains why snakes, though land animals, have no feet.

IA.8 708a9–20

The cause of footlessness in the snakes is *both* that nature does nothing in vain, but in every case attends to what is best for each thing among the possibilities, preserving the proper substantial being of each, and its essence; *and further,* and as we've stated previously, none of the blooded animals can move by means of more than four points. For from these ⟨two premises⟩ it is apparent that none of the blooded animals that are disproportionately long relative to the rest of their bodily nature, as are the snakes, can be footed. For on the one hand they cannot have more than four feet (since in that case they would be bloodless), and on the other hand having two or four feet they would be pretty much completely immobile – so equipped, their movement would necessarily be slow and useless.

NP* is here stated in its full form, and is cited as a part of a complex cause of the lack of limbs in snakes, the other part being the principle that restricts the blooded animals to four limbs. The second premise restricts the possible limb number to two or four (given that the number of feet must be even, which Aristotle has argued for in 708a21–b21). The first is used, in combination with a premise that identifies the crucial aspect of a snake's nature, its unusual length relative to its overall size, to argue that, since two or four limbs would be useless to snakes, they must have no limbs. Once again, a premise to the effect that a snake's nature does not provide it with useless appendages (within the range of possibilities for blooded animals) is needed to establish this conclusion.

VI CONCLUSION

When William Harvey, in 1628, lists the considerations that led him "privately to consider if [the blood] had a movement, as it were, in a circle," the second item he mentioned was the one with which we began:

I . . . considered the symmetry and size of the ventricles of the heart and of the vessels which enter and leave them (since Nature, who does nothing in vain, would not purposelessly have given these vessels such relatively large size) . . . (Harvey [1628] 1963, 57)

As in Aristotle, the principle **NP** is a premise in an argument, and one that is required by Harvey's teleological explanations. The vena cava, pulmonary veins and arteries, and aorta, are indeed large and symmetrical. But what leads Harvey to use this as evidence for a circulatory hypothesis about the motion of the blood is the supposition that there must be a purpose for the size and symmetry of these structures. This is another example of the form of Harvey's teleological reasoning that so impressed Robert Boyle, as reported in his *Disquisition on the Final Causes of Natural Things:*

> I asked our famous *Harvey,* in the only Discourse I had with him (which was but a while before he dyed): What were the things that had induc'd him to think of a *Circulation of the Blood?* He answer'd me, that when he took notice that the Valves in the Veins of so many several parts of the Body, were so plac'd that they gave free passage to the Blood Towards the Heart, but oppos'd the passage of the Venal Blood the Contrary way: He was invited to imagine that so Provident a Cause as Nature had not so Plac'd so many Valves without Design: and no Design seem'd more probable, than That, since the Blood could not well, because of the interposing Valves, be sent by the Veins to the Limbs; it should be Sent through the Arteries, and Return through the Veins, whose Valves did not oppose it course that way. (Boyle 1688, 157)

However, while the premise used in these cases is **NP,** and it is indeed a first principle of Harvey's reasoning, the reasoning is *inductive.* Harvey is using **NP** to lead him from observations of the biological systems he is studying to a specific teleological hypothesis which would explain those observations. As with Aristotle, the explanation is teleological in form: The circulatory hypothesis would explain the observations as consequences of appropriate design for such a cardiovascular system, which, guided by **NP,** Harvey supposes them to be. In short, **NP** is a first principle in Harvey's biological studies, as it is in Aristotle's; and it often plays the role of an explicit premise in Harvey's reasoning. But, at least on certain occasions, Harvey uses this principle in a way not in evidence in Aristotle's biological treatises, as a *constraint on hypothesis formation.*

This doesn't mean that Aristotle didn't also use this principle to guide his investigations. Indeed, in at least one passage he seems to say that he does.

> The principal reason for investigators not speaking well about these matters is that they lack experience of the internal parts, and fail to assume that nature in every case acts for the sake of something; had they inquired for the sake of

what respiration belongs to animals, and had they investigated this question in the presence of the parts involved, i.e. the gills and lung, they would quickly have discovered the cause. (*On Respiration* 3 471b24–29)

Aristotle is typically chary when it comes to telling us how he arrived at his explanations. Harvey, by contrast, like many of his contemporaries, is self-conscious about the methodology of 'the new science'. Thus the study of his use of **NP** adds a piece to the puzzle of the role of Renaissance Paduan Aristotelianism in the development of an experimental and hypothetical approach to nature in the 16th and 17th centuries. Harvey accepted the principle that Nature does nothing in vain, or without a purpose, and explicitly used it as a regulative constraint on the search for acceptable explanations. He did not, it seems, do so in vain, but rather because it served his purposes well, given the possibilities.

NOTES

1. See Cunningham 1985, 195–222; Bylebyl 1979, 334–370. It is noteworthy that the title of this work, and of his later masterpiece on generation, both stress that they are universal studies of the animals, not studies restricted, as medical works would be, to humans.

2. Gotthelf 1987a, 187 and note 50 briefly discusses this principle as one of two 'posits' regarding formal natures that underwrite the teleological explanations in *PA* II–IV, and suggests provocatively how these may play a role in understanding the concept of 'hypothetical necessity' in Aristotle's biology. As a broad discussion of the 'axiomatic' structure of the *PA*, Gotthelf's discussion can be seen as laying the groundwork for the present discussion. There are some further suggestive comments about the philosophical role of this principle in Gotthelf 1997.

3. Cf. chapter 8. Cataloguing these passages is a challenge. I count 79 passages in *PA* alone which use nature as a subject of a verb of agency, in an explanation for why an animal of a certain kind does or does not have some feature. Of these, there are seven that use the precise formula that nature does nothing in vain, or superfluous, which I will label **NP**. But there are a number of more specific principles which play similar explanatory roles, about nature when possible using a distinct organ for distinct functions, but when not possible doubling up functions (683a20–26), assigning parts to animals that are alone, or best, able to use them (684a28), making the part for the function, not the function for the part (694b14–15), putting the more honorable part in the more honorable place (665b20), or using a part for more than one function (688a24).

4. In the biological explanations we will be studying, the principle that nature does nothing in vain or superfluous is often used without the added positive assertion that it does what is best given the possibilities. I will use **NP** to designate the negative assertion, and **NP*** to designate the combined statement. It is noteworthy that **NP** is used on its own to explain the *absence* of features for one reason or another thought to need explanation; while, as the last sentence in the passage we are about to

discuss suggests, **NP*** is typically, though not invariably, used when a part is shown to be present because it is better that the animal have it. A suggestion will be made below about this difference.

5. Cf. *APr.* I.3 25b14–18; *Ph.* II.5 196b11–32, 7 198b5–9.
6. I don't here have space to deal with a further complication with this notion of 'best among the possibilities', a complication raised in Gotthelf 1997, 7 and note 15. This complication has to do with whether the features explained by explanations positing **NP*** are thereby necessitated, in the ways one expects in canonical Aristotelian explanations. The problem arises because Aristotle sometimes contrasts explanations by reference to what is better with explanations by reference to what is necessary (e.g. *PA* I.1 640a35–6; *GA* I.4 717a12–22).
7. I discuss this issue in detail in chapter 8 above.
8. Cf. *Resp.* 10 476a13, where he says simply that we see that nature does nothing in vain.
9. Here a number of caveates should be entered. First, while there is rather pointed use of various forms of the verb *hypotithēmi* in our two passages, the noun based on it, *hypothesis,* is not. Second, in a perfectly straightforward sense, axioms are also 'supposed' in demonstrations, and so the use of this language does not rule out **NP*** being seen as an axiom. I give reasons in favor of its not being so viewed, and it is worth noting two counters to those reasons suggested to me by Malcolm Wilson. Against the argument that axioms are completely general, while **NP*** is restricted to natural science, he notes that the Equals Principle is referred to by Aristotle as a 'common principle' (cf. 76a41, 77a27–32), and **NP*** would seem to be as unrestricted as the Equals Principle. Though never termed an axiom, the Equals Principle is mentioned in conjunction with the laws of non-contradiction and excluded middle in these passages.

   Wilson also counters my argument that since **NP*** is actually a premise in arguments in *PA,* this counts against it being viewed as an axiom. Again, he notes that the Equals Principle is often a premise in Euclid, and if it can be an axiom, so can **NP***. Note that the form of these two objections is identical – the apparent conflict between what is said generally about axioms and the laws of logic, on the one hand, and the claim that the Equals Principle is a 'common principle', on the other, is used to weaken the conditions on being an axiom to the point where **NP*** could be an axiom. Supposing, however, either (a) that Aristotle is confused about the Equals Principle, or (b) that he has a wider concept of 'common principle' than of 'axiom', such that there are certain things that can be true of both, or (c) that there are crucial differences between the status of the Equals Principle in mathematics and the status of **NP*** in natural science – supposing any one of those possibilities, and it remains reasonable to view **NP*** as a supposition, rather than an axiom.
10. McKirahan, Jr. 1992, chapter 3, 36–49.
11. Cf. Jonathan Barnes 1975a, 103–104 (1993, 99–101). In *APo.* I.9 76a17–19, and I.10 76a32–33 Aristotle also says that it is not possible to demonstrate the proper principles of each genus. In chapter 10 he again distinguishes common from proper principles. The examples of proper principles are "Line is such and such," "The straight is such and such," and they are said to refer to things that "are assumed to exist, concerning whose proper attributes the science studies" (76b3–5). Now this last certainly *includes* the subject genus of the science, mentioned a few lines later

as one of three things every demonstrative science is concerned with (76b12–16); but there is no reason to think it is restricted to this. Finally, about suppositions, Aristotle continues to distinguish them from definitions on grounds that they assert that something is or is not the case (76b35–39); but he adds that they are also distinguished from definitions in that they are either asserted as universal or as particular (77a1–4). This latter claim gives the distinct impression that they, unlike definitions, are capable of being premises in demonstrations.

12. One further wrinkle in Aristotle's account of suppositions should be noted. In *APo.* I.10, Aristotle, in distinguishing dialectical from demonstrative contexts, notes that the term 'supposition' can be stretched to include suppositions that are 'proveable', but since the student accepts them as obvious, need not be proven. But I take it this is a footnote in the account, not an essential feature of it, a footnote intended to explain why in practice many provable theorems are used in the sciences without proof.

13. And once Aristotle seems to say as much explicitly: "No demonstration takes ⟨as a premise⟩ that it is not possible to simultaneously assert and deny, unless it is required to prove the conclusion in this manner." (77a10–12)

14. Cf. *Ph.* II.8 198b24–33, *GA* V.8 788b10–29.

15. So too outside of *PA: IA* 11 711a1–6 (why bipeds other than birds, man in particular, have no wings), *GA* II.4 739b16–20 (why females don't emit seed), *GA* II.5 741b2–5 (why females don't generate on their own), *GA* V.2 781b22–28 (why seals lack external ear shells), *Resp.* 10 476a10–16 (why animals with gills don't have lungs and vice versa).

16. Or, since seals are viviparous quadrupeds (well, in a manner of speaking – see *PA* IV.13 697b1–8), and most of these have external ear shells, why don't they (cf. *GA* V.2 781b22–28)?

17. The argument here assumes that neither *Ph.* II nor *PA* I are actually providing scientific explanations starting from first principles, but rather are providing philosophical justification for those principles. Thus I am accepting Aristotle's actual views about the special principles of specific sciences, that one does not provide scientific demonstrations for them, but can justify them on philosophical grounds (cf. *APo.* I.9 76a17–25, 12 77b4–8; *Ph.* I.2 184b26–185a20). An extremely difficult question is what science it is that the principles I am discussing are the principles *of.* Aristotle typically refers to the science of nature, and doesn't seem to mark off the science of animals as a distinct science. For example, in the preamble to *Meteorology* I (338a20–339a9) while investigations of animals and plants are mentioned, they seem to be treated as parts of an investigation of nature; and *Meteorology* IV.12 closes its discussion of the material construction of uniform parts by looking forward to discussing the specifically biological ones, then the organs composed of them, and then the entire organisms (390b20–22), that is, by looking forward to what we refer to as 'the biological works'. Even *PA* I, which does seem to have animals primarily in focus, describes itself as establishing the principles of natural inquiry (639a12–16), an inquiry focused on the things that come to be and pass away by nature (644b22). I here hazard the speculation that in the hierarchy of the 'natural sciences', Aristotle, unlike us, would see the study of animals and plants as most fundamental, precisely because the formal and final cause operate there, and in such a way as to direct the material and efficient causes

toward goals. Thus, while when we think of identifying a particular science with natural science, it is physics, when Aristotle does so, he thinks of the study of animals and plants.

18. That it is an etiological account, i.e. one in which goals and functions of structures are ineliminable features in the causal explanations of the coming to be and presence of those structures, I cannot here argue. For the case, see Allan Gotthelf 1987b, 241. To make such a claim, of course, one must define etiological accounts in the general way just stated, rather than as coupled to a specific scientific account of the way in which goals and functions are causes.

19. In *IA* 7 there is a puzzling attempt to explain the connection between these two differentiae, a connection that elsewhere is explanatory bedrock. See Gotthelf 1985b, 27–54, esp. 43–50. This much is clear, at least – being blooded limits to four the possible number of 'limbs' (or motion points, the term Aristotle uses so that even the fins of fish and touch points of snakes (cf. *IA* 7 707b6–9) can be included).

20. Cf. *PA* IV.10 686a25–31 (which parallels our passage in a number of respects), and Gotthelf 1985b, 43–44.

III

Teleological Explanation

> The major reason for the failure to speak correctly about respiration is a lack of experience with the internal parts *and a failure to grasp that nature in every case acts for the sake of something.* Had my predecessors been seeking *what respiration was present in animals for* and been investigating this among parts such as gills and lung, they might have discovered the cause with ease.
>
> Aristotle, *On Respiration* 3 471b 24–29 (emphasis supplied)

Research done during the last twenty-five years of the twentieth century has provided a new level of understanding of Aristotle's use and philosophical defense of teleological explanation. (Allan Gotthelf has provided two valuable guides to this research, on page 172 of the "Report on Recent Work" appended to Balme 1992; and in his contribution to Hassing 1997, "Understanding Aristotle's Teleology.") My own work on this subject has focused, not directly on Aristotle's philosophical defense of teleology, but on his understanding of the relationship between teleology and chance, on the contrast between Platonic and Aristotelian teleology, and on the reaction of Aristotle's friend and collaborator Theophrastus to Aristotle's teleology.

The first chapter in this section, "Teleology, Chance, and Aristotle's Theory of Spontaneous Generation," tackles a rather deep puzzle in Aristotle's biology – how could Aristotle consistently believe that a diverse collection of different kinds of organisms was spontaneously generated? That he did believe this is beyond doubt – the *Generation of Animals* even provides a general model for their production. And yet on the usual understanding of his concepts of chance and spontaneity, as presented and defended in *Physics* II.4–6, nothing that occurs regularly in the course of nature can be spontaneous.

This problem had been written about twice in the 1960s, once by David

Hull and once by David Balme. Both had pointed out the apparent inconsistency, and both had failed to resolve it. My paper is an advance on those, but I now think it an instructive failure. The essay that convinced me of this was Allan Gotthelf's "Teleology and Spontaneous Generation: A Discussion" (Gotthelf 1989, in Kraut and Penner 1989).

In doing the research for that paper I uncovered fascinating treatments of Aristotle's discussion of chance and teleology in the ancient Greek commentators. "Aristotle on Chance" begins with the core puzzle that worried them – Aristotle seems both to *contrast* teleology and chance *and* to say that chance events are in the class of teleological events. One modern textual critic wished his philosophical solution to this puzzle into existence by changing one letter in the Greek text, so that Aristotle said that chance events were among those that *might* have been for the sake of something. The solution I offer has received interesting development and critical commentary in Lindsay Judson's "Chance and Always or for the Most Part in Aristotle" (in Judson 1991, 73–100). Some of its themes are also developed in Susan Meyer, "Aristotle, Teleology and Reduction" (Meyer 1992). The key to a solution lies in recognizing that for Aristotle what happens 'by chance' or spontaneously is the *coincidental* result of two or more natural processes directed at some *other* result. Thus the result is not the goal for the sake of which the component processes occurred. This is compatible with the outcome being necessary given the intersection of the component processes.

A thorough review of Aristotle's biology makes it clear that, though he grants teleology priority when it is present, he is no Dr. Pangloss, seeing every natural process as for the best. Nevertheless, he does occasionally stretch for a teleological explanation when one would wish he had not. By examining carefully Theophrastus' cautions about teleology in his *Metaphysics,* I was able to conclude that he had a good sense of where Aristotle had overstepped his own limits on the use of teleological explanation.

This paper has had a significant and valuable impact on the current revival of interest in Theophrastus (cf. van Raalte 1993; Gottschalk 1998, 285–288; Sharples 1998, 272–274, in van Ophuijsen and van Raalte 1998; and Hughes 1988, 68, note 12, in Fortenbaugh and Sharples 1988).

In two cases, however, the argument has been badly misunderstood (cf. Laks, Most, and Rudolph, in Fortenbaugh and Sharples 1988, esp. 224–233; and Repici 1990). Despite my citing the texts in which Aristotle himself indicates that not all biological phenomena should be accounted for teleologically (see p. 261, and notes 12–14), and despite my stress on the absence of such explanations in many of Aristotle's biological texts, I have been accused of seeing Aristotle as a panglossian teleologist. Because these authors assume this to be my view, they further assume that I take Theophrastus to be criticizing Aristotle for being a 'universal' or 'panglossian' teleologist.

I am thus delighted to be able to reprint this essay, so that what I am *actually* saying will be readily available. My view was, and *is,* that Aristotle is not as clear and explicit as he could be on how to determine when teleological explanation is inappropriate, and that Theophrastus is drawing our attention to the problem by pointing to some problematic examples.

These authors also attempt to argue, against me, that none of Theophrastus' examples echo Aristotelian examples. Dr. Most is forced to argue this, since any other claim would conflict with his view that Theophrastus' *Metaphysics* was written before Aristotle's *De Partibus Animalium.* This relative dating strikes me, and a number of other scholars, as extremely unlikely. As for the strictly philosophical question of the closeness of fit between Theophrastus' problematic cases and Aristotle's questionable uses of teleological explanation, I leave that for the reader to decide.

There is one further misunderstanding, in Repici 1990, that may be cleared up. She argues that it is strictly impossible for me to render Theophrastus' argument as one regarding the limits of teleology, on grounds that "whatever the meaning of *horos* may be, it apparently never stands in Aristotle's usage for a *subjective limit* to be *set to* things or *to* an explanation. Instead, it always connotes the *objective limit* which is *in* things (must be found *in* them) and then *assumed*" (1990, 188). I, of course, do not mention 'subjective' limits in my paper. I agree completely with Professor Repici that for Aristotle such 'boundaries' or 'limits' are objective. As I say myself (p. 261), they are "conditions under which teleological explanations are and are not appropriate."

As for Aristotle's usage of the term *horos,* the following passage from *De Partibus Animalium* I.1 is definitive:

> So it is clear that for the investigation of nature, too, there should be certain standards *(horoi),* such that by referring to them one can appraise the manner of its explanations, apart from the question of what the truth is, whether thus or otherwise. (639a12–15)

This passage introduces a chapter devoted to developing such standards, essentially a set of rules for judging natural investigations. When Theophrastus talks of *horoi* for the use of teleological explanation, it is surely a use of the term analogous to the one above: principles *we apply,* but principles *objectively based in the way the world is.*

Aristotle's teleology, like so much of his natural philosophy, is a *via media* between a reductive materialism defended by a variety of Ionian natural philosophers and the natural theology of Plato's *Timaeus*. To understand its distinctive character it is important to see the ways in which it differs from Plato's teleology by divine design. That is the rationale for including the last chapter in this section, "Plato's Unnatural Teleology" – unnatural because Plato argues that the 'natural' world is an artifact, the product of a divine *craftsman* realizing his purposes in its order and harmony.

The contrast is instructive for those of us who live in a post-Darwinian world. The natural theology to which Darwin responded can be traced directly back to a Platonic ancestry. It has often been assumed that, since Darwin's theory undermines natural theology, he must have been an opponent of teleology – contrary to everything he ever said on the subject (see Lennox 1992, 1993). The contrast between Plato and Aristotle demonstrates clearly that a student of the living world can remove the teleological baby before throwing out the theological bathwater.

10

Teleology, Chance, and Aristotle's Theory of Spontaneous Generation

I

Aristotle's theory of sexual generation is recognized as an exemplary application of his theory of causality to a specific natural occurrence (or alternatively, as the source from which the theory was generated).[1] Yet Aristotle readily admitted that many living things were *not* sexually generated, but arose 'spontaneously';[2] including all testacea, many insects, certain fish, and (analogously) certain plants. It has been claimed that, unlike his theory of sexual generation, Aristotle's account of spontaneous generation is inconsistent with his metaphysical doctrines of causation and chance.[3]

Briefly, the alleged inconsistency is this. *Physics* II.4–6 presents a theory of things generated by chance and spontaneity. Such productions are argued to be nonteleological, contrary to what is both natural and comprehensible to man, unusual, and due to incidental causes. Yet in the *Generation* and *History of Animals,* Aristotle argues that entire genera are generated spontaneously. In *GA* III.11, in fact, we are treated to a theoretical, that is, causal, account of such productions. And in *Parts of Animals,* teleological explanations of the parts of such creatures – the testacea and insects, for example – abound.

The question to be faced is this: Why does Aristotle call the working-up of creatures out of inanimate materials 'spontaneous'? Assuming that the theoretical account of the *Physics* predates the concept's use in the biological works,[4] it is reasonable to expect its biological use to reflect that theoretical account.

I shall argue that for the most part it does. A careful understanding of the interaction among Aristotle's concepts of teleology, chance, and spontaneity is, however, necessary before a general harmony between theory and practice can be established. I will argue that an observable causal pattern is at the

229

core of Aristotle's doctrine of teleology and that certain kinds of events – even the nonbiological production of organisms – might occur quite regularly and yet fail to exemplify that pattern. This chapter argues that it is the absence of teleological causation, in things that might have been so caused, that is the interesting core of Aristotle's theory of chance and spontaneity. It further concludes that the spontaneous production of organisms is a case in point.

II

Before we begin to pursue the philosophical issues involved in resolving the difficulties raised by Aristotle's account of spontaneous biogenesis, it will be helpful briefly to compare Aristotle's accounts of sexually and spontaneously originated generation.

In sexual generation, the female contributes *all* the bodily material (a constituted segment of the menstrual discharge) of the organism to be (*GA* 727b31–33, 729a11, a32, 730a16, a20, a27, 740b24). During copulation the male emits semen that possesses an inherent power (*GA* 727b16, 729b5, b27, 730a14, 740b35) or source of change (*GA* 724a11, b13–14, 730a27, 740b25) for the production of form, that is, for the organization of the female material into an organism of a specific kind.[5] Semen operates not by adding material for growth (736a27) but as the agent that transmits the inherent warmth of the semen's *pneuma* to the material within the female. This warmth is analogous to the heat of the sun, which ripens[6] and fosters growth (*GA* 729b27, 734b30, 736b33–737a18, 739a11, b23, b28). This warmth of the *pneuma* is the immediate source of the changes that eventually result in a new organism. In a manner unclear from the texts we have,[7] the natural, or life-promoting, heat operates according to an order or pattern (*logos* [734b33]) that reflects the organization of the parent organism. This power of orderly heating is able to initiate an organized set of changes in the female's material, changes eventuating in and directed toward an offspring formally identical to the parents. Aristotle insists that sexually initiated biogenesis is *for the sake of* its outcome, a fully functional organism of a specific kind. Furthermore, he insists that what such processes are *for* is is responsible for them (see esp. *PA* I.1 639b12–642b4).

What does Aristotle mean by saying that sexually originated generation is *for* its outcome, or that that outcome, a functioning organism of some kind, is responsible for the generative process? I propose that to grasp the meaning of the doctrine we must take seriously Aristotle's insistence on a close analogy between processes originated by a craftsman and processes of a biological kind originated by the male parent (*GA* I.22 730b8–32, II.1 734b20–735a29, II.4 740b25–741a4).[8]

In the crafts, the artist works on his material by means of his manipulation of the tools of his trade. But the movements of those tools are not random: they are guided by the principles of the craft Aristotle identifies as the form of that craft's product.[9] The arts of medicine or architecture exist, as knowledge of what it is to be healthy or a good dwelling, in the craftsman's soul (cf. *Metaph.* Z.7 1032b1ff.; *GA* I.22 730b16; *PA* I.1 639b16–21). To state the doctrine generally: the artist, by means of his knowledge of the form of the product of his craft, governs the activity of his tools so that his materials come to exemplify that form. In this way, that form is responsible for the process that eventuates in its exemplification in certain materials.[10]

Aristotle insists that natural biogenesis exemplifies the same pattern of causality. The semen operates like a tool of the parent. Its heating and cooling activity possesses a pattern, and that pattern is determined by the male parent's form. In parallel with craft production, the male parent, by means of having the form he does, determines the pattern of the heating contributed by the semen. This heat acts in its orderly way, and on the appropriate species-specific material, so that it comes to exemplify the appropriate form. This form is thus ultimately responsible for the process that results in its being the organization of a specific organic body.

This pattern, which is exemplified in both craft and biological production, I shall refer to as *formal replication*. It generates *teleological* explanations, explanations of processes by means of their outcomes, because the form at the end of such a process *is* the form responsible for the process's having just the structure and direction necessary to proceed to it.

Aristotle claims that in spite of the lack of intentionality in biological generation, "the relation between the latter and the earlier stages is the same in nature and in the arts" (*Ph.* II.8 199a18–20). The pattern for a specific result has been built into the semen's heat, unifying the various heatings and coolings by a *logos*.[11]

However, there is one critical disanalogy between artistic and natural generation: "For the art is source and form of what comes to be, but in another; whereas the movement of nature is in what is coming to be, issuing from another nature which has the form in actuality" (*GA* II.1 735a2–4). The immediate efficient cause (i.e. the pattern of heat) of natural changes is inherent in the material that changes, not constantly imposed from without as in artistic production.

The teleological nature of the process does *not* prevent Aristotle from describing the process itself as analogous to the movements of "spontaneous marvels" (*GA* 734b9–17, 741b9–15):[12] "The parts of the embryo already exist in the matter potentially, so that when a source of change comes about, the rest follows as a connected series, just as in the case of the spontaneous marvels" (*GA* 741b7–9). The points this analogy stresses are:

a. What goes on in the embryo goes on in the absence of an external agent (unlike his other favored analogy for embryogenesis, the craftsman creating his product) (see *GA* 734b9–17).
b. An initially tiny source of change is perfectly capable of generating a large complex of changes (see *MA* 701b1–9).
c. The embryonic material is not totally amorphous, but like the wheels and pulleys of a mechanical toy, its 'parts' have a potential for certain changes, awaiting that first initial infusion of warmth to initiate differentiation (see *GA* 734b11–12).
d. Embryogenesis is a continuous series of changes, not a collection of discrete alterations. In fact, Aristotle wants to claim that each biological development is *one organized* production, like writing a novel or building a house. This does not exclude the possibility of describing subprocesses of this development independently of their place in it, although such descriptions are not ultimate explanations (see *GA* 789b3–8).

That sexual biogenesis, looked at in isolation from its causation, can be characterized as 'spontaneous' is revealing. It suggests that, if we are to locate the difference between spontaneous and sexual generations, we must look beyond descriptive features of the processes themselves.

III

GA III.11 constitutes Aristotle's only theoretical account of spontaneous generation and comes during his discussion of the generation of testacea. According to this chapter, some testacea give off a substance from which new creatures like them are produced (mussels and whelks, for example), but this is akin to a botanical 'budding' process and is *not* a case of sexual generation. "The generation of all the other testacea is spontaneous" (*GA* 762a9). The following passage gives us the general pattern for such cases.

> Animals and plants come into being in earth and in liquid because there is water in earth, and air in water, and in all air there is vital heat; so that in a sense

all things are full of soul.[13] Therefore living things form quickly whenever this air and vital heat are enclosed in anything. When they are so enclosed, the corporeal liquid's being heated, there arises as it were a frothy bubble. The differentiae which determine whether the kind is more or less honorable are determined by the organization of the vital principle in the enclosure. And both the places and the enclosed material are causes of this organization. (*GA* 762a18–27)[14]

The 'foaminess' of a liquid indicates the presence of *pneuma* in it, and semen is characterized by the possession of imperceptible foamy bubbles, according to *GA* II.2. The creatures spontaneously generated are ranked in terms of 'nobility', just as are those biologically generated (*GA* II.1 732a24–733b23). In the biological cases, this ranking depends on the internal vital heat characteristic of the species, and this is likely the force of the passage just quoted as well. But the causes of the sort of vital heat that gets enclosed and produces the species characteristics are said to be two: the location of enclosure and the material that becomes enclosed.

Presumably the exact amount and strength of the *pneuma,* and consequently the vital heat, that an enclosure would 'capture' would be a function of the locale where it took shape. This suggestion is born out by the various accounts of spontaneous generations in the *Historia Animalium* (e.g., *HA* V. 539a18–26, 547b18–22, 548a7–19, 551a1–552b25, 557b1–13; VI. 569a24–b22). The pattern of development is a function of the amount of *pneuma* relative to earthen and liquid material in the enclosed mix. That mix, which can change depending on climatic and geographic variables, will determine whether the enclosure comes to be a sea urchin or an oyster. The vital heat that is productive of a species member is thus *not* derived from anything one in form with that species member. To put it another way, there is no description of the course of such a process that identifies it as *for* the production of a specific kind of organism.

IV

The various descriptive differences between the two sorts of generation give rise to crucial differences in the types of explanations appropriate for these two biological phenomena. Recall first that before formal-final causal explanations of a developmental process can be given, there must be a *formal identity* between the source of that process's efficient causes and the process's outcome. As no such identity is available between either the source of the enclosing heat, or the locale, and the kind of creation that emerges as the terminus of spontaneous generation, two of Aristotle's four types of causal explanations will not be appropriate in these cases.

Next, note that circumstances *external* to the actual developing 'embryo' are said to determine the 'form' of the end product. It is *not* the heat of the *pneuma* that determines form, but external contingencies.[15] The 'embryo' does not, in spontaneous generation, have the "source of *form*" in itself. Thus, in explaining such developments, one must point to determinants beyond the nature of the developing organism itself.

Finally, the lack of any apparent causal relation between the heat enclosing *pneuma* and the heat productive of an organism suggests that the relationship is one of mere coincidence.

Two texts in Aristotle with clear theoretical intent contrast processes characterized by what I have termed formal replication with processes not so characterized. In both, the latter type of process is termed spontaneous.

> It is this way too with things that are formed by nature. For their seed produces just as do the things that work from art – it has the form potentially, and that from which the seed comes is in a sense the same in name. . . . *Those can come to be spontaneously like the ones considered above, of which the matter can be moved and changed by itself* in the way in which the seed characteristically moves and changes it. (*Metaph.* Z.9 1034a34–b1, 1034b5–8; emphasis supplied)

> Now there are some things whose producing agent preexists resembling them,[16] for example the art of making statues: *for they* [i.e. statues] *do not come to be spontaneously.* And the art is the definition of the work without the matter. (Balme trans., *PA* I.1 640a28–32; emphasis supplied)

Both passages insist on a fundamental kinship between natural and artificial productions: in each, there is a synonym – that is, resemblance in name and definition – between the source of the immediate efficient cause of production and the product. As Aristotle sometimes puts it, the source of the potentially X's becoming actually X is what *is* actually X: "When asking how each thing comes to be, it is necessary to grasp – to make it our first principle – that what comes to be by nature or art comes to be *by* something which is in actuality *from* that which is such potentially" (*GA* 734b19–22). Each passage *also* insists that where processes occur that resemble these, but where the formal replication pattern is not present, these processes are *spontaneous*. In fact 640a30 seems to imply that the absence of a preexistent resembling cause is sufficient grounds to term a process spontaneous.

The importance of the synonymy of cause and effect for distinguishing natural generations from those that are spontaneous is also the theme of a difficult text in *Metaphysics* Λ.

> Next note that each substance (for both natural things and others are substances) comes to be from a thing with the same name. For everything comes to be either by art, nature, chance or the spontaneous. Now art is a source [of generation] in another, nature is a source [of generation] in itself (for a man generates a man), while the remaining causes are privations of these. (*Metaph.* Λ.3 1070a4–9)

This passage, like many in the corpus, has the appearance of notes for a lecture, which leaves much content to be filled in.[17] Nonetheless, I think its meaning is tolerably clear. Art and nature differ in that one is an external, the other an inherent, source of change in specific materials. Both produce something identical in form, however. Spontaneity and chance are privations of art (in the case of chance) or of nature (in the case of spontaneity). What they specifically lack, in each case, is any essential connection between source of change and product of change. That is, the model of formal replication is missing.

This interpretation of the crucial differences between sexual and spontaneous generation has two controversial implications.

1. Merely having a unified, directive process that is determined to lead to a certain outcome is not sufficient to characterize a process as formally or finally caused.
2. Aristotle's view of spontaneous processes ought to be contrasted, not with (or merely with) *regular* and *orderly* processes, but with those that exemplify formal replication.[18]

It is now time to look at these implications more closely, and to offer and defend a new interpretation of Aristotle's theory of chance and spontaneity that explains why Aristotle calls the biological processes and outcomes discussed above 'spontaneous'.

V THE HULL-BALME DILEMMA

On the basis of a particular understanding of Aristotle's theory of spontaneity, his biological account of spontaneous generation is paradoxical. To quote David Hull:

Aristotle usually calls an event or thing spontaneous if it fails to occur always or for the most part, but in the case just cited (*HA* 551a13) and in numerous others, the changes occur regularly whenever the appropriate material is present and the conditions are right.[19]

Hull's *coup de grace,* as he thinks, comes when he asks:

If lower animals can partake in the eternal and divine without the help of efficient, formal, and final causes coinciding, why not all species? And if there is no need for the efficient, formal, and final causes to coincide in any species, nothing stands in the way of accepting evolutionary theory.[20]

Hull thus voices two objections to Aristotle's overall account of spontaneity: (1) Spontaneous processes are alleged to be irregular and sporadic, but typically spontaneous outcomes (such as flies) occur with irritating regularity. (2) If formal-final causation is not needed in *these* explanations, why are they *at all?*

The second objection can be answered on the basis of the preceeding discussion. Formal-final explanations are intimately connected with the sexual-generation model. Explanation in terms of a form as the end for the sake of which a process is taking place are applicable only where that end product is identical in form with the efficient cause of the generation. If the end product is identical, *qua* functional organization, with what initiated the process, it is legitimate to say that the nature of the entire process is 'caused' by its end.[21] To put it bluntly, what prevents Aristotle's abandonment of formal-final causality in sexual generation is *the very nature* of sexual generation; while what prevents their use in the realm of spontaneous generation is that Aristotle simply *could not find* this formal-replication model in certain cases.

Hull's first point is repeated by a more sympathetic commentator, David Balme:

The doctrine [of spontaneity] as presented in *Phys.* and *Met.* together has three aspects relevant to the present enquiry.
(1) Spontaneity is unusual: The occurrence of natural ends happens [irregularly] and is therefore not [natural].
(2) Spontaneity is random: it is caused incidentally, by chance, as the byproduct of some other action. . . .
(3) Spontaneity brings about the same ends as are normally caused by purposeful act or by nature.

[This doctrine is] completely at variance with the treatment of spontaneity in A[ristotle]'s actual zoology.[22]

According to the zoology, spontaneity is *not* unusual (contra 1), is *not* random and indefinite (contra 2), and *does not* bring about animals normally due to natural (i.e. sexual) generation (contra 3).[23] The inconsistency, as Balme sees it, is all-pervasive.

In order effectively to meet these difficulties it must be shown that there is a plausible interpretation of Aristotle's theory of spontaneity that renders it consistent with his biological account of spontaneous generation.

VI

In *Physics* II.3 Aristotle has outlined his theory of causality, a theory devoid of any element of chance or spontaneity. In chapter 4 Aristotle faces up to the facts that it makes sense to refer to certain events as due to chance or spontaneity and that many of his predecessors have used these concepts in their accounts of nature. Part of Aristotle's scientific methodology involves taking such reputable opinions seriously, and chapter 4 of *Physics* II is a perfect example of his working dialectically on these "opinions of the many and the wise."

What, then, can people mean when they say something is due to chance and spontaneity? Aristotle sides with those (probably Leucippus and Democritus)[24] who believe both in a universal necessity *and* in chance occurrences (196a11–17).[25] Nonetheless, his predecessors failed to give a consistent theory of chance (196a16–17, a19–20, a35-b1), especially when they made use of this concept at crucial points in their cosmology (a20–24, a33–34). Generalizing a disappointment of Socrates',[26] Aristotle accuses a number of pre-Socratics of postulating a principle to explain the orderly nature of things, only to abandon it in specific explanatory contexts – usually where it is most required (196a17–35). They rely on spontaneity where things are most orderly and deny it where it most clearly occurs (196b1–5).[27]

Chapter 5 aims to give a general characterization of chance, applicable in the domains both of human action and of natural processes. Aristotle's approach is to focus on the characteristics of events typically described as 'by chance'. The chapter opens with a lengthy analysis of "generations," which can be represented by Table 10.1. Aristotle begins by noting that nothing found in boxes 1 or 2 would be said to be due to chance. He next insists that it is possible that some generations are both unusual and for the

Table 10.1

Generations	Always or usually in the same way	Not always or usually in the same way
For the sake of something	1	3
Not for the sake of something	2	4

sake of something – namely, whichever could have been done by thought or by nature. This is the general class of processes found in box 3.[28] Processes found in box 1 involve true teleological causation, that is, where the result of the process is that for which the process took place.

In box 3 are processes due to incidental causes, that is, where what results from the process was *not* the cause of the process. In these cases the *actual* cause of the process was only incidentally related to the outcome we are seeking to understand, while that outcome is not explanatory of the process at all. These processes are nonetheless prima facie teleological for three reasons. First, they have the *appearance* of being goal-directed. There is a process with a complex set of steps leading to an end (compare the characterization of what is 'by nature' at 199b15–17). Second, that sort of outcome is *typically* a goal of processes leading to it. Finally, what is undergoing the process typically does so for the sake of some end.

Nonetheless, while they are "among things that come to be for the sake of something" (*Ph.* 196b33, 197a6), these processes are not for the sake of what results; that is, they are *not* teleological.[29] The following example is used to present the theory in a concrete form: "Thus the man would have come for the purpose of getting back the money when his debtor was collecting contributions, if he had known; in fact, he did not come for this purpose, but it happened concurrently that he came, and did what was for getting back the money" (Charlton trans., *Ph.* 196b33–36). Aristotle goes on to stress that (1) he didn't go there always or usually, (2) the end of recovery was not among the causes in him, and (3) the end of recovery is among the things chosen and due to reasoning (*Ph.* 197a1–5).

Aristotle has in mind a class of results that are *typically* due to forethought and choice. When a member of this class of results occurs *without* forethought and choice – when the result is not a goal of the agent's – it is said to be due to an incidental cause. These cases, we say, are by chance. If this is Aristotle's theory, it is clear that incidental causation in such cases refers to the absence of teleology where it is characteristically present.

A number of passages in these chapters, however, seem to suggest a somewhat more subtle interpretation, apparently first suggested by Porphyry (see note 29, above). After considering two other examples, Aristotle gives a *general* characterization of results that are spontaneous: "Hence it is clear that events which (1) belong to the general class of things that may come to pass for the sake of something (2) do not come to pass for the sake of *what actually results* and (3) have an external cause, may be described as 'from spontaneity'" (Hardie and Gaye trans., *Ph.* 197b14–20). This could mean that among the class of processes that in an *unqualified* way are for the sake of something, there are some that are *not* for the sake of what in fact results. In the human sphere, for example, we may allow that the action that happened to result in collecting a debt was for *something,* without thereby saying it was for the result or that the result was *the* goal of the action. Commenting on this example, Aristotle notes: "And the causes of the man's coming and getting the money (when he did not come for the sake of that) are innumerable. He may have wished to see somebody or been following somebody, or avoiding somebody, or may have gone to see a spectacle" (*Ph.* 197a15–17).

Note that all of the incidental processes that led to collecting the money mentioned are goal-directed actions. What is stressed is that there is no causal-teleological connection between the actual result and the action that produced it. This, by the way, is what Aristotle believes is at the root of the popular belief that chance productions are outside of rational accounting. If the result of an action was not the agent's goal, there will be an unlimited number of potential candidates for its actual cause, and the list of possible causes will be different in every case in which that result occurs incidentally.

Thus Aristotle seems to allow that the incidental production of something typically the goal of its own production may be by means of actions aimed at *other* goals. There does not, however, seem to be any reason to see this as a *requirement* of spontaneous generation. For in other cases it is enough to say that results of the sort that generally are the "final cause" of their own productions in some cases are not. In those cases, the result is "due to spontaneity or chance," and the process is a spontaneous or chance process.

Aristotle's mode of presentation makes it seem as though the criteria that spontaneous processes be due to incidental causes and occur neither always nor usually are quite independent of each other. In fact, I now want to argue that spontaneous productions occur neither always of necessity nor usually, because incidental causes, by their very nature, cannot produce the same effects always or usually.

As Aristotle characterizes it in *Physics* II.3, an essential part of scientific method is discovering what it is about the nature of a substance that allows it regularly to produce a certain *sort* of result (*Ph.* 195b22ff.). Thus it would be inappropriate to say that the pale person or the musical person or Polyclitus is the cause of a statue, because it is not in virtue of being white, being musical, or having the constellation of peculiar characteristics denoted by the name "Polyclitus" that this individual is responsible for a statue. It is in virtue of the possession of a kind of skill that he is so.[30]

This doctrine plays a crucial role in the discussion of chance and spontaneity. Aristotle tells us that just as flute playing or being pale may be incidental causes of a house though only house-building is *properly* so called, so many reasons for going to the market may be incidental causes of collecting on a debt, though only when collecting on a debt is the goal of going can we say that the result explains going to the market (*Ph.* 197a14–17).

The felt difficulty in carrying over this notion to Aristotle's theory of chance is that here the causation is teleological.[31] Thus it is only when one can say that the result was the cause of the process leading to it – and thus indirectly its own cause – that one has a *proper* causal account. When the result occurs, but was not responsible for generating the process leading to it, the actual cause of the process, qua productive of that result, is incidentally related to it. The incidental cause will typically, in human cases, be some *other* goal.

Insofar as the result is the incidental effect of the attempted realization of another goal, it will not characteristically be produced *in this way*. Thus, as Aristotle opens his account of chance and spontaneity, he contrasts chance events with those that "come to be always *in the same way* or usually" (196b–10). Insofar as they are due to *incidental* causes, they will not occur with any regularity *in just that way*.

Throughout chapters 5 and 6, Aristotle has stressed that chance events are to be found both in the realm of human agency and in the natural realm (196b21, 198a6–7). In chapter 6 he introduces a distinction between chance and the spontaneous, making spontaneity the inclusive notion for chance events in either realm, and restricting chance to the domain where human choice is operative (see *Ph.* 197b7–36).[32]

Unfortunately, the examples Aristotle gives to show that natural productions can be due to spontaneous causes are confusing rather than illuminating. Not that they do not fit Aristotle's description of spontaneous occurrences – they do.[33] They fail, however, to illuminate the truly significant role this concept plays both in Aristotle's theory of nonsexual generation and as a characterization of the views of Democritus and Empedocles on biogenesis and cosmology generally (e.g., at *Ph.* 198b30, 196a25, 26, 34, 196b3; *PA* I.1 641b23, 29). We are left then, with the problem of how this account of spontaneous generation adequately characterizes the processes so described in Aristotle's natural science.[34] Let us review the essential ingredients of a spontaneous generation according to the *Ph.* II.5–6 account.

1. The process must be such that it might have come to be for the sake of something. 'Might have' in this context is shorthand for the facts that the result is typically a goal, that the subject of the process is typically goal oriented, and that the process proceeds to the result with the appearance of being goal directed.
2. The result must not come about always or usually in the way it in fact did in this case.
3. The cause must be incidental, that is, the result must *not* be the goal for which the process took place.

It is clear that a proper understanding of teleological processes in nature is crucial to understanding how certain natural processes can be described as spontaneous. In sections 2 and 4 above, I argued that to describe biological development as for the sake of its end was to take cognizance of the fact that such processes are the means by which a form is *reproduced.* The end of the process is responsible for it because it is one in form with the agent of that process.

I have argued that spontaneously generated organisms lacked, in Aristotle's understanding, this formal identity between productive agent and end product and thus were *not* teleological. Now let us read this account of teleology into the three key conditions that must be met in order that a process be 'spontaneous'.

1'. Spontaneous generations must be the sorts of things that might have been due to the agency of something identical (formally) with them.

Now a good deal turns on the force of 'might have been' (*Ph.* 197a35, 198a6). In the examples we are given in the *Physics,* the results occasionally produced incidentally, and thus due to chance, are typically produced teleologically – by the goal's being a cause in the agent (*Ph.* 197a1). In the cases characterized as spontaneous in the biological works this is *not* the case: sea urchins are *never* produced by other sea urchins.[35]

However, self-maintaining, perceptive beings – organisms – *are* typically produced by formal replication, and their development exemplifies the directedness it does for this reason. Thus, Aristotle's point may be only that biological development is typically for its end, but in these cases it is not. If so, condition 1' is met at a very general level only.

2'. Spontaneous generations must not come about always or usually in the same way.

Here the detailed examination of Aristotle's theory of spontaneous generation, in combination with the view that 2' must be understood within the context of incidental causation, yields an interesting result. Spontaneous biogenesis *is* outside of what occurs always or usually, notwithstanding the *frequency* of spontaneous production. Aristotle's general account of spontaneity is that, where the result of a process was not its goal, it is not likely always or usually to be produced in the same manner. Thus, while many species of organisms are spontaneously generated with great regularity, they do not come to be always or usually due to the same cause.[36] Thus, as Aristotle characterizes natural processes, these do not qualify: "A thing is due to nature, if it arrives, by a continuous process of change, starting from source in itself, at some end. Each source produces, not the same thing in all cases nor any such thing, but always something proceeding to the same thing, if nothing impedes it" (*Ph.* 199b15–18).

3'. The cause must be incidental, that is, the result must not be the goal for which the process took place.

It is clear that Aristotle sees this as the fundamental differentia of spontaneous processes. It is *equally* clear that those processes described as spontaneous in the *Generation* and *History of Animals* exemplify incidental causation in the appropriate sense. For what I have argued is that the core of incidental causation of processes described as by chance is that the result is not responsible for the process leading to it. And only where processes exemplify formal replication can the result of a process be said to be responsible for it. Thus the production of a specific sort of organism out of some material due to the combined effects of climate, ratio of *pneuma* to earth in the material, the sun's heat (or warm winds or currents), and so on, is paradigmatically spontaneous.

VII

Between the 17th and 20th centuries the theory of spontaneous generation was, species by species, shown to be false.[37] Nonetheless, I want to conclude by suggesting that Aristotle's concept of chance in nature has an analog in contemporary biological theory.

The synthetic theory of evolution by natural selection is, in part, an explanation of functional adaptation. It can explain why a particular population tends to possess certain characteristics. Arguably, such explanations are teleological.[38]

What evolutionary theory cannot explain teleologically is the infrequent but regular appearance of mutations. These 'errors' in the genetic program are typically said to arise 'at random' or 'spontaneously'. When evolutionary biologists talk this way, they do not mean to convey the idea that mutations occur causelessly. Rather, such descriptions are offered within the context of the typical case, where a structure's repeated production in a population can be explained by the function it serves for an organism of that kind in its normal environment. That is, to refer to such mutations as 'random' or 'spontaneous' is to say that they cannot be accounted for in terms of the adaptive requirements of the organism.[39]

This, as we have seen, is Aristotelian in spirit. To see the characteristics that can be explained teleologically as the norm, and to identify the production of similar structures by similar mechanisms that cannot be explained teleologically as 'random' or 'spontaneous', is the essence of his doctrine.

Yet there is a fatal flaw in Aristotle's application of the concept of spontaneity to species and genera that are *typically* well adapted.[40] His critique of Empedocles' and Democritus' nonteleological accounts of biological development turns on the absurdity of supposing that such highly organized, typically well-adapted entities could arise regularly 'by chance' or 'spontaneously'.[41] His own doctrine seems well suited to an *occasional* such production. But to imagine entire kinds being produced regularly, and being well suited to their specific sort of life, seems as unlikely as anything Empedocles might have said.

Thus, the central outstanding problem with Aristotle's doctrine of spontaneity in nature is *not* that it violates his overall theory of causality, *nor* that the chance production of organisms by nonreproductive means would not be appropriately termed "spontaneous" according to his general doctrine. The basic problem is that to hold that every member of many species of insect and testacea are so produced would seem to be open to the following critique: "The things mentioned, and all things which are due to nature, come to be as they do always or for the most part, and nothing which is the outcome of luck or an automatic outcome does that" (*Ph.* 198b35–198a1).

I have argued that the *essence* of Aristotle's concept of natural spontaneity is the incidental, that is, nonteleological, production of a result that might have been teleologically produced. There is nothing in such a concept that rules out the regular occurrence of events. But Aristotle insists that an outcome can be labelled spontaneous only if it *normally* is produced teleologically. Thus, if some sea urchins are produced spontaneously, we should expect a much larger number to be *re*produced. The lack of a contrasting set of *re*produced species members told against the hypothesis that many testacea and insects were spontaneously produced. At the same time, the apparent absence of formal replication told against such processes being for the sake of their results. Since the absence of formal replication where one would expect it is the fundamental criterion of spontaneity on Aristotle's account, he opted for labelling these creatures spontaneous. There is no doubt that he would have been excited with the discovery that formal replication is as pervasive among invertebrates as among vertebrates.[42]

NOTES

1. There will not be space here to discuss the general issue of the logical and chronological relationships between Aristotle's *Physics* and his biological works. I shall assume that the *Physics* is earlier, and I will be arguing that the theories of explanation, teleology and chance argued for in *Physics* II.3–9 are used more or less systematically in Aristotle's study of living things. This view has been defended in a general way in Lloyd 1968; Preus 1975; and Balme 1962a, 91–104.
2. When discussing Aristotle's theory of nonreproductive generation, I shall render *automaton* 'spontaneous'. When Aristotle's theoretical account of *tuchē kai to automaton* in *Physics* II.4–6 is discussed, I will use the term 'chance' for the general phenomena under discussion. The reader is here forewarned that the same Greek term is used in the one context and under philosophical scrutiny in the other.
3. By Balme 1962a, 96–97; and David Hull 1967–68, 2: 245–50.
4. If there is such a thing as an uncontroversial claim with regard to the dating of various parts of the corpus, this is one. Of course, given the various catalog references to different treatises composed of parts of our *Physics,* one cannot presume a dating for the *Physics* per se. For our purposes it is enough to note the clear reference to *Ph.* II at *Metaph.* A.3 983b1, and the likelihood that *Metaph.* Δ.2 and 4 are borrowed from *Ph.* II.1 and 3 (rather than the other way around). See the discussion in Ross 1955b, 1–19; and in Mansion 1946, 1–34. Mansion, however, has a complex theory of the development of Aristotle's theory of chance (hazard), which I discuss in note 28 below.

5. The concepts 'power' and 'source of change' are used interchangeably in *GA*, following the definition of 'power' provided at *Metaph.* H.1 1046a10–19.
6. One of Aristotle's usual verbs for the effect of the semen on the material is *pettein*, which derives from both an agricultural usage meaning 'to ripen' and a culinary usage meaning 'to brew' or 'to cook' (see Bonitz, 590a45ff.).
7. For some conjectures, see Balme 1972, 157; Gotthelf 1987b, in Gotthelf and Lennox 1987, 215–22; and Nussbaum 1978, 158–64. More general discussions are cited usefully in nn27–35 of Nussbaum's essay.
8. As Gotthelf notes (Gotthelf 1976, 252), a casual reading of this analogy might lead one to suppose some analog of consciousness in organic development. What I shall argue is that this is not an aspect of the analogy between art and nature at all.
9. See Balme 1972, 95, 153. Balme argues convincingly that Aristotle is comparing the *craft* with the parent's *nature* or form, not the craftsman with the parent. Balme also reads *Ph.* 199b28, "The art too does not deliberate," as *literally* about the art and not about the craftsman. It seems easier to read it (as does Charlton 1970, 125) in the light of *EN* 1112a34. The point there is that deliberation comes into play only when we consciously have to plan out the steps to some goal. In the established arts, the practitioner need not deliberate how to achieve his desired end.
10. Marjorie Grene has raised questions (private communication) about the consistency of such a doctrine with *Ph.* II.1 193b13–14. "It [nature] is not like doctoring, which has as its end not the art of medicine but health." But all that this passage demonstrates is that the form replicated in craft production is *not* the form of the craft but the form of the product. Nonetheless, "the art is the account of the product without the matter" (*PA* I.1 640a31–32). Note that, at *Ph.* III.2 202a9–12, Aristotle distinguishes between the agent of a change and the form it bears, which is the source and cause of the change.
11. See references in note 7, above.
12. The automata are discussed in more detail in *MA* 701b1–13, where a detailed structural-functional analogy is drawn between them and the locomotive physiology of mammals. A. S. L. Farquharson's notes to 701b2, b4, b5, and b7 give all the background needed (with a wealth of references) to allow one to picture the toys Aristotle is discussing. (Smith and Ross 1912, vol. 5.) See the Pseudo-Aristotelian *Mech.* 847b1–848a38; Brumbaugh 1968; Nussbaum 1978, 50–51, 347–48; Nussbaum 1976, 146–52; and Gotthelf 1976, 242–43n30.
13. Perhaps a reference to Thales, as Pierre Louis suggests (1961, 131n1). See *de An.* I.5 411a8ff. A passage in *HA* (577b11–13) talks of "those materials which have life" in the context of spontaneous generation. David Balme has suggested (private communication) that this text is suspect, while P. Louis (Louis 1964–69) has suggested that all that Aristotle intends is that these materials possess nutrients required for life and growth. The "one-liner" on spontaneous generation at *Metaph.* Z.9 1034b5–7 says that matter has the capability of moving itself in these cases.
14. I have here followed Arthur Platt's translation (Smith and Ross 1912, vol. 5) except for the last two sentences.
15. There is a parallel in this respect to what Aristotle says about plant seed, at *GA* II.4 738b34–37. He refers there to the land as the provider of the matter and body for the seed and argues that a seed deposited in a foreign land will produce a plant of a different form, reflecting its atypical environment.

16. There are no textual grounds for Peck's bracketing of 640a27–33 (Peck 1955). As Balme notes (1972, 86), the passage is in note form but answers a possible objection to the preceding argument and so has obvious point.

17. Note the omission of most of the definition of art and nature, the unexplained reference to "the others" in 1.6, and the opening of the passage ("Next note that . . ."), of which Ross comments, "This phrase . . . is one of the clearest indications . . . that Aristotle is jotting down notes for a treatise (or lecture) and not writing a treatise in its finished form" (1924, 2: 354, 1069b35n; see 1071a4n). Ross's notes (354–55) are generally helpful on this passage.

18. At this point it is worth dealing with two obvious objections to my account of teleological causation in Aristotle.

    The first would argue against insisting that an *aitia* be strictly interpreted as a factor *responsible* for a change and thus would object to the view that teleological *explanations* involve the identification of processes the results of which *are responsible for* their occurrence (e.g., Sorabji 1980, 155–74; Nussbaum, 1978, 74–88; Charlton 1970, 112–13). There are a number of responses to this objection. First, Aristotle typically uses all the vocabulary of causation in discussing the coming to be and existence of biological structures – normally insisting that where one can say "*X* for the sake of *Y*," *X* came to be on account of the end (*GA* V.8 789b6). Second, there is good reason to think that the modern notion of 'cause' as some change in the conditions external to the event to be explained, which is necessary, sufficient, or both for the occurrence of that event, is not even correct as an account of Aristotelian *efficient* causation. There is every reason to think that Aristotle offers us a more robust account of what a cause is, one designed to account for spatially and temporally continuous processes of a highly complex nature.

    A second objection might allege that the account offered here *reduces* final causation to efficient causation (Sorabji 1980, 166–71). The involvement of the form in three of Aristotle's four causes leads *him* to a *sort* of reduction, of course (*Ph.* II.8 198a25–27; *PA* I.1 641a26–28). yet he does *distinguish* three types of explanation in these passages, and I suggest that if we focus on what is being explained, it is clear why. Nature *qua* form can account for two quite different things. It can account for a composite's being what it is (see *Metaph.* Z.17) and it can account for something's coming-to-be in a certain way. In the latter cases, the form is responsible as that for the sake of which the coming-to-be takes place. To account for the organization of the complex efficient causes involved in a teleological process, we need to identify the *parent's* form. But at a more general level, i.e. in explaining why the coming-to-be takes the course it does, the identity between the form at the end of the process and the form of the agent is central. To the extent that one must establish this identity between agent and end for this type of explanation to go through, it is a distinct form of explanation from one which *simply* focuses on the agent of the change.

19. Hull 1967–68, 247. See also Hull 1967, 309–14.

20. Hull 1967–68, 247.

21. The analogy between Aristotle's explanation and that of contemporary molecular regulatory genetics has not gone unnoticed. See the interesting comments by Stent 1971, 3; and Stubbe 1972, 33–40.

22. Balme 1962a, 96–97.
23. Ibid., 97–99.
24. See Leucippus, Fr. 2 in Kirk, Raven and Schofield 1983, 420n1; (*DL* IX.45). Leucippus's "Nothing comes about in vain, but everything from reason" is very much in the spirit of Aristotle's principle that "Nature does nothing in vain, but always the best" (*IA* 704b14–16; see *PA* 658a9, 661b24–25, 687a15–16; see *Bonitz* 1870, 247b5–10). Whatever their differences, the atomists and Aristotle are in agreement that the occurrence of certain events by chance does not imply uncaused events.
25. Sorabji (1980, 1–15) takes Aristotle to be saying that if a result can be described as due to an incidental cause, then it is uncaused. What Aristotle quite clearly says (*Ph.* 196a11–15) is that saying "*X* is by chance" and "*X* has some cause" are utterly compatible. Such events are a breach of a *very special* causal pattern, but not of causation per se. (Sorabji attempts to avoid this, and other, objections on pp. 23–25.)
26. See *Phaedo* 97b–99d, and Aristotle's comment at *Ph.* II.8 198b13–16.
27. Similar criticisms are made in *PA* I.1 641b18–24; *Cael.* II.3 287b25, 289b21–27.
28. Mansion 1946, 292–314 argues that two different, though compatible, accounts of chance are reflected in these paragraphs and elsewhere. Mansion takes Aristotle's statement that what occurs neither always nor usually is due to chance as a *definition* of one sense of chance. I take it to be the first of three necessary conditions that together define chance.
29. I shall avoid discussing a number of difficulties raised by Aristotle's account of chance in these chapters, but one in particular *must* be faced. At *Ph.* 196b23–24, Aristotle parenthetically remarks, "Whatever might have been done due to thought or to nature is for the sake of something" (and see 197a35, 198a6). He repeatedly refers to chance events as "in the things which come to be for the sake of something" (196b33, 197a6). He *also* insists that chance processes do *not* come to be for the sake of their results (196b34, 197a16, 18, 30, 199b21–22). Finally, he says that chance results are those that might have been done by thought or nature (198a6). Thus he appears simultaneously committed to saying (i) that chance processes are for the sake of something and (ii) that they are not for the sake of what results.

    Simplicius and Porphyry represent the two most plausible ways out of this dilemma: "The 'for the sake of which' is observed not because the process *is* for the sake of *something*, e.g., for the sake of marketing, as Porphyry says, but because it might even have come to be *for this end*" (Simplicius, *in Ph.* 366.27–29). These issues are the subject of "Aristotle on Chance," chapter 10.
30. See Sorabji 1980, 5–6.
31. Simplicius (*in Ph.* 377.15–378.3) argues that the same action, e.g., digging, may be a proper cause of one thing (planting) and an incidental cause of another (finding treasure). His analysis avoids, however, the question of how one makes this distinction. Clearly it entirely depends on which result was conceived by the digger as what his digging was *for.*
32. *Ph.* II.6 also replaces the condition of *incidental* causation with that of *external* causation (197b20). Some commentators, notably Ross (1955b, 524), have tried to read the distinction between 'internal' and 'external' causes at 197b35–37 as a

distinction between two types of spontaneity, and they therefore suppose that the definition offered in chap. 6 does not cover spontaneous biological generation. But there is a much more plausible reading of this passage. *Ph.* 197b32–37 reads: "[That which is due to the spontaneous] is most separate from that which is due to chance in the things which come to be by nature. For when something comes to be contrary to nature, we do not say it came to be due to chance but more due to the spontaneous. But there is also this difference; the cause of the one is external, the cause of the other is internal." With Simplicius (*in Ph.* 353.1–12), Themistius (*in Ph.* 36.16), and Charlton (1970, 110–11) I take the contrast in 11.35–37 to be between the *truly* spontaneous and what is contrary to nature, e.g., monsters (cf. 199b1–7).

By saying that the cause of spontaneous processes is external, Aristotle is saying that the cause of such a process's leading to some result is *not* that result itself but some other thing (see Philoponus, *in Ph.* 288.23–28; *Metaph.* Δ.30 1025a28–29). This applies equally to spontaneous biological development and human chance, where Aristotle has already explained that "the end, namely recovery of a debt, was not one of the causes *in* the man" (197a1–2; see Philoponus, *in Ph.* 275.12–13).

33. On the lack of relevance of the examples in *Ph.* II.6, see Charlton 1970, 109–10.
34. I shall here restrict my discussion to the general account of spontaneous generation in *GA* III.11.
35. As Balme (1962) notes.
36. As Aristotle notes in *Ph.* II.5, if the result was not what the process was for, an indefinite number of things might have brought it about. It is unclear how Aristotle might have reacted to the suggestion that in cases where the identical sort of creature is produced the material conditions must, at *some* level, be identical. What is clear is that such an identification would not yield a cause identical in *any* sense with the result of the spontaneous process.
37. See Farley 1977; Mendelsohn 1976, 43ff.
38. For such arguments, see Wright 1976, Ayala 1970, 1–15.
39. "The forces that give rise to gene mutations operate at random in the sense that genetic mutations occur without reference to their future adaptiveness in the environment. . . . it did not arise as an adaptive response but rather proved to be adaptive after it appeared" (Ayala 1978, 14–29).
40. That they are is shown clearly by the discussion of the organs of various testacea at *PA* IV.5–8.
41. Gotthelf 1976, 244–49. My account of Aristotle's teleology owes much to Gotthelf's. I was led to the formal-replication model of teleology originally by troubling over three difficulties in Gotthelf's paper. First, it is not clear to me how that account can distinguish between nonteleological, spontaneous processes and teleological, reproductive ones. Second, I have doubts that Aristotle held that reproduction involves "*an* irreducible potential for form." The relevant texts suggest an integrated *series* of potentials, the crucial factor being the formal pattern possessed by the series. Finally, it remains unclear to me how 'the end for the sake of which' is *responsible for* the end-directed process on Gotthelf's account.
42. Many people provided helpful comments on earlier drafts of this chapter. It is a special pleasure to acknowledge the valuable comments of Jim Bogen, Alan

Bowen, Gail Fine, and an anonymous referee. Their perceptive criticisms contributed in important ways to the argument's final form. Part of the research was supported by a faculty research grant from the University of Pittsburgh, which I acknowledge with gratitude.

11

Aristotle on Chance

Aristotle's technical concepts of chance and spontaneity play three crucial roles in his philosophy of science. In his accounts of his predecessors' views on scientific explanation, Empedocles and Democritus are singled out as invoking chance and spontaneity where Aristotle insists teleological explanation is required.[1] In his own theory of explanation, certain biological processes are characterized as 'spontaneous'.[2] And in his ethical writings, chance plays a crucial role in determining responsibility for an action.[3] It would seem, then, that a proper grasp of his considered doctrine of chance, which is worked out in *Ph.* II.4–6, is central to the evaluation of a number of areas of his philosophy. Unfortunately, the development of that doctrine contains a number of difficulties. The central ones turn on the relationship between processes which are due to chance and those which are for the sake of something. The purpose of this discussion is to resolve these difficulties.

The argument of *Ph.* II.4–6 makes the following claims:

(1) Whatever might have been due to thought or to nature is for the sake of something (196b23–24; cf. 197a35, 198a6).

(2) Chance events are "among the things that come to be for the sake of something" (196b33, 197a6).

(3) Chance processes are not for the sake of their result (196b34, 197a16, a18, a30, 199b21–22).

(4) Chance processes might have been due to thought or nature (198a6).

Although (2) may be ambiguous, (1) and (4) imply that:

(5) Chance events are for the sake of something.

Now if one takes (5) to mean that the something which a chance event is for the sake of is some result of that event, then (5) appears to contradict (3). At

250

the same time, it seems clear from numerous discussions in the corpus that (1) does not represent Aristotle's doctrine of teleology: it is not what *might have been done* by thought or nature which is for the sake of something, but what *is* done by thought or nature.

There are two plausible strategies for dealing with this problem. The first is to reinterpret (1) to bring it in line with standard Aristotelian accounts of teleological outcomes. This would block the inference to (5) and avoid any chance of contradiction. The other, first suggested (as we shall see) by Porphyry, is to say that chance events are for the sake of *something,* but what they are for the sake of is *not* what results.

Proposition (1) occurs as a parenthetical aside near the end of Aristotle's preliminary enumeration of the conditions which must be satisfied before a process can be referred to as "by chance" or "spontaneous." Aristotle has claimed that a necessary condition for being by chance is that a generation be outside that which occurs always or usually. He then goes on:

> Of things which come to be some come to be for the sake of something, some not (and of these some are according to forethought, some not, but both are among things which come to be for the sake of something), so it is clear that among those things outside the necessary and usual there are some which may be for the sake of something. *But whatever might have been done by thought or by nature is for the sake of something.* And when such things come about by accident, we say they are by chance . . . (196b17–24; emphasis added)

The 19th century German philologist Torstrik was vexed enough by this account of what things are for the sake of something to change *prachtheiē* – might have been done – to *prachthēi* – is done.[4] This brings (1) into line with Aristotle's standard account of teleological explanation, thereby restricting the subjunctive formulations to descriptions of chance events. This move is, however, utterly *ad hoc.* There is no textual basis for these changes, and the ancient commentators were reading the same puzzling texts we are. Such a desperate attempt to deal with the problem should remain an unsatisfying last resort.

On the other hand, the suggestion of Ross and Wieland, that Aristotle's teleology is only "*de facto*" or *als ob,* goes against the spirit of Aristotle's attempt to contrast chance processes with those which are truly teleological.[5]

Simplicius' commentary on this passage (366.27–29) preserves not only his own way out of this dilemma, but also Porphyry's. These two ancients offer the two most plausible interpretations of the passage in question. In his commentary to the proposition I have been calling (1), Simplicius is considering the question why chance processes have the appearance of being goal directed. He rejects a suggestion made by Porphyry, namely, that they appear goal directed because they are.

> Something's being for the sake of something is observed not because the process is for the sake of something, e.g., for the sake of marketing, as Porphyry says, but because it might even have come to be for this end.[6]

Porphyry's suggestion seems to be this. The man who happens to collect on a debt goes to the market place for a purpose, though not to collect on a debt. Perhaps he was there in order to take care of a legal matter, or for a festival. Porphyry thus attempts to avoid our dilemma by noting a distinction between claims (1) and (4), on the one hand, and (3) on the other. Claims (1) and (4) generate,

(5) Chance events are for the sake of something.

while the third states that

(3) Chance events are not for the sake of their results.

Porphyry, then, argues that Aristotle's theory of chance is that while chance processes are for the sake of something, they are by chance because they are not for the sake of the result they bring about. It is this unexpected result of a teleological action which is said to be by chance.

Simplicius' view, however, implies that chance processes have the appearance of being goal directed because the result is something which might have been done for that result. This implies that claim (1) is a characterization of a loose sense of 'for the sake of something', covering processes which are of a kind which are normally goal directed, but in certain cases are not. Those cases are differentiated from genuinely teleological ones by the fact that they have 'incidental causes' (196b23–29).

There are three pieces of evidence which one suspects led Porphyry to his view.

(a) Aristotle persistently uses the phrase "among the things which come to be for the sake of something" in his characterizations of chance processes. One might suppose this phrase picks out processes which are generally for something, and thus would include those which are for something other than what they produce (i.e. chance processes).

(b) The list of the incidental causes of collecting on a debt by chance given by Aristotle at 197a17–18 are all goals other than the result achieved.

That is, while I might have collected a debt by chance, I had some other reason for being where my debtor happened to be.

(c) The definition of spontaneity in chapter 6, which is a crucial text, can be read in a way which supports this interpretation.

Let us consider each of these points. The use of 'in' in the phrase "in things which come to be for the sake of something" needn't signify that something so described is one of the class of things which are for the sake of something. The Greek *en*, which I am translating 'in', is much broader in signification than this. At any rate, the issue is how one takes 'for the sake of something' in these passages; so we can't derive support for either interpretation from this turn of phrase.

With respect to (b), it must be admitted that the incidental causes mentioned by Aristotle are other goals. But this need not always be the case. Suppose you were compelled to be at some location and, having been taken there, discovered the woman of your dreams. You didn't go there for *any* purpose, but you would have, had you known whom you would meet. It is also important that Aristotle's account be extendable to natural cases. And it is unclear how spontaneous natural processes could be said to be for ends other than those which result.

But there is another more serious objection to using the evidence noted in (b) to support Porphyry's interpretation. The evidence cited in (b) above in fact says that such processes really are due to thought or nature, not that they might have been. But we are trying to understand why Aristotle says that things which might have been due to thought or nature are for the sake of something, and this end is not furthered by simply saying that they *are* due to thought. Porphyry's understanding of Aristotle's concept of chance processes seems inadequate to account for the subjunctive in claim (1).

The strongest evidence for Porphyry's reading is the definition of spontaneity (the wider concept, covering chance in both the human and natural realms) in *Ph.* II.6.

> Hence it is clear that events which are among those things which come to be without qualification for the sake of something, when they do not come to be for the sake of the result, and which have an external cause, are due to the spontaneous. (197b14–20)

The crucial distinction, for the purpose of this discussion, is that between processes without qualification for the sake of something and processes not for the sake of the result. Spontaneous processes are said to be both. On Porphyry's reading 'without qualification' signals that such processes are for something, and the requisite qualification is then specified: they are not for the sake of their results.

Though on its own this passage suggests Porphyry's solution to our dilemma, that solution faces insuperable difficulties. These difficulties can be clearly seen by paying careful attention to the closing lines of *Ph.* II.5.

> Both chance and spontaneity are incidental causes, as we said, among the things which may come to be neither without qualification nor for the most part; and among these, whichever might have come to be for the sake of something. (197a32–35)

Here Aristotle claims clearly that it is what might have been for the sake of something that is by chance. This suggests that Simplicius' solution, rather than Porphyry's, is the correct one.

The difficulty in the way of accepting Simplicius' solution is that it forces us to take proposition (1) seriously. But this proposition seems to sanction the description of processes which are not causally tied to their outcomes as for the sake of their outcomes. This solution also requires us to accept the implication that, in some sense, chance processes are for the sake of something. Simplicius' solution, then, forces us to take account of *all* of the textual evidence, and brings with it all the attendant problems.

Simplicius' original insight was to note that Aristotelian chance processes have to be descriptively like those which normally achieve the end achieved either by thought or by nature. I would like to pursue this idea by suggesting that Aristotle is willing to describe such processes as for the sake of their outcomes. In such cases, however, 'for the sake of something' doesn't carry the causal force that it does when applied to processes where the goal is essentially (rather than incidentally) related to the process leading to it.

To defend such an interpretation, I wish to consider the crucial passage of *Ph.* II.5, in which Aristotle attempts to present us with an example of a chance event, and then notes its philosophically relevant features.

> Thus, the man would have come for the sake of recovering the money when his debtor was collecting contributions, if he had known; in fact, he did not come for the sake of recovering the money, but he happened to come and to do this for the sake of collecting the money. (196b34–36)

This passage presents numerous difficulties of interpretation which I shall gloss over. The central philosophical problem is making three claims in this passage consistent. They are:

(i) He might have come for the sake of debt-collecting.

(ii) He did not come for the sake of debt-collecting.
(iii) He happened to come and to do what was for the sake of debt-collecting.

Bonitz[7] excised the phrase "for the sake of collecting" in the last line, and one can sympathize with him. However, the best manuscripts and commentators have it, and I believe we can make sense of the passage with it. To do so, we need to consult Aristotle's notes on this example, which run from 196b36 to 197a8. In summary, Aristotle takes the example to show that:

1. The person in question does not usually, or necessarily, go to the market for the purpose of collecting on debts (196b36–197a1).
2. The end result, collecting on a debt, was not among the causes in him (197a1–2).
3. Recovering money owed to one is the sort of thing typically achieved by intelligent action (197a2, a5).

Point 3 tells us that debt-collecting is the sort of thing that could, and normally would, be the result of goal-directed, intelligent action. This also helps us to understand claim (i) – 'might have' has the force of 'could, and normally would have'. However, point 2 reminds us, collecting on a debt was not in this case a part of the explanation of the events that led up to it. Unlike truly teleological processes, the end result of this process was not that for the sake of which the process took place. This explains the meaning of (ii) in the example.

Aristotle's comments on his example give us a means of rendering (i)–(iii) consistent. For one way of understanding the claim that someone did not come to p for the sake of X (ii), but happened to be at p and did what was for the sake of X (iii), is this. He carried out all the actions required to achieve X, having done so he achieved X, and yet X was not the goal of his actions (not a cause in him). And while achieving X is not the reason why he normally goes to p, X is the sort of thing which typically is achieved by purposeful action.

The outcome is related incidentally to the process which leads to it in the following way. Just as the proper characterization of the efficient cause of a home's being built picks out the sort of agent whose actions characteristically lead to completed houses – i.e. the builder – so the proper characterization of the final cause of a process picks out that result which was the goal of the process. If the result being scrutinized was not the goal of the process leading to it, then it is related to that process only incidentally.

However, if the process is one which might have been properly for that goal, if the end result is capable of being the proper goal of the process, and if the process does in fact achieve that end result, there is sense to saying, as Aristotle does, that the process is, by accident, for the sake of that result.

Let us now return to the original problem. We noted that Aristotle asserts:

(1) Whatever might have been due to thought or nature is for the sake of something.

and

(4) Chance processes might have been due to thought or nature.

These assertions imply

(5) Chance events are for the sake of something.

But (5) seems to violate Aristotle's insistence that chance processes are *not* for the sake of anything (3).

The suggestion I have made is that Aristotle is willing to describe chance processes as for the sake of their results provided certain conditions are met. When he says they are for the sake of something without qualification, but not for the sake of what actually results, I suggest he means this: the result was not responsible for (not a cause of) the process that lead to it;[8] nonetheless, the result was valuable for the agent, and was the sort of thing that is typically achieved by goal-directed activity.

To the suggestion that there are two senses of 'for the sake of' in Aristotle, the obvious response is to wonder why we aren't explicitly told about it. I will thus close by indicating what positive evidence there is for the view that two different sorts of processes are described as 'for the sake of something'.

First of all, there are passages which tell us about two sorts of things which can be described as that for the sake of which.[9] These passages are usually taken to contrast the beneficiaries of processes and the actual goals of processes.[10] Only one of these is causally related to the process which leads to it. The favored example among the commentators is that health is 'that for the sake of which' of medicine as its goal, while the patient is 'that for the sake of which' as its beneficiary. One of Aristotle's favored examples of a chance result is the restoration of health by unintended means (e.g., a chance change in the weather).[11] In this case, we could say the change was for the good of the patient, *if* the change brought about just that physiological effect the doctor would have had as his goal.

It seems plausible to suppose that there will be processes – just those Aristotle says are by chance – which achieve beneficial results and yet are not goal directed. That is, there will be a non-causal sense of 'for the sake of something' corresponding to the non-causal sense of 'that for the sake of which'.

A second bit of support for this view derives from the *Nicomachean Ethics*. Ross has pointed out that *EN* 1111a5, 1135b12 and *EE* 1225b2 describe actions as for the sake of some result even though that result was not a goal of the agent.[12]

Finally, there is a natural and reasonable distinction to be made here, as is suggested by a contemporary dispute on the same issue. In his *Teleological Explanations*,[13] Larry Wright insisted that a consequence could be a function only if that consequence was the reason why what has that consequence exists.[14] A number of philosophers have properly objected to this, on ground that structures can come to have functions which have nothing to do with why they are there.[15] To legislate against this very natural use of 'function' seems utterly uncalled for. Nonetheless, it is true that to attribute a function to something is to at least raise the possibility that it *explains* the structure in question. Discovering that it achieves good things for its possessor quite by chance does not, however, lead one to stop saying that this is its function. Likewise, evolutionary biologists distinguish between useful traits which are *adaptations* and those which are not, where the former are *explained* by their functional contributions to fitness and the latter are not.[16] Such a distinction closely resembles Aristotle's between good results achieved by chance and good results which were goals of the changes that produced them. There is every reason to suppose he felt the need for such a distinction as acutely as contemporary philosophers and biologists. The terms of Aristotle's teleological vocabulary are ambiguous in much the way that our concept of 'function' is, and the passages referred to in note 9 indicate that Aristotle was sensitive to this ambiguity.

Let me conclude by reviewing the original problem, the difficulties in the way of a Porphyrean solution, and the solution I am recommending.

The problem originates from Aristotle's apparently conflicting desires to sharply distinguish chance processes and outcomes from teleological ones, and to describe *certain* chance processes in teleological terms. Porphyry's solution, a subtle and ingenious one, was to distinguish between the actual goal of a process and its outcome, and to see chance processes as those which achieve outcomes other than those toward which they were directed. There was a fundamental difficulty with this solution. It insists that all chance processes are due to thought or nature (in a causal sense), whereas Aristotle only makes the weaker claim that they might have been. It thereby restricts the range of processes which can be described as chance in a way which conflicts with Aristotle's scientific and ethical use of the term.

The solution I have suggested, a descendant of Simplicius', is that Aristotle has a causal and a descriptive sense of 'for the sake of', and that chance processes are for the sake of their results only in the noncausal sense. It is these which 'might have been due to thought or nature', while *truly* teleological processes *are*. The form of an object of craft, the good perceived as achievable by action, or the nature (form) of a sort of organism are all aspects of the world which Aristotle argues are typically responsible for the processes which produce them. When just these sorts of things are produced and yet are not responsible for the processes which produce them, they are by chance.[17]

NOTES

1. *Physics* II.8, *Parts of Animals* I.1, *Generation of Animals* V.8.
2. *Generation of Animals* III.11.
3. *Nicomachean Ethics,* 1111a5, 1135b12; *Eudemian Ethics* 1226b2; cf. Sorabji 1980, 227–256.
4. Torstrik 1875, 425–470.
5. Ross 1955b, 518; Wieland 1962, ch. 16, 271–272.
6. Diels 1882.
7. Bonitz 1862.
8. A detailed defense of this analysis can be found in chapter 9 above and is suggested in Charlton 1970, 106–108.
9. Cf. *de An.* II.4 415b1–3, 415b20–21; *Ph.* II.2 194b1; *Metaph.* Λ.7 1072b2–4.
10. E.g., Simplicius, *in de An.* 119.22–37; Philoponus, *in de An.* 269.26–270.10; Ross 1961, 228; Hicks 1907.
11. Cf. *Metaph.* Z.7 1032b22–32; *PA* I.1 640a27–28.
12. Ross 1955b, 518.
13. Wright 1976.
14. Wright 1976, 78–79.
15. Achinstein 1977, 341–367; Boorse 1976, 70–86.
16. E.g., Lewontin 1978, 114–125; Williams 1966, 3–20, 251–273.
17. I would like to take this opportunity to thank Professor E. M. Curley and an anonymous referee for many valuable suggestions which led to improvements in the argument of this chapter.

12

Theophrastus on the Limits of Teleology

One of the most distinctive features of Aristotle's philosophy of science is its attempt to defend the use of teleological explanations in accounting for a variety of natural phenomena without assuming some sort of consciousness involved in their causation. Such a position attempts to steer a middle course between the Scylla of the reductive materialism of certain pre-Socratic natural philosophers and the Charybdis of a Platonic demiurge. Recent discussions of Aristotle's teleology thus rightly stress its naturalism, and argue that it bears up rather well under philosophical scrutiny.[1] These discussions develop interpretations based on certain paradigmatic uses of teleological explanation in Aristotle, especially those that explain biological generation as for the sake of its end and biological structures as for the sake of some functional contribution to the organism's being or well-being.[2]

Anyone involved with such an enterprise is aware of certain teleological explanations which place serious roadblocks in the way of a fully naturalistic account of Aristotle's teleology. Four of these types of explanation are:

1. The use of the premise that nature does nothing in vain, but always what is best among the possibilities for each organism.[3]
2. Explaining some arrangement or structure being as it is, because it is more honorable thus.[4]
3. Explaining a certain kind of organism's existence, or its behavior, as due to the needs or requirements of some other kind.[5]
4. Explaining reproduction as being for the sake of "participating in what is everlasting and divine."[6]

I do not mean to suggest that such explanations are not, once we understand Aristotle's overall project, consistent with those paradigm cases already mentioned. My own view is that they are. What I want to draw attention to in this chapter is that at least the first two of these modes of explanation were already bothering Aristotle's friend and successor, Theophrastus. That is, there is nothing peculiar to our contemporary way of looking at these explanations which makes them seem problematic. They appeared equally so to someone as close to their source as anyone could be, as I believe the last chapter of Theophrastus' *Metaphysics* 'fragment' demonstrates. Section II examines Theophrastus' worries about Aristotle's use of the first sort of explanatory pattern, and section III examines his qualms about the second.

I

A provocative beginning can be made by drawing attention to a current controversy in evolutionary theory. A popular explanatory strategy among orthodox neo-Darwinians is to construct a design or 'engineering' model of the optimal adaptive strategy for dealing with a certain niche, given certain assumptions about the functional requirements of the organism in question. This model is then compared with the actual structures of the species or genus in question. If the fit between model and structure is a good one, that is taken as presumptive evidence in favor of the hypothesis that the structure is an adaptation, a product of selection for that niche.[7]

Such a strategy has certain inherent dangers. To mention three: First, we nearly always have prior knowledge of the actual structure for which we are developing an optimal design. This leads to the ever-present danger of allowing this knowledge to bias the construction of the model in the direction of the actual. Second, there is no guarantee that a good fit between model and structure tells us anything about its causation. We require independent methods of judging whether a structure is a product of selection, and such methods are hard to come by. Finally, there is such an array of possible subsidiary hypotheses available in evolutionary theory that one might be tempted to bring a rough fit into neat alignment with the real world by invoking a few of these in an *ad hoc* manner; an epicycle here, a change in diameter there, so to speak.[8] None of these problems shows that the use of optimality models will necessarily lead to such methodological pitfalls, but they do show that the dangers are especially great and the excesses especially tempting.[9]

Theophrastus is not, of course, concerned with finding criteria for testing claims about natural selection. But since it can be said without anachronism that Aristotle made vigorous use of optimal design strategies in his explanations of biological structures and behavior, it is legitimate to wonder whether Theophrastus' remarks about final causality in *Metaphysics* 9 are concerned with issues about the scope and legitimacy of such explanations. It is my hypothesis that that is indeed one of his principal concerns.

Twice during this discussion Theophrastus tells us what he is attempting to do. Here is the second of those manifestos:

> We must attempt to find a certain limit (*horos*) with respect to those things which are for the sake of something and which have an impulse toward what is better, both in natural things and in the entire cosmos. (11b25–7)[10]

What we take our task to be here depends on how we interpret the notion of a *horos*. Antecedently, it might mean either that which all teleological processes tend toward (so that our task would be to determine what sorts of things can legitimately be considered goals), or some limitation on the legitimate range of teleology (in which case our task would be to find conditions under which such explanations would be legitimate). The way to determine this issue is to examine carefully the cases of explanation which Theophrastus finds troubling. By so doing we ought to gain some insight into what it is he is looking for.

There is an important limitation on this task, dictated by the nature of the text we are working with. The whole of Theophrastus' *Metaphysics* has, as Ross notes in his introduction,[11] all the marks of an *aporetic* text, akin to *Metaphysics* B of his teacher. It is thus designed to raise problems in need of investigation, and is not to be read as a positive investigation, driving toward a solution to those problems. Nonetheless, this should not be taken to mean the discussion will be a mere logical exercise. One normally raises problems that do present legitimate blockages to theoretical advance. While it is unlikely that we shall learn much about Theophrastus' considered views on the nature of teleology from this discussion, we shall learn a good deal about the sorts of teleological explanations that he found problematic.

It is also true that we inherit a legitimate problem in this respect from Aristotle. There are texts which seem to tell us that certain ranges of phenomena are outside the scope of teleology,[12] and texts within the scientific works, both biological and nonbiological, in which teleological explanations are virtually absent.[13] And yet his few attempts to characterize the nature of such explanations make it difficult to give a general account of how one is to determine whether some phenomenon must be accounted for teleologically.[14]

I propose, then, that we interpret Theophrastus' call for the setting of a *horos* on those things that are for the sake of something as a call for an explicit account of the conditions under which teleological explanations are and are not appropriate. The test of this proposal will be its ability to account for the details of *Metaphysics* 9.

This chapter opens with a list of natural phenomena which seem to violate the maxim that "all things are for the sake of something and nothing is in vain."[15] Theophrastus suggests (11a2–3) that a consideration of such cases leads one to question whether such explanations should be put forward "in all cases without qualification," and indeed raises doubts about whether "nature in all cases desires the best" (11a5–6). Some cases that seem to be troublesome for such panglossian assurances of universal design are:

1. the ebb and flow of the sea (10a28–10b1);
2. dry and moist seasons (10b2);[16]
3. male breasts (10b8);
4. the female emission (10b9);[17]
5. beards, and hair where it has no function (10b10–11);
6. the size of the deer's horns, and the troubles this causes (10b11–14);
7. certain forced and unnatural phenomena such as heron copulation[18] and insects that only live for a day[19] (10b14–15);
8. the nutrition and generation of animals, "for these are for the sake of nothing, but they are coincidental and due to other necessities" (10b16–19).

In addition, Theophrastus questions explanations of the windpipe's position and the location of the blending of the blood by reference to what is honorable (11a8–13). I shall establish in what follows that, in many cases, Theophrastus is discussing specific explanations to be found in Aristotle's scientific corpus. I shall argue that Theophrastus has a keen eye for explanations in which Aristotle appears to have sought a final cause, but prudence would have counseled otherwise. By examining such cases, Theophrastus challenges his readers to articulate the grounds of their uneasiness about these cases.

My first task is to establish the claim that Theophrastus is directing our attention to actual Aristotelian explanations. In the section immediately following, I do so for cases 3 and 6 from the list of 'panglossian' explanations given above. In section III, I do the same for explanations of the placement of the windpipe and heart by reference to what is honorable. In section IV, emboldened by success, I make some speculations as to why the other panglossian cases appear on Theophrastus' list. In some of those cases, while we lack actual explanations, we have clear Aristotelian texts that discuss the phenomena in question. In other cases, the relationship to any specific Aristotelian material is, at best, not obvious. I conclude in section V with some remarks about whether this chapter of Theophrastus' *Metaphysics* sets a trend within the Peripatetic tradition which led to extreme skepticism about the legitimacy of teleological explanation. If it did, it is a watershed document that deserves more attention than it has received so far.

II

Case 3

At *PA* IV.10 688a12 Aristotle takes up a consideration of the chest or front part of the trunk. He argues that in viviparous quadrupeds the breasts are not located on the chest because of its narrowness, which in turn is explained by indicating the locomotive requirements of quadrupeds. In man, things are otherwise:

> In the case of humans, however, due to the breadth of the chest and the need to protect the area around the heart, this place is fleshy, and the breasts become distinct; and these are fleshy in the case of males for the reason given, while in the case of females nature uses them as well for another function, as we often say it does; for it stores the nourishment for the offspring there. (688a18–25)[20]

Aristotle does indeed often argue that the nature of the organism will make use of an organ for two different functions.[21] Typically, one function is treated as primary. In the case presently under consideration, the primary function of the fleshy nature of the chest in general, and breasts which are fleshy in particular, is to protect the area around the heart, which in Aristotelian physiology is the central location of sensory integration and the origin of locomotion. When Aristotle says "nature uses them *as well* for *another* function" at 688a23, he stresses the fact that the female breasts are viewed as *primarily* for this function, and only secondarily as repositories of nourishment.[22]

However, Aristotle is not entirely consistent on this point. At 692a12, the breasts are carefully defined as receptacles – vessels for milk. This fact is then used to explain why creatures that are not nourished by milk lack breasts. Aristotle nowhere faces the puzzling question of why males have nipples, if only the females use their *mastoi* for nourishing offspring. He seems not to have asked why nipples should be required to carry out the chest's primary function.

It is precisely these questions to which Theophrastus is drawing attention. If one is going to define breasts as Aristotle does at 692a12, why do human males have them, equipped with nipples, no less? In this respect, they seem to be quite without a function as male organs. On the other hand, should Aristotle retreat to claiming that their primary function is the protection of the area of the heart, he is left with a functional explanation that fails to account for the specific nature of the structure in question: for nipples are not necessary for the performance of this protective function. Theophrastus has pointed to an explanatory anomaly in Aristotle's biology. As it stands, and no matter which aspect of Aristotle's explanation we stress, male nipples appear to be 'in vain'. Theophrastus' intent, it will be recalled, was to question the universality of the claim that nature does nothing in vain, but always the best among the possibilities. Were one to attempt an adaptationist explanation of male nipples today, one would be similarly open to charges of panglossian functionalism.

Case 6

Aristotle discusses horns among animals with no upper incisors and multiple stomachs in a variety of texts.[23] Theophrastus' remarks seem to make a direct reference to the discussion of *PA* III.2, where Aristotle says:

> Now in those animals to which the emanation of the horns is naturally useless, nature has provided in addition another defense; for example in the case of deer, speed (for the magnitude of the horns and their great number of branchings are more a difficulty than an aid) . . . (663a8–11)

Aristotle's general functional explanation of horns is that they exist for the sake of defense and attack (662b27–28). He denies that deer use them for this purpose, however, which makes him wonder why they have them at all. Indeed, not only are they useless – their density and structure make them a positive nuisance. In discussing the question why, if female deer have no horns, they lack teeth just as if they had horns, he responds:

> . . . one reason is that they [male and female] are both of the same nature, that is, horned; and while horns are excluded from females because they are useless, they are useless to males as well. However, males are less likely to be harmed by them because of their strength. (664a5–8)

The background explanation to the above remarks is complex. Aristotle argues that to be horned is to have a systematic set of hypothetically necessitated structures.[24] Horns are possessed for the sake of defense; the construction of horns uses up a good deal of the bony material used in other viviparous quadrupeds for teeth.[25] Thus they lack upper incisors. This lack of chewing teeth necessitates a more complicated digestive system with, for example, multiple stomachs. Aristotle appears to argue that, even though the female is without horns in the case being considered, the *species* of which she is a member is by nature horned. Thus the rest of the species properties follow from her being a sort of deer.

A peculiarity of the horns of deer is that they are solid, and therefore heavier than horns in other animals.[26] Their greater relative weight is the factual basis for both a teleological and a 'mechanical' explanation for the shedding of horns in deer. By lightening the stag's load it is useful, and it occurs of necessity because of their weight.[27] To suggest that the stag's losing his horns is advantageous, and to further explain this loss teleologically, is to carry the notion that they are not only useless but harmful to its logical conclusion. It also raises a serious question as to why this kind of animal is horned at all.

Aristotle can easily deal with characteristics that are not functional by seeing them as necessary consequences of the possession of functional defining characteristics, or even as species with developmental anomalies relative to some broader class.[28] But two of his premises in this case prevent him from falling back on these strategies. First, he insists that it is of their nature that deer are horned; second, horns are not merely useless in this case, they in fact are harmful. It is simply unprecedented for Aristotle to attempt a teleological explanation for an aspect of an organism's nature which is, at the same time, a harmful appendage that it is better off without. So, in a chapter devoted to specifying what horns are for (662b28, 663b21) it is the benefit of losing them that is the best he can do in this case.

Again, Theophrastus is on target. He has drawn the attention of serious students of Aristotle's science to an argument that forces us to ask whether a teleological explanation is out of place. To indicate a case in which one of an animal's defining features is not merely useless but harmful encourages us to puzzle as to when teleological explanations are inappropriate. In this case we must question the general teleological account of horns, for the deer establishes that they are not in all cases for defense. How do we then explain their presence when they are functionally harmful? And perhaps we must question the extent of nature's functionalism if it first provides a kind of animal with a harmful structure, and then the mechanisms periodically to shed that structure. Pointing to such cases is a fine method of forcing us into *aporia* about teleology.

III

I now proceed to consider the second sort of explanation listed in the introduction, namely, that in terms of the more honorable. It is found repeatedly in Aristotle's biological writings and gives pause to current commentators, as it did to Theophrastus. In regard to this second kind of explanation there will be an added benefit in looking at Theophrastus' comments, for they provide a test of adequacy for present-day interpretations. The sort of explanation in question has the form "X is p rather than q, because it is more honorable thus." Theophrastus offers some examples:

> . . . for wherever the better is possible, there it is in no way lacking; for example, the windpipe is in front of the gullet, for it is more honorable;[29] and the best mixture takes place in the middle of the heart, because the middle is most honorable; and similarly whenever something is said to be for the sake of order. (11a8–13)

It should be noted straightaway that Theophrastus sees these as specimens of a single kind of explanation which accounts for some state of affairs by noting the good it achieves, but on the face of it they are quite different. Moreover, what appears most objectionable in these explanations is precisely that they *do not* tie what is to be explained to some biological good that this feature serves for the organism. As Geoffrey Lloyd argued some years ago, there is the appearance here of the a priori evaluation of a certain location as 'honorable' independently of any biological good that might be achieved by the placing of a structure there.[30]

Again, Theophrastus is referring directly to Aristotelian arguments. While the esophagus is the avenue of food to the stomach, it is not necessary for the sake of nourishment, but rather because the lungs require a common tube which then divides. It must be of some length, which entails there being some distance between mouth and stomach. There must, then, be some passage, in order that the food which enters the mouth may proceed to the stomach, and this is the esophagus (664a12–31). Fish, as Aristotle points out, lacking lungs, and thus a windpipe, have no need of an esophagus.[31]

The windpipe, however, is placed in front of the esophagus, though this causes problems – for food to be swallowed, it must first pass over the windpipe (664b3–6).[32] Nonetheless, the windpipe *must* be so placed: it is for the sake of breathing, which is for the sake of regulating heat in the area around the heart. The heart is, in Aristotle's theory, the source of locomotion and perception, and it is the direction of an animal's locomotion and perception which determines the part to be labeled 'front'. Thus, if the heart is to be properly situated for its cognitive and locomotive roles, it will of necessity be located in a forward position. The lung (Aristotle considered it one bifurcated organ) must be there as well, for it is there in order to cool the area around the heart by means of the inhalation of *pneuma*. Again, breath is conveyed to the lungs by the windpipe, so it must be forward as well. So runs the argument from 665a9–21. But if the windpipe is located in front of the esophagus, the need for the epiglottis becomes apparent: ". . . and this is why nature has doctored up the poor placement of the windpipe, manufacturing the so-called epiglottis" (665a8–9). Aristotle concludes the discussion with the following grand generalization:

> Generally, the better and more honorable is always, wherever nothing else greater impedes it, in the case of the upward and downward more in the upward location, of the forward and behind more in the forward location, and of the right and left more on the right (665a22–26).[33]

If Theophrastus is referring to this discussion, we need first to ask whether he has the reference of 'more honorable' correct. Could not Aristotle be referring to the placement of the heart, which has just been said to be in the front necessarily?

Given that Theophrastus was an intimate of Aristotle for a number of years, it seems unlikely he is confused. The above conclusion is a generalization, but it certainly sanctions any claim that, with respect to any two structures related as before/behind, the more honorable one will be in the forward position, all things being equal. And he has just been discussing the problems created by the windpipe being in front of the esophagus. Aristotle's understanding of the cardiopulmonary system requires that the windpipe be in the front of the neck, even if this requires some doctoring up on nature's part. On the other hand, Aristotle seems to be arguing not that the windpipe is where it is because that *place* is honorable, but rather that what is more honorable is generally found in such and such a place. This is not what Theophrastus has led us to expect. To determine what is going on here we need to fill in some of the theoretical context of these evaluative references to locations. Theophrastus has once more forced us into a thorny thicket of problems.

First, it is useful to remind ourselves of Aristotle's biological concepts of direction, alluded to in our passage. "Upper" and "lower" are defined by function,[34] not by cosmic orientation. Those parts are upper which are toward the mouth and lower which are toward the opening through which waste is excreted (*Juv.* 468a2–4; *IA* 705a30–b6). Therefore, plants "have the upper lower and lower upper" (*PA* 686b35); that is, their functional upper is cosmologically lower. Only in humans is the part which determines what is biologically upward also the cosmologically upper part (*PA* 656a13; cf. *Juv.* 468a5, *IA* 705a30–1).

Likewise backward and forward are defined in virtue of the location of an organism's perceptual faculties, front being determined by the orientation of the sense organs (*Juv.* 467b31–34, *IA* 705b10–13), and right and left are defined by the location of the origin of locomotion – this will be right (*IA* 705b17–18).[35] He concludes his discussion of these basic principles of the investigation of animal motion with the following comment:

> But it is also reasonable that the origins are from these parts; for the origin is honorable, and that which is above, forward, and to the right is more honorable than that which is below, behind and to the left. Yet it is also well to speak of these matters the other way around; to say that it is due to the sources being in these positions that these positions are more honorable than the opposing parts. (706b11–16)[36]

Two quite different inferences are sanctioned here. The first inference is:

Being honorable belongs to what is above, forward and on the right.
Being honorable belongs to the origin of organic activities.
Being above, forward and on the right belongs to the origin of organic activity.

This argument may be a conscious echo of *Timaeus* 45a:

> And the gods, holding that the front is more honorable and fit to lead than the back, gave us movement for the most part in that direction. (Cornford translation.)

As Aristotle states it, the inference is faulty. But the force of it is this: If certain placements for bodily organs are intrinsically more honorable than others, and if those organs which serve as the origins of organic activity are honorable, they ought to be placed in the intrinsically more honorable position. Under the guiding assumption of a nature (or a demiurge) which does nothing in vain, this does seem a plausible pattern of reasoning.

Aristotle, however, goes on to suggest a pattern of reasoning which allows us to give an explanation for *why* we call certain regions of the body more honorable than others. He appears to want to provide a rational grounding for some hitherto unargued assumptions. *This* explanation goes:

Being honorable belongs to anything which is an origin of organic activity. The origins of organic activity belong to certain positions more than others. Thus, being honorable belongs to certain positions more than to others.

Thus, as opposed to the *Timaeus,* movement in the forward direction is not a result of the front being honorable – rather, we predicate 'honorable' of the front *because* that is the direction of locomotion and perception. Aristotle is not denying the truth of the view that the front is more honorable than the back. But he is suggesting that this is not a basic principle of biology; it is to be explained by principles having to do with basic organic activities.[37]

Let us now return to the location of the windpipe. Its location is ultimately a function of the position of the heart. From a teleological perspective, the problem raised by the position of the windpipe is that it is potentially quite harmful. Being in front of the esophagus, food must pass over it when we eat. The epiglottis looks to Theophrastus (and even Aristotle's own language suggests this) like a Rube Goldberg device which, prima facie, could have been rendered entirely unnecessary by the placement of the esophagus in front of the windpipe. In the aftermath of this worry comes our passage about honorable things being placed forward, upward, or to the right. The fundamental issue for an Aristotelian is this: Why is the windpipe positioned in such a way as to require the epiglottis, when, by being behind rather than in front, it could have solved the engineering problem in a simpler way. Aristotle's answer goes something like this. The heart's forward position is explained by its being the fundamental organ of perception and locomotion. It is therefore determinative of what part of the body shall be called 'forward'. Its 'honor' (and that of the other perceptual and locomotive organs) accounts for our willingness to call the forward part of the body 'honorable'. By extension, the windpipe is where it is because of its relation to the heart; it is for the sake of the heart. The esophagus, on the other hand, while a part of the digestive system, plays no digestive role – it is only there because of the windpipe, which is hypothetically necessary. And "the better and more honorable wherever nothing else greater interferes . . . is always in the more forward rather than the more backward position." That is, the better and more honorable organ (the windpipe) is in front of the less valuable organ (the esophagus), because of its functional tie to the organs which lead to our labeling a certain part of the body front.

The *ceteris paribus* clause ("whenever nothing else greater interferes") in Aristotle's formulation can be explained by reference to a passage in *PA* II.14. Aristotle is attempting to explain the relatively greater amount of hair on the backs, as opposed to the underbellies, of viviparous quadrupeds. What is initially surprising is that nature does *not* offer greater protection to the more honorable part (the belly being functionally front, i.e. where the heart is). "Now in quadrupeds the back needs more protection, while the forward parts, though more honorable, are without hair because quadrupeds are bent over" (658a19–21). Here is a case where "something greater interferes" – all things being equal, nature protects the more valuable parts, but when they need protection less than some other parts this tendency is overridden.

In the explanation we have been scrutinizing, the windpipe might have been behind the esophagus, but only on the condition that it would have been biologically worse off the 'honorable' way. Had we no epiglottis, for example, surely nature would *not* have situated our windpipes in front of the esophagus just to insure that the entire cardiopulmonary system were in the biologically most honorable place. It could easily run the windpipe down behind, and extend it over to the lungs on the way down. But given the possibility of an epiglottis (or its analogue in birds and reptiles), there is no reason not to place it in the more honorable place; and nature "is always a cause of the better from among the possibilities."

Theophrastus has again been a useful guide to Aristotle's actual difficulties. His objections to these arguments were not based on their importation of prebiological value judgments into the biological realm but on their *ad hoc* panglossian flavor. "Nature does nothing in vain? But you yourself say the windpipe is placed most carelessly, so that nature must 'doctor things up' with the epiglottis. Even at that, as you note, we still choke on our food from time to time." There is force to this objection, but it is the objection of one biologist to another.

As we have seen, Theophrastus also raises an eyebrow at the suggestion that blood is blended in the middle of the heart because that is the most honorable place for this to occur. Aristotle does indeed refer to the heart as the origin of the blending of the blood (650b29, 686a9) and discusses the purity of the blood in the middle cavity of the heart (667a3–6, *HA* 496b10),[38] but to my knowledge he nowhere in the works now extant develops these ideas in the precise direction suggested by Theophrastus. We can, however, imagine such a development from the following aspects of the theory with which we are provided:

1. It is better, where possible, for the heart to have three cavities, in order that there can be a single, common beginning for the two distinct systems of blood vessels (666b33–34).
2. Where the heart has three chambers, the blood in the middle chamber is finest (*HA* 496b10).

3. The heart is properly located in the middle, front, upper portion of the body, as equidistant from the senses and limbs as possible (665b18–21; 666a13–17).[39]
4. Nature establishes what is more honorable in the more honorable place, where nothing more important prevents it, which is why the heart is where it is (665b20–21).

Thus, Aristotle certainly believed a blending occurs in the heart, that there is a middle cavity, and that its contents are the most pure or fine. It is thus not unlikely that he also said that the blending that occurs there is best, and it occurs there because it takes place in the most honorable location. That is, this is the best place for such blending to take place, and therefore nature has arranged for it to occur there.

Once again, Aristotle is arguing from assumptions about what biological functions are 'most honorable', about which location would be biologically best for the organ which performs those functions, and from his biologically based premise that nature always goes for the best among the possibilities.

And this is just the way in which Theophrastus treats them. He does not object to their normative nature but to the premise that because an arrangement is best, nature will invariably arrange things thus:

> For even if desire is thus, still it is surely apparent that that which neither obeys nor is receptive of the good is great, or rather is much greater. For that which is alive is quite a small part of things, while that which is lifeless is boundless. And among those which are living, their existence is brief, even if it is better. (11a13–18)

That which has desire may go after the good (at least the apparent good), but this is a small part of the natural world. To tighten up this objection, Theophrastus would need to point out that even among things that are alive, many processes are not under the direction of *orexis* (development, for example), and thus not obviously directed toward and invariantly productive of the good. As it stands, it seems only to rule out claims that non-living processes are for the best and happen because it is honorable thus, and the arguments he has been critical of up to this point have made no such claims. They have, however, extended teleology beyond the realm of *orexis,* and that is to extend it beyond the realm where it obviously has a place into a realm in which its place needs to be argued for. Moreover, Theophrastus seems to be pointing out that one must not simply assume that a part's being in a certain location guarantees that it is in the best possible location – this must be shown, not assumed. At first glance, many of Aristotle's arguments appear to make the assumption that the arrangement that is found in nature *must* be the best possible, and it is the task of the biologist to figure out why this arrangement is best. (This is, of course, a valuable heuristic, and lies behind the engineering analysis-optimal design arguments in evolutionary biology – it becomes dangerous only when it is forgotten that it is a heuristic for design hypotheses and not a shortcut to well-confirmed theories.) Take the case of the placement of the windpipe. Aristotle wants to argue that its placement is *poor, and yet* more honorable than any alternative. Perhaps he would have been better off withholding teleological explanation in this case. I have suggested the lines of reasoning that may have led Aristotle to the conclusions he reaches, but Theophrastus' point, understood as a means of raising problems of a methodological sort, is that we have little to go on beyond our uneasiness as a means of determining when such explanations are inappropriate.

IV

Theophrastus' rapid, shorthand list of examples of things inappropriately said to be for the sake of something (10a28–10b20; see above, section I) is rather like Aristotle's list of objections to the theory of Forms at 990b9ff. of his *Metaphysics*. Taken on their own, the examples are nearly meaningless. I have attempted to place them within the context of a debate within the research program of the Lyceum, and to identify the examples as based on actual explanations in Aristotle's biology. In this section I briefly point to the context of other items in this list, though I do not have the space to consider these cases in detail. In all of the following cases my suggestions for a context are far more speculative than those previously made, about which I am reasonably confident.

Case 2

A text in *Physics* II.8 is thought by some scholars to suggest an Aristotelian teleological explanation for fair weather in the summer and rain in the winter.[40] Others feel this is not required by the text and is unlikely, given Aristotle's account of seasonal changes and rainfall elsewhere.[41] I have no desire at present to enter into this debate, but that the text is open to this sort of controversy may account for "dry and moist seasons" being on Theophrastus' list.

Case 4

The female emission (*proesis*) may refer to two quite different phenomena – either the vaginal discharge, which Aristotle mentions at *GA* I.20 728a1–6, or the menstrual discharge (*ta katamēnia*). Aristotle uses the term *proesis* for a variety of physiological emissions, including that of the *katamēnia,* but not of the vaginal discharge, which he refers to as an *ekkrisis.*[42] One other piece of evidence in favor of the *katamēnia* being referred to here is that the referent must be something that could plausibly be given a teleological explanation by Aristotle, and that explanation could also plausibly be called into question by Theophrastus.

The menstrual fluid seems ideally suited, for it is both a residue that is produced of necessity and with a good deal of random variation in its production, which nature nonetheless uses "for the sake of the better and the end" (*GA* II.4 738a10-b9).

Case 7

As I have already noted, the references to the day fly and the copulation of herons are almost certainly to phenomena noted in Aristotle's biology. The day fly could be described as contrary to nature for two quite distinct reasons. Aristotle gives them both in *HA* I.5: "It is peculiar not only in virtue of its way of life, whence also it gets its name, but also because, while being four footed, it is winged as well" (490b1–3). The latter combination of traits violates a universal principle of Aristotle's theory of motion, namely, that bloodless creatures are either footless or have many (i.e. more than four) feet. He avoids this problem by shifting in this passage to talking about points of motion and noting that the day fly has eight of those (four feet, four wings).[43] At any rate, Theophrastus' mode of expression ("as the day fly lives") suggests it is the shortness of their life that is the characteristic to which he is referring. The less said about the copulation of herons, the better.

Case 8

Finally, let me consider that case which is most difficult to understand as a reference to Aristotle. It is 10b16–19 and may be translated as follows:

> And the greatest and most obvious case concerns the nourishing and generation of animals; for these are for the sake of nothing, but are coincidences and due to other necessities. For, if they were for the sake of animals, they ought always to be uniform and unvaried.

Part III. Teleological Explanation

Aristotle's theory of generation is rightly thought to lie at the heart of his teleology. Up to this point, Theophrastus has been challenging explanations that stretch teleology to the breaking point, but this comment seems to challenge the very idea of teleology as Aristotle understood it. And again, if biological generation, among natural processes, is not "uniform and unvaried" – what is?

I cannot pretend to understand what is going on here, and offer only two brief speculations in order to suggest lines of inquiry. It may be possible, first, that Theophrastus is raising the problem of variability in the results of animal generation which is played down in Aristotle's *GA* I-II but comes to the fore in *GA* IV–V. There, variations in the 'temperature' of the semen's heat and the 'coolness' of the menstrual discharge are used to explain the vast range of variation that characterizes the offspring of a given kind of animal. Are these coincidental and due to other necessities? And is this the sense in which generation is not uniform and unvaried? To see a conflict between these explanations and teleology, Theophrastus would need to misunderstand Aristotle on a fairly basic level; on the other hand, if he is only raising issues for a dialectical resolution, there is a more charitable way of taking these comments.

What of nourishment? As with all these examples, the central problem is the telegraphic style – Theophrastus' text gives us no help as to what he might mean by nourishment at 10b17. However, there is a text of Aristotle which does concern a teleological explanation for the existence of nutrients. It appeals to the 'nature does nothing in vain' principle under scrutiny here and is thought by modern commentators to be prima facie dubious on Aristotelian grounds.[44] I am referring to a well-known text in *Politics* I.3, which suggests that plants are for the sake of animals and the other animals are for the sake of man, especially for the sake of nourishment.

> If, then, nature does nothing incomplete or in vain, it is necessary that nature has made all these things for the sake of human beings. (1256b21–22)

This passage uses the very axiom which is under examination in *Metaphysics* 9, and suggests a dubious teleological explanation which is at odds with Aristotle's usual teleological account of why various plants and animals come to be. I am not suggesting that this explanation cannot be made consistent with Aristotle's biological theory when seen in the proper perspective, only that it is worthy of skepticism and dialectical examination.[45]

This suggestion has many problems, the worst of which is that, however much this passage may conflict with Aristotle's biology, it is *not* because the biology suggests that plants and animals come to be 'coincidentally'. The conflict is between an apparent cosmic teleology in the above passage, requiring a design capable of ordering kinds for the sake of other kinds, and Aristotle's typical view that it is the activities constituting an organism's life which are the goal of development and what the various structures produced are for.

Theophrastus has deftly picked out passages in Aristotle that reveal him struggling to explain a structure teleologically, when he might have been better off admitting that such explanations are inappropriate. Has Theophrastus provided a clear statement of criteria for determining the limits of teleological and normative explanations in science? Not explicitly, and given that this is just a problem-raising discussion, that is not surprising. However, when the examples Theophrastus provides and the questions he raises are focused upon, two limiting principles suggest themselves:

1. Neither regularity of production nor universality of a trait within a species is sufficient to establish that something exists for the sake of some end.
2. At some point in the construction of an adaptationist explanation (which no one to this day has very carefully specified) the number of subsidiary hypotheses required outweighs the explanatory value of the design story.

As with a detective novel, there is a point at which the twists and turns of the plot pass beyond the point of being ingenious yet plausible, and become farfetched and implausible. The male's breasts, the deer's horns, the location of the windpipe, have, we might think, passed over that line, however hard it may be to define precisely.

Metaphysics 9 suggests a debate going on during Aristotle's life and after his death about the limits and logic of teleological reasoning. Theophrastus' *De causis plantarum*, a book one might expect to be filled with teleological reasoning, is remarkably devoid of it.[46] We know from Lucretius that an attack on teleology in biology was an aspect of Epicurus' thought.[47] And if we are to trust the reports of Diogenes Laertius and Galen, and suppose that Erasistratus was closely tied to the Peripatetics, we may see the move to rein in teleology in his work. For Galen is often angered by the fact that a man of such obvious gifts could from time to time lapse into saying that certain organs have no function.[48]

The twin concerns of biologists who use teleological explanations permeate Theophrastus' discussion. On the one hand, it is temptingly easy to invent a story which will explain why what one finds to be the case is best. On the other hand, we are always too hasty in concluding that things which are not understood have no function. Galen's anger with Erasistratus is often well founded. And though no one has determined their value, I stubbornly refuse to do away with my tonsils. What was, and is, needed is a methodology for the testing of adaptationist claims. Like so many debates in the history of scientific methodology, the one over the limits and value of teleological explanation in biology is as alive today as it was when it began – in the fourth century BC. And there is evidence that some of Aristotle's admirers, like some of Darwin's, had a tendency to throw the teleological baby out with the panglossian bathwater.[49]

NOTES

1. Sorabji 1980, ch. 10; Nussbaum 1978; Balme 1972, 96ff.; Gotthelf 1987b, 226–54; Cooper 1982, 197–222; and chapter 9 above.
2. The relationship between the functional explanation of existent structures and the teleological explanation of development is a complex issue; see Gotthelf 1975, 277–90.
3. Bonitz 1870, vol. 5, 836b28–37. Bonitz lists twenty-three statements of this principle, and the list is not complete.
4. *PA* II.14 658a22–24, III.3 665a9–26, III.4 665b18–21, III.10 672b21.
5. *PA* IV.13 696b28–32, *Pol.* 1256b15–22; cf. Balme 1972, 96.
6. *De An.* II.4 415a22–b8; *GA* II.1 731b24–732a1; cf. Balme 1972, 96–97.
7. Cf. Beatty 1980, 532–61; and Smith 1978, 31–56.
8. The critics' points are well put in Lewontin 1979, 387–405; Gould and Lewontin 1979, 581–98; Smith 1966, 16–18; Williams 1966, intro. and ch. 9.
9. Lewontin 1978, 125.
10. Lest the reader put too much interpretive stress on the singular 'a certain standard', the earlier passage says 'certain standards' (11a2).
11. Ross and Fobes 1929; cf. Zeller 1962, German ed. 1897, 354n2.
12. For example, *Metaph.* H.4 1044b12 says that eclipses are not for anything; *Ph.* II.7 198a18 says the same of mathematical entities; *Ph.* II.5–6, 8 rules out chance processes; and *PA* III 677a16–18 notes that certain residues are necessary by-products of processes that are for the sake of something. Thus, while these residues presuppose processes that are for the sake of something, they are not themselves for the sake of something.
13. One has to keep in mind the aims of any given work, as well as its subject matter, in dealing with this issue. *HA* is virtually devoid of teleological explanation, but that is because it is virtually devoid of explanation, not because teleological explanations of its subject matter would not be appropriate. On the other hand, the *Meteorology* and *On Length and Shortness of Life* are explicitly explanatory and shun teleology. (The remarks at *Mete.* IV.12 389b28–390b1 refer to material elements of organisms, as is clear when the passage is compared with a similar one at *PA* II.1 646b11–19. Cf. Furley 1983, 73–93.
14. On which see Gotthelf 1975, 1–89.
15. *Metaph.* 9 10a22–23. It is worth noting that Aristotle never speaks this loosely; it is always *hē physis* that does nothing in vain and acts for the good. Given Aristotle's persistent use of the definite article in stating this fundamental biological principle,

I take it to be the nature of this or that sort of thing which does what is best in the circumstances. Thus processes or structures not under the control of the nature of the thing would not turn out best, unless by chance.

16. Theophrastus generalizes from these first two cases to any where what is to be explained are "changes, destructions, and generations which are first one way and then another" (omitting, with Ross, *ē tinos hai prochōrēseis* at 10b1).

17. Either the menstrual blood or the vaginal discharge; I incline to the former. See below, section IV.

18. The account of this subject in *HA* IX.1 609b23–25 is quite fantastic: "Among the herons, one species mounts and has intercourse only with great difficulty. For it both screams and, so they say, drips blood from its eyes when it copulates; and it lays its eggs awkwardly and painfully." As so often with such claims, Aristotle reports it, but uses 'as they say' to remain neutral on its veracity.

19. Discussed at *HA* I.5 490b1–4, *HA* V.19 552b20–23. See below, section IV.

20. Notice that the nipple, which seems clearly designed for the nutritive function, is not mentioned here.

21. E.g., 658b33–659a36 (on the elephant's trunk); 671b1–3 (kidneys); 684b1 (lobster claws); 687a15, b5 (hands); 689a6–8 (penis); 690a1–5 (tails); 691b8–9, 25 (crocodile's upper jaw).

22. I am assuming that the presence of another function is not intended to exclude the first function. I assume this because (1) women's hearts also require protection, and (2) in this pattern of explanation Aristotle typically assumes that the second function is additive in this way. So, for example, the fact that the elephant uses its strange nose as a hand does not change the fact that it is a strange nose (658b33–659a36). An exception is the lobster's claw (684a31–b2), but it is anomalous in a variety of respects. That there are primary and secondary functions is clear from the way such cases are described, i.e., "x is for a, but nature makes use of it for b."

23. *APo.* II.14 98a17–19; *PA* III.2 663b33–664a8, 14, 674a31–33; *HA* II.17 507a33–507b12.

24. That is, if they had horns, they would have these other features. The direction of the implication is crucial, because Aristotle knew that having few teeth and many stomachs did not imply that a creature was horned. Cf. *HA* II.1 501a14; *PA* III.14 674a31–34.

25. *PA* III.3 663b25–664a3.

26. *HA* II.1 500a6–7, *PA* III.3 663b12–13.

27. *PA* III.3 663b14.

28. Cf. *PA* IV 677a16–18. All moles can be viewed as developmentally anomalous viviparous quadrupeds (*HA* I.9 491b28–34), as can seals (*HA* 566b27–567a14).

29. Here 'more honorable', as the next example shows, refers to the placement of the windpipe, not to the windpipe itself.

30. In 1966, 52–61. Cf. *Timaeus* 44d–45b – clearly the origin of Aristotle's use of this language.

31. Cf. *PA* III.3 664a3–36.

32. Aristotle's vocabulary can be confusing. When necessary, he distinguishes the larynx from the trachea; but he often uses the two terms interchangeably. I have followed Peck's expedient of using *windpipe* as a translation for both to avoid

confusion. Theophrastus is discussing the trachea, though he uses the term Aristotle uses for the larynx.

33. Cf. *HA* I 494a20–b1; *Juv.* 476a16–b10; *IA* 4–5.
34. Cf. *IA* 4705a28–b5; *PA* IV 680b14–15, 683b19–25.
35. As so often, the language is Platonic but has been given a naturalistic meaning by shifting the basis of evaluation to biological considerations. This radical shift in basic assumptions seems insufficiently appreciated in the work of Lloyd and in Byl 1980, 210–51.
36. Michael's reading of this passage (*in MA* 144.5) is confused, and makes the argument appear circular. Cf. Preus 1981, 155–56.
37. I have a better understanding of this argument thanks to the persistent proddings of Malcolm Schofield and Richard Sorabji during the Theophrastus Project Conference at the University of Liverpool, March 1982. This does not imply their agreement with my current interpretation.
38. As Ross notes (*Theophrastus, Metaphysics* 72). Compare *HA* 496a20–23, *De somno* 458a17. For a recent account of the problem of determining the reference of *hē mesē koilia,* see J.R. Shaw 1972, 335–88.
39. Cf. *Juv.* 468b32–469b8. Lloyd's account here (*Polarity* 53) is, I think, confused. He quotes *PA* 665b18ff. That the heart is in the middle in virtually all animals that have one, except man. As it is never thought of by Aristotle as on the right, *PA* 666b6 cannot be considered a special argument to show why it is not on the right in man, as Lloyd suggests. The right side is more honorable, because the more honorable is whichever side originates motion – but this normative claim can only be used to explain facts having to do with motion.
40. Cooper 1982, 29–30, nn22, 23; Furley 1985, ch. 12.
41. Nussbaum 1978, 94; Gotthelf 1976, 240–42; Charlton 1970, 122–23.
42. This admittedly proves little, for virtually everything that Aristotle refers to as a *proesis* he also refers to as an *ekkrisis,* so that which term was used in a given instance is probably a matter of chance. Cf. *Bonitz* 1870, 228b20–28, 637b44–53.
43. This is consistent with *De incessu*'s basic assumptions, which are that blooded things move on four points or less, and bloodless on more (704a10–13; 707a16–20). However, the dayfly does seem to be evidence against the claim of *IA* 14 713a 27–29 that "of the bloodless creatures that have feet, none are four-footed, all are many-footed." Professor Balme has suggested that this is evidence that the dayfly was not part of Aristotle's experience when the *IA* was composed.
44. For two very different attempts to deal with it, cf. Cooper 1982 and Balme 1972, 96.
45. Cf. *PA* IV 696b26 on sharks' mouths; and see Balme 1972, 96.
46. Professor Luciana Repici-Cambiano's comments on this paper in Liverpool stressed Theophrastus' use of analogy between plant and animal structures and of the principle of reproduction according to type, in order to indicate that Theophrastus was working within the teleological tradition of Aristotle. I have two responses to this. First, I don't wish to deny that Theophrastus allows for teleological explanation in principle – only that, at least in botany, he is careful and guarded in their use. Second, the use of the above patterns of reasoning, in the absence of the patterns A for the sake of B, or A on account of the better, does not establish her point. For Lucretius uses the patterns she mentions and denies the legitimacy of teleological explanation.

47. Cf. *De rerum natura* 4 823–858.
48. Cf. *De naturalibus facultatibus* 1.13, 35 and 2.4, 91, 132; *De usu partium* 4.15 and 5.5 (Kühn 1965). Galen consistently accuses Erasistratus and even more his followers, of paying lip service to the principle that nature does nothing in vain, while in practice denying that various organs are for the sake of anything. On the connection between Erasistratus and the Peripatos, cf. I.M. Lonie 1964, 426–43; *DL* V.57.
49. This is a revision of a paper first presented at the University of Texas, Austin, in 1982 and at the Theophrastus Project Conference at the University of Liverpool in 1983. I would like to thank all those with whom I discussed this topic on those occasions, and especially Michael Frede, David Furley, Luciana Repici-Cambiano, Malcolm Schofield, and Richard Sorabji. I would also like to thank Geoffrey Lloyd for written comments, and the Trustees of Harvard University for providing support (through a fellowship at the Center for Hellenic Studies) so that revisions could be made.

13

Plato's Unnatural Teleology

In a number of later dialogues, Plato contrasts two sorts of accounts for features of the natural world.[1] One would account for the pattern of the visible world's changes by invoking chance, spontaneity or blind necessity, *and nothing else,* as the responsible force. The other insists that an intelligent maker or craftsman is the truly responsible agent. Plato encapsulates the former well in these lines from the *Laws:*

> Fire, water, earth and air, all of them they say are by nature and chance, while none of them is by craft. And again, the bodies made from these, earth, sun, moon and stars they say have come to be due to these [nature and chance], being entirely without soul. Each one, moving about among each of the others by chance of its power, hot to cold, dry to moist, soft to hard, and all what-soever have been blended by the blending of opposites according to chance from necessity, by which has been concocted a harmony which is somehow fitting. (889b1–7)

He has the Athenian endorse the latter view three pages later:

> And so judgement and foresight, wisdom, art and law would be prior to hard and soft, heavy and light; and the great and primary works and actions just because they are primary, would be those of art; those of nature and nature

The first half of this chapter was originally a paper delivered to the APA Pacific Division meeting in 1981. An earlier version of the whole was read at The Catholic University of America, November 1983, and at the Princeton Ancient Philosophy Colloquium, December 1983. The final rewriting benefited from many discussions and suggestions on those occasions, most especially from David Furley's commentary at Princeton. I was also helped by the written comments of Alexander Nehamas and Joan Kung; and from discussions with Anne Carson, Mary Louise Gill, Areyh Kosman, and Deborah Roberts. The penultimate draft was written while enjoying the varied pleasures of a junior fellowship at the Center for Hellenic Studies.

herself – this very thing which they mis-name – would be secondary, having its
origin from art and intelligence.

The idea of the "natural" world as *unnatural,* as the product of a *technē,* is a
stable feature of Plato's later thought and had momentous consequences for
the history of natural philosophy.[2] Robert Boyle, a leader among the British
'mechanical philosophers' of the seventeenth century, looked back self-
consciously to Plato in his *Disquisition on the Final Causes of Natural
Things:* "The provident Δημιουργός [craftsman] wisely suited the fabric of
the parts to the uses, that were to be made of them: as a mechanic employs
another contrivance of his wheels, pinions, etc., when he is to grind corn with
a mill."[3]

The world viewed as the product of a good and benevolent craftsman was
one of two aspects of Darwin's formal education at Christ College, Cambridge which he looked back upon with approval (the other also had Greek
roots, Euclidean geometry). Indeed, as one traces the numerous versions of
Darwin's argument for natural selection, from its first formulation in the late
1830s through the last edition of *On the Origin of Species* one sees a palpable
struggle to free himself from the implications of this picture.

The tradition begins in a familiar passage in the *Phaedo,* and so shall we.
In it, Plato provides two models of explanation which he clearly feels are
preferable to those put forward by the "natural investigators." One of these
types of explanation is teleological in nature; the other uses forms as *aitia* of
coming to be and being. The *Phaedo* self-consciously announces Socrates'
failure to develop the former, and to integrate it in any way with the latter.
During Plato's middle and later period there is a persistent exploration of a
model of skillful craftsmanship, a major theme in the *Gorgias, Cratylus,
Republic X, Timaeus, Statesman, Sophist,* and *Philebus.*[4] The central question of the second half of this paper is this: to what extent, and in what ways,
do these explorations help Plato develop a more integrated theory of scientific explanation?

I

In an allegedly autobiographical digression, Socrates tells of his initial enthusiasm for, and ultimate rejection of, certain attempts to provide explanations of generation and destruction. These accounts had consequences that
left him confused even about things he had once thought he understood. At
this point he reports, "I now rashly adopt a different method, a jumble of my
own, and in no way incline toward the other" (97b6–7, Gallop trans.).[5]

Socrates appears about to introduce the safe (100d8, e1) but simple-minded (100d3–4) form of explanation, which, however, is not introduced until 99b4. What interferes is an apparently parenthetical discussion of Anaxagoras, which puts off the presentation of the method of explanation by hypothesized forms for two pages.[6]

I want to look carefully within the parentheses with two primary questions in mind. What can this passage tell us about Plato's views on what a fully adequate account of a feature of the natural world should look like? Second, what is the significance of the placement and style of this passage for an evaluation of other discussions of teleological explanations in the Platonic corpus?

Professor Frede has pointed out that this passage exploits a distinction, integral to the moral/legal contexts in which it arose, between *to aition,* the agent responsible for a state of affairs, and *hē aitia,* that in virtue of which the agent is responsible, which may be called the reason why.[7] In legal contexts, this would be the distinction between the accused and the basis of the accusation. The doctrine attributed to Anaxagoras is that intelligence is the agent responsible for orderly arrangement and all else.[8] But at 97c6ff., discovery of the reason why each thing comes to be, is, or passes away as it is or does is said to depend on discovery of why that particular arrangement is *best.* Likewise, Socrates had hopes that after Anaxagoras had said whether the earth was flat or spherical,[9] he would set out in detail the *aitia* and the *anankē* of it, which would be a matter of showing that it is better to be this sort of thing (97d8–e3; cf. 98b1–4, 99c5–7). Throughout *Phaedo* 97–99 intelligence is the responsible agent, while a certain state of affairs' being good (better, best) is said to be the reason why the agent brings that state of affairs about. Further, accounts that make reference to intelligence and the good are contrasted with 'mechanical' explanations – the former provide the true explanation, though their ability to bring about appropriate states of affairs is dependent on the operations of the relevant physical processes.[10]

The operative presupposition that accounts for this distinction comes out clearly in the following comment, revealing to us the nature of Socratic expectations for Anaxagorean *Nous:* "For I never supposed that someone who said these things to be ordered by intelligence would offer any other cause for them than that these things are best just as they are" (98a7–b1). Let me encapsulate Socrates' presupposition in the following formula:

P   If intelligence bestows a certain order on something, that thing has that
    order *because* its having that order is best.[11]

What this hypothetical formulation is intended to stress is the conceptual link in Socrates' thinking between intelligent agency and the explanatory efficacy of goodness. Only intelligent agents bring about certain states of affairs *because* they are good, though good states of affairs may arise by chance. Aristotle encapsulates the point neatly in a fragment of the *Protrepticus:* " . . . something good might come about by chance; but in respect of chance, and insofar as it results from chance, it is not good" (Fr. 11).

Aristotle, like Plato, will only allow the good outcome of a process to explain it if that good outcome was somehow responsible for the process. They differ, of course, over the issue of whether an intelligent agent is the only sort of agent that can initiate changes for the sake of a goal. But they agree, I would argue, that some such agency must be involved if explanations by reference to the goodness of the outcome are to be legitimate.[12]

P maintains the *to aition/hē aitia* distinction in the following way. Intelligence, conceived of as productive of a certain state of affairs, is its *aition.* That state of affairs, *identified as best,* and therefore as the outcome desired by intelligence, is the *aitia* – the reason why – for that production.

Confirmation that *P* adequately captures Socratic presuppositions on this subject comes from an examination of his distinction between true causes and the things without which they wouldn't be such. Socrates chooses a timely example to explain the distinction. Why does he remain in Athens, though he is about to die? It isn't a matter of constraint – he could easily flee.[13] Nor is it simply a matter of pointing out that, given the way his bones and sinews are arranged, he could hardly do anything else. No – he has an opinion that remaining is *good,* and he has chosen to remain because it is good that he do so.

Socrates' characterization of the 'careless' account of his actions also stresses the role of both intelligence and 'what is best' in the preferred account.

> If someone were to say that without having such bones and sinews and what-ever else I have I would not be able to act on my judgements, he would speak truly; but to say that I do what I do *because* of these things, and do these things with intelligence, but not by means of the desire for what is best, would be an extremely careless account. (99a5–b1)

Commentators regularly note the carelessness Socrates finds in saying "*be-cause* of bones etc." and "*with* intelligence."[14] But they ignore the fact that Socrates is drawing it to our attention that these accounts fail to make reference to the desire for what is best. But the above remarks stress its importance. In fact, that those who offer the careless account leave this out altogether may explain why they are so careless as to reverse the true order of priority between intelligence and physical systems. By leaving out of account what is best, they ignore the fact that intelligence is intentional; and thus they will fail to realize that the crucial agency involved in bringing about *this* state of affairs (i.e. the one that is best) is not the physical processes involved, but intelligence.

This passage is not, of course, putting forth a theory about human action per se, but rather a perfectly general thesis about causal attribution. This is made clear as Socrates goes on to note that various theories of why the earth came to be and remains where it is make the same error of taking the physical preconditions of its becoming or remaining where it is to be the actual cause. Later, in a craftily hypothetical mode, Socrates claims to have been persuaded that *if* the earth is a sphere and in the heaven's center, *then* the mere uniformity of the heavens would insure its remaining.[15] Such an account in no sense *competes* with the teleological; rather, it provides the appropriate answer to the question, by what means does intelligence accomplish this good?

The radical discontinuity between the Anaxagorean excursus and the rest of the exploration of the *aitia* of generation and destruction is clear, and clearly self-conscious.[16] At the same time, there is no obvious shift in philosophical motivation. The entire discussion in 96a6–106c9, where its results are applied to the issue of the soul's immortality, is governed by the requirement that a general examination of the reason for coming to be and passing away (95e9–96a1) be carried out. Throughout, Socrates is concerned with answers to the very general question, why does each thing come to be, pass away and exist (96a9–10)? To have such an answer is to know what's really responsible for each thing (96a9). It is this knowledge he pursues in natural investigation (96c7–97b7),[17] in the book of Anaxagoras (97b8–99c6)[18] and in the idea of form-participation (99d1–105c11).[19] There is no hint that the question has changed, nor that different types of answers will be required either for different domains or for different questions. The sort of *aitia* hoped for in Anaxagoras, and those which occupy Socrates' attention from 100a onward, both attempt to substitute for, and avoid basic problems of, explanations provided either by common sense or by the 'natural investigators'. Both lay out stringent, though different, Socratic constraints on what can legitimately be said to be responsible for a state of affairs.[20] Both provide preferred responses to *dia ti* questions. And most importantly, the method which makes use of hypothesized forms is introduced as a "second best voyage in search of the explanation" (99d1),[21] implying a single search for adequate explanations in general.[22]

Thus we are left with a continuous background of explanatory concerns and motives, yet two radically different accounts of explanation each with its own claims to superiority. Faced with this fact, commentators have tended to polarize around two extreme positions. At one extreme is the view that the teleological parenthesis is of no significance to the rest of the dialogue; on this view, Socrates' claim that the hypothetical use of the theory of forms is a 'second best' is a bit of characteristic irony.[23]

Opposed to this are attempts to find, hidden away in the *deuteros plous*, (second best voyage), teleological explanations of some sort.[24] Neither strategy works. There is no evidence for the latter position.[25] Against the former, one needs to consider the following facts in the context of the characteristic care taken by Plato over the structure of a dialogue. First, the philosophical intelligence of the Anaxagorean excursus, in combination with Socrates' impassioned expression of the need and importance of explanations which make use of intelligence motivated by the good, speaks for its importance. Second, the intrusiveness of the passage appears clearly intentional. Third, even as Socrates "takes to the oars," he criticizes those who say nothing about "the good or binding, that genuinely does bind and hold things together." And, with hindsight, of course, we know that the developed use of the theory of forms for various philosophical purposes did not lessen the importance of teleological explanations in Plato's system.

I propose to take at face value both the continuity of concern to find a general explanation for the world of generated things throughout, and the clearly flagged intrusiveness of the Anaxagorean discussion. Once one does so, very natural *comparative* questions arise, questions concerning the relative virtues and shortcomings of various forms of explanation.

There is, for example, a clear preference for intentional/teleological explanations in certain explanatory contexts. Repeatedly (97c2, c5, 98a7, 98b15, 98c1, 99c1, 99c56) Socrates formulates his vision of a noetic *aitia* as an explanation for the *order* that we find in the world. *Nous* is an *ordering* cause, and chooses a certain order because it is best. This is not in itself surprising, in that this was just the role Anaxagoras himself claimed for Intelligence. Yet, no such concern is in evidence during the discussion of the safe, simple explanations which explain something's coming to be beautiful by participation in Beauty Itself. And this would appear to be an inevitable shortcoming of the safe form of explanation. The appearance of an order and pattern in the world's comings and goings is left inexplicable. Or, to put it in a manner Aristotle was fond of, given the theory of form-explanation in the *Phaedo*, we will *still* need a theory of why things come to have the features they do as and when they do.[26]

It remains true throughout Plato's philosophic life, in dialogues as diverse as the *Republic, Timaeus, Philebus,* and *Laws,* that intelligence is invoked to explain the *order* and *unity* in a potentially disordered and dis-integrated world. This is one clue to the centrality of craftsmanship as a metaphor for divine intelligence.

Another curious difference between the two sorts of explanations presented here can be brought out by following a clue quietly dropped by Socrates early in the discussion of form participation explanation.

When Socrates turns to the *deuteros plous,* he seeks agreement from his interlocutors that certain things exist 'in and of themselves'. Among these things he includes 'good' (100b6). This is not surprising, of course. This passage relies on the easy agreement obtained at 65d6 for the theory of forms, and Beauty and Goodness were among the forms mentioned there as well. But immediately after discussion of a theory in which the goodness of a state of affairs was said to be its *aitia,* a theory reluctantly abandoned by Socrates, the use of the good to exemplify the *deuteros plous* has curious implications. Here is a theory in which the good, or alternately participation in it, is once more said to be an *aitia.* It is instructive to see why Socrates (rightly) doesn't see this as a substitute for his preferred teleology.

What can be explained about a thing by citing its participation in the good itself on its own? Only this, that it happens to be good. But Socrates had much grander hopes for a theory which used Mind bringing about various arrangements because they were good. In each case, goodness ought to account, not only for the *goodness* of a state of affairs, but *also* for that state of affairs itself – that is, we ought to be able to say, citing its goodness, why intelligence brought *that* about (e.g., brought it about that the earth is a spherical thing).

By comparing the two examples Socrates has given us of how the good can be an *aitia,* we have isolated a crucial ingredient in the intentional/ teleological accounts. Explaining by goodness is not just one more explanation of a feature (namely goodness) possessed by a number of particulars: it is a way of explaining why particulars possess the *other* feature they do. We may wonder, then, whether Plato ever considered form participation as an adequate account of why a particular or sort of particular can be said to have some feature or other.

Now the issue of how goodness is related to other features of the world is one which Plato explores from a variety of directions. The *Republic's* analogy between the sun and the good is one such exploration,[27] the *Philebus* in its entirety is another. Whether these explorations constitute a linear development or are mutually consistent I am not prepared to say. But they all in their way deal with the issues that arise when the Anaxagorean excursus and the *deuteros plous* are treated as components of a single discussion of explanation.

I wish to consider the *Timaeus* as another such exploration. Its affinities to the Anaxagorean component of the *Phaedo* have been noted since ancient times, at least in a general way: a craftsman uses his intelligence in order to produce a good order.

But with a more fine-grained picture of the *Phaedo*'s account of explanation, a richer understanding of this relationship is available to us. In particular, the *Timaeus* appears to develop a theory of explanation in which the distinction between forms and the world of perceptible particulars is an aspect of an intentional/teleological account of the world. The first and second voyages have been united. If this appearance is not deceptive, such unification must entail an account of 'good intentions' explanations and how they relate to an account of the perceptible world in terms of separate forms. Many questions are opened up: what has happened to the distinction between true causes and the means of their operating? Are forms, Intelligence and the good *all aitia?* What is the nature of the good the Demiurge is seeking to achieve? Behind these questions, I shall argue, is one basic one: Why did Anaxagoras' *Nous* become a divine *craftsman* in Plato's later thought? A reasonably detailed answer to this last question goes a long way toward explaining the differences between the account of scientific explanation in the *Phaedo* and in the *Timaeus*.

II HUMAN CRAFTSMEN

The *Cratylus* considers the giving of appropriate names to be a craft. The good rhetorician is, according to the *Gorgias,* just like other craftsmen (503e1). The maker of good laws is a practitioner of statecraft, a craft parallel in many ways to weaving (*Statesman,* passim). And, as we've seen, the divine intelligence which is responsible for our world having the character it does is also a craftsman (*Republic* VII 530a6, X 596c4, *Laws* X 889–906, *Timaeus* passim, *Philebus* 26e5, *Sophist* 262b5–c4). The *Republic* is already toying with the idea that the natural world is the product of a craftsman;[28] and the later dialogues consider it wrongheaded to treat the products of nature as anything other than *craft* products. Looking carefully at what Plato imagines to be involved in the production of a craft product is thus an integral part of understanding his philosophy of nature.

As a focal text, we can do no better than this characterization of the craftsman in the *Gorgias:*

> Come now, the good man who speaks with a view to the best, surely he won't speak at random, but will look to something? He will be like all other crafts-men; each of them selects and applies his efforts looking to his own work, not at random, but so that what he produces will acquire some form. Look for instance if you like, at painters, builders, shipwrights, all other craftsmen – whichever one you like; see how each of them arranges in a structure whatever he arranges, and compels one thing to be fitting and suitable to another, until he composes the whole thing arranged in a structure and order. (503d6–5404a1; Irwin trans., with modifications)

This protean passage makes note of five distinctive features of the crafts-man's activity.

1. Craftsmen proceed by looking to a paradigm, a form, an idea or product.[29] Indeed, to use the language of paradigm and likeness or imitation is simply to use the natural language of craftsmanship. But lest we imagine the image of 'looking to' as pictorial, it is relevant to recall that the 'that-which-is k' locution *is* substitutable for any of the above names of the craftsman's intensional object. The requirement that a craftsman look to a paradigm insures that his activity is, as the above passage stresses, orderly rather than random. It does not by itself insure that his actions produce the best possible product: the *Cratylus* warns against using a faulty paradigm (389b1–3), and the Demiurge of the *Timaeus* fortunately looked to a timeless rather than a changing paradigm (28a6–b2).

2. There is no suggestion that such copies or imitations would arise *without* the activity of the craftsman. Nor is it suggested that the form or paradigm of the craft product is an *aitia* of its likenesses, copies or imita-tions. Within this model – that is, when the dialogues recount a discussion of craftsmanship – the language of communion and participation to describe the relationship between what-k-is and the many (sorts of) k's is absent. The *Cratylus* likes "placing the form in the materials" (389c1, 389c6–7, 389c9, 389d9–6, 389e1–3, 390b1–2, 390e3–4); above, we have the craftsman placing, sometimes compelling, things into proper order, which entails hav-ing a certain form. In every case, being like a form is not something which just happens; it is the result of a goal-directed productive activity. Inter-estingly, the goal is never to make a good copy: making a good copy is a means to accomplishing some (other) good.

3. If a craftsman must *look to* a paradigm (which may simply mean that he must know *what it is* that he is making), so must he *work with* materials.

Becoming a likeness of a form is not like becoming warm through being acted on by a very hot object; it is a matter of materials being structured, organized, and arranged. An unorganized, disintegrated plurality is compelled to become "a whole thing arranged with structure and order." In fact the quoted passage introduces a section of the *Gorgias* in which Socrates suggests that just as it is the physician's task to restore or maintain structure and order in the body, so it is the good rhetorician's task to restore or maintain structure and order in the soul (504a3–e4).

This element of the craft model recalls that it was in contexts where the order that prevailed in nature required explanation that Socrates found the idea that Intelligence was its cause and the goodness of that order its goal so compelling. Socrates spoke reprovingly of those who forgot that "the good and binding truly does bind and hold things together" (99c5–6). This reproach echoes throughout the *Republic*. At 462b1–2 we are told that the greatest good for a state (whatever it turns out to be) "binds it together and makes it one, its greatest evil is whatever fractures it and makes it many instead of one." The breakdown of the good *polis* begins when it becomes two rather than one (551d5). The fact that justice is each part of a state or soul doing what is naturally its own is always a theory about the harmonious order achieved by a soul or *polis* being just.[30]

4. This speaks (briefly) to the issue of the nature of the *ordering* materials receive. But it is equally central to craftsmanship that it is constrained by the fact that it is an activity of ordering *materials,* and these materials are a given, in two distinct ways. First, the nature of the craft product constrains the *choice* of appropriate materials: knives must cut; making them of soft or crumbly material won't do. Second, whatever material is used has a nature of its own: the craftsman cannot do anything he likes with his material, but only what it is capable of being compelled to do. The *Cratylus* compares the namegiver to a smith or a carpenter. If one is to produce names, awls, or shuttles, one must use letters, iron or wood (whether Greek or foreign is not relevant). If one is given letters, iron or wood, only certain sorts of copies can be made (387e–390b5). In neither case need we imagine that the given determines a *unique* choice; but it *constrains* the craftsman's choices and actions considerably.

5. Finally, Socratic discussions of craftsmanship present a curiously ambiguous attitude toward the good intended by the craftsman's work. This ambiguity results from a distinction which periodically emerges within the craft model between the person who *directs* or *oversees* production and the producer himself. The maker of a shuttle produces an instrument for the weaving of other things; it is the *weaver* who will know what a good shuttle

ought to be like, and will use this knowledge to guide the actions of the carpenter. In such cases, the user of the instrument is said to have knowledge of what a good instrument consists in and to direct the builder (*Republic* X 601e–602a, *Cratylus* 390b–d).[31]

The idea that the good achieved by craftsmen is instrumental is at times subordinated to a quite different notion of goodness, one which "faces in the other direction," so to speak. The very existence of the craft product, because it represents the triumph of order, unity, proportionality, and harmony over their opposites in a given domain is viewed as a good in itself. In the passage with which we began, for example, the usefulness of the craft product is not discussed, for the production of a good soul, the focus of the discussion, is not measured by its instrumentality but simply by the unity and harmonious order of its parts. This counterentropic concept of goodness is relative to the random, uncoordinated dis-integration that would exist in the absence of the goal-directed intelligence of the craftsman — relative perhaps to that world of universal flux which, surprisingly, the world we live in is not.[32]

III DIVINE CRAFTSMEN

It is not news that the *Timaeus* fulfills the fondest wishes of the Socrates of *Phaedo* 97–99. But while this is often noted in a general way, the comparison between the hope and the fulfillment is seldom looked at in detail. I now propose to do just that. The first order of business is to establish that *Phaedo* principle *P* is in place, and to explore the rich theory of causality in the *Timaeus* against the background of the *Phaedo*. Then I wish to explore in some detail the influence of the model of craftsmanship just discussed on the role of intelligence in the *Timaeus*.

Timaeus opens his portion of the feast being served up to Socrates by stating the reason why the framer of the entire universe did so.

> For the god, wishing all things to be good and nothing to be bad in so far as possible, took over everything which was visible – not at rest but moving in a discordant and disorderly manner – and led it from disorder to order, judging this to be in all respects better. (30a2–5)

The explanatory role of the good to be achieved by a state of affairs coming about is no longer expressed in the language of *aitia*. The *aition/aitia* distinction is reserved for the divine craftsman and the necessary motions of materials or for propositional accounts of their respective causal functions. The typical explanation has it that the divine craftsman uses or persuades various unintelligent cooperative materials to bring about a certain state of affairs, *in order that* some good is achieved, or *for the sake of* some good.[33] Thus the divine craftsman of the *Timaeus* acts with intelligence, and what is thus brought about does so *because* that state of affairs is good – the best, given the possibilities. Which is to say, *Phaedo* principle *P* is at the heart of Timaeus' plausible story about the cosmic likeness of the Living Thing Itself.

Whereas the aetiological role of the good in the *Timaeus* is virtually always expressed by prepositional phrases or final clauses expressing purpose, there are two sorts of causal agents reference to which is taken to be essential to a fully adequate explanation of any stable feature of the world. At 68e7–8 these explanations are referred to as the divine *aitia,* which makes reference to intelligence or the craftsmen of beautiful and good things as a cause (46e4, 48a2), and the necessary *aitia,* which makes reference to whatever produces in a random and disorderly fashion in the absence of the divine *aitia* (46c7, d1, e6, 76d6). The latter are cooperative causes used (46c9, 76c6) and ruled through persuasion (48a2) by the former.

The compatibility of these two "agencies" and the consistency of the idea of a "necessity" which can be ruled and persuaded and which is equated with chance when not so ruled was persuasively argued some years ago by Cornford, followed by Morrow, and others. The brilliant Epilogue to *Plato's Cosmology* reveals the extent to which such ideas were a legacy bequeathed to Plato rather than inventions of his own.

But Cornford, in attempting to avoid the idealism of earlier accounts of this distinction, was misled into positing two *realms* in the cosmology of the *Timaeus* corresponding to this distinction, and Vlastos has recently followed.[34] These authors both imagine that the Demiurge had a better world than the one actually produced in mind, that the inherent powers of the world's basic constituents were recalcitrant, that the Demiurge was thus forced to compromise with his ideals, and as a consequence this world has an irreducible realm where necessity reigns, unpersuaded by intelligence.

The first premise in this argument is crucial, for it determines what will count as evidence for the others. If one postulates a world quite different from our own as the Demiurge's goal, then the world we see necessarily falls short, and one might look to the distinction between the two *aitia* as an account of this.

But in none of the statements of the Demiurge's aims is any goal mentioned other than to bring the maximum order and perfection possible to the materials at hand. And it is consistently maintained that this is achieved.

The central portion of Timaeus' story, concerning what occurs of necessity, drops copious hints that necessity is an aspect of every part of the Demiurge's construction and that within that construction it is always a servant of intelligent ends.

First, the random flux described in the language of chance and disorder is explicitly described as what the Receptacle, *absent intelligence,* would be like (48b, 53b, 69b). This suggests that, even as an account of the physical world, Plato could not buy the ontology of radical flux described at *Theaetetus* 179–83 and *Cratylus* 440. It is rather an ontology of a world uncontrolled by intelligence working for the good.

Second, "the productions of necessity" rely throughout on intelligent design. Only the random traces of the elements would occur in the absence of intelligent design, but much, much more than that is described at 46a–c and 53c–68d.

Finally, the description of the necessary powers and properties of the physical world as *sunaitia* is a give-away.

The necessary causes in the *Timaeus* are *always* the inherent necessities possessed by the materials at hand, used or persuaded by divine intelligence "to lead the greatest part of the things that come to be to the best" (48a3). It is important to stress that it is the material necessities that are doing the leading here. The role of intelligence is clearly circumscribed. Plato does not conceive of intelligence as superimposing *other* sorts of activities on a recalcitrant matter with its own — intelligence uses those very material powers, insuring that they shall *work together* for the best result.

> When our creator made our heads shaggy with hair, he used the aforementioned causes, while reasoning that this rather than flesh ought to be the covering around the brain for the sake of protection. . . . (76c5–d1)

Among "the aforementioned causes" is the necessary behavior of thin skin when acted on by the heat and moisture necessarily emanating from the brain. "Nail was crafted by these agents, but due to the most responsible reasoning for the sake of the fashioning of the things which were to be later" (76d6–8). Here then are the necessary results of drying on a compound of sinew, skin, and bone.

Such passages indicate clearly that a *sunaition* is the physical *agency by means of which* intelligence achieves good ends. Plato doesn't conceive of *nous* superimposing other kinds of activity on those of matter, but as insuring certain specific interactions will take place among all those possible, namely, just those which will cooperatively produce the best possible cosmos. His model is of a reasonable counsel who accomplishes his ends by *persuading* various agencies to operate cooperatively, according to a plan, for some end.

These explanations recall the relationship between the master weaver (and by analogy the statesman) and the subordinate craftsmen in the *Statesman* who are referred to as cooperative causes (281d11–e10). They are described as "that without whose attendance the ruler of each of the arts would never produce" (281e2–4), words which again recall the *Phaedo*'s notion of "that without which the cause would not be a cause." The statesman, and the true weaver, act by directing and commanding their subordinates.[35] This image, perhaps borrowed from the world of craftsmanship, captures well the nature of the relationship between reason and 'necessity'. Viewed independently of the *guidance* and *coordination* of intelligence these active materials are 'wandering' causes (48a7), producing in a disorderly manner whatever chances to occur (46e5). Without intelligence, only fleeting traces of the four elements would appear, and then only by chance (69b5–c2). But *they* are capable of being persuaded to produce the best order possible (46c7–8; 48a1–4). To call them *sunaitia* is to describe them as operating and interacting according to a plan which is, however, not their own, much like the *productive* craftsmen are guided in their work by the *directive* craftsmen.

The chief methodological message of the *Timaeus* is that, of any feature of the physical world we must ask two distinct questions, and seek out two distinct 'becauses': (i) What are the physical interactions required to produce this result? (ii) What is the good for the sake of which these physical processes are *cooperating* to produce this result?

> . . . he [the Demiurge] made use of causes of this sort as subservient, while he himself contrived the good in all things that come to be. We must accordingly distinguish two kinds of causal account, the necessary and the divine. (68e4–7)

> We must speak of both kinds of causes but separate those which, with intelligence, are craftsmen of fine and good things, from those which in the absence of foresight, produce their sundry effects at random and without order. (46e3–6)

The *Timaeus* recommends that *we*, as far as possible, distinguish these two sorts of explanation. But this is a recommendation concerning how best to understand the world, not an account of distinct aspects of the world's makeup. These passages do *not* picture a layer of the operations of the world where necessity is unconstrained, not does it distinguish, as Prof. Vlastos suggests, between triumphs of "pure teleology" and compromises between teleology and necessity. Precisely, it characterizes a world which, at every level of structure, is the product of necessary physical interactions ordered and coordinated for the sake of some good.

The *Timaeus* thus develops the teleology of the *Phaedo* in rich and complex ways. In contrast with the *Phaedo,* however, the *Timaeus* never describes *forms* as causes. This, and the introduction of a third element in Plato's ontology, the Receptacle, are directly attributable to Plato's use of the image of divine craftsmanship, an image absent from the *Phaedo.*

A common image used by commentators to characterize the Receptacle in the *Timaeus,* though not one used by Plato, is the image of the mirror.[36] The things which come to be are images of the forms, reflected in the Receptacle. This image is dangerously misleading, for it ignores the fact that anything which has a stable enough existence to be named at all is *constructed* by intelligence (69b3–c2). Thus there is no sense in which the world we perceive is due to simple reflection. Plato *does* describe a precosmic activity in the Receptacle (52d2–53c2; 69b5–c3), which involves mere chance occurrences of traces (*ichnē* 53b2) and characters (52d6) of the four elements. This suggests that, without intelligent guidance the receptacle may, somehow, participate in the two basic sorts of triangles out of which the elements are constructed.[37] But what is crucial for Plato is that the world is *not* such an indeterminant and nameless flux, though if intelligence were not present it would be. In so far as 'space' has the character of the rational, ensouled, mathematically structured and stable organization that it does, it is due to intelligent persuasion.

Participation, then, understood as a relation between copy and paradigm in virtue of which the copy may bear the name of the paradigm, is not something which occurs independently of an intelligent agent aiming to achieve some good.[38] Thus the explanation of some feature of our world in terms of its likeness to a paradigm is, in the *Timaeus,* only an aspect of the nature of intelligent production, not worthy of independent identification as a cause. As we were led to expect by our brief look at Plato's human craftsmen, paradigms within a craft model are not *aitia.*

But again, everything which comes to be does so from necessity by some cause; for in all cases it is impossible for there to be a generation apart from a cause. Now, whenever the craftsman, looking to that which is always the same and using some such paradigm, fashions the *idea* and capacity of it, everything

thus completed is from necessity beautiful. But whenever he looks to a generated thing, using a generated paradigm, what is thus completed is not beautiful.
(28a4–b1)

Out of an extensive list of questions this passage raises, the one I wish to
focus on is why it is stressed that the craft product will only be beautiful if the
divine craftsman uses a changeless model.[39] This is not justified in our
passage, and on a certain interpretation of what it is the craftsman hopes to
achieve, it is unjustifiable. For if he simply wants to make a living thing, and
has no desire to make it changeless, why should it matter whether the
paradigm is changeless?

The same question can be raised about arguments that the copy must be
single and unique (30c2–31b3), that air and water are needed to make the
body one and insoluable (31b4–331), that all movements but one are to be
removed from it (34a2–6), that it be made if not eternal without qualification, at least an everlasting likeness (37b6–d8). As David Keyt has noted,
such arguments seem to confuse copying the form of *living thing* with
copying *the form* of living thing, *qua* form. Any paradigm has properties *qua*
paradigm that it would be "mad" to instantiate in one's copy – houses, as
copies of blue prints, should not be made of blue paper.[40]

The consistent stress of the above arguments in the *Timaeus* on producing
a copy with these "formal" features make us doubtful that such a criticism
understands Plato's motives. The assumption of this criticism is that the goal
of the Demiurge is to produce a living being (or living beings). This assumption is false. What the Demiurge aims to do, as we've seen, is to bestow
maximum unity, order, and persistence on his materials, because this is, in
itself, *good* for those materials. The *means* of achieving this is to copy the
form of Living Thing in these materials. Reconsidered in this light, the
Demiurge is, at least from an economic point of view, sane.

One can achieve *this* sort of goodness only by looking to the changeless
paradigm, for only it truly instantiates those features you strive for in your
model. Your copy must be, if possible, unique,[41] for a number of related
reasons, all given by Plato.

First, the form of living things is pictured as a genus/species hierarchy
(30c5–6, 39e3–40a7). If the god made two animals, each would be a 'part'
and thus a copy of one *sort* of living thing, but not a copy of Living Thing
itself. "Now we must never suppose the maker composed the world of those
things which are in the form of parts – for nothing akin to the incomplete
could ever come to be beautiful – but of that of which the other animals,

295

individually and by kinds, are parts" (30c3–6). Thus Plato views making "two or a plurality" of living things as akin to making copies of subkinds of Living Thing. But an obvious alternative view seems possible – why could the craftsman not work with many distinct parcels of material, providing each parcel with copies of *all* the living things, and therefore a complete copy in the relevant respect? Plato's response to this alternative is parallel to his response to supposing that there are two *forms* of Living Thing. On what grounds do we claim that more than one copy has been produced? Each of these "parcels of material" contains the same four kinds of living thing, under the same (generic) kind. If there were two islands that possessed the same four species of the genus finch, no biologist would argue that we had two finch kinds and eight distinct species.

A response to this argument carries me to my next point. One might say that Plato has to admit the possibility that a good craftsman could construct two animals, at least insofar as they are spatially differentiated, even if they are of one kind.[42] But this is false, because a good craftsman is out to unify and organize his material to the greatest extent possible, and this would not be accomplished by the construction of two formally identical but materially distinct universes. It must never be forgotten that the materials of *this* craftsman make up the *entire* visible flux. If it can become one, unified, bound together whole, it will be better than if it remains to whatever extent a plurality.

Which introduces a third reason for the Demiurge's monomania – a composite body, if acted on from without, can be destroyed. An antidote to this possibility is to produce one, self-contained physical system, as the Demiurge is craftily aware (33a–b).[43]

Briefly consider the other Demiurgic activities, remembering that the goal of the informing process is not in the first instance to make a living thing, but to provide maximum unity and harmonious structure (mathematically conceived throughout) to the visible and tangible world. Take the puzzling account of why the world body consists of just the four elements. Fire and earth are introduced as implications of our world's visibility and tangibility. Air and water, however, are provided with a very different explanation. "But two things alone cannot be satisfactorily united without a third; for there must be some bond between them, drawing them together. And of all bonds the best is that which makes itself and the terms it connects a unity in the fullest sense and this is naturally effected best by a proportion" (31b8–c4). The three-dimensional nature of the cosmos requires a four-term proportion and thus (with some work), air and water are explained.[44]

Notice that two goods are effected by the creation of precisely four elements. One is that the visible and tangible plurality becomes a *unity;* the second is that it becomes *indissoluble,* except by the one who bound it together. This is the beginnings of a world that is as far from the randomly shifting flux of the *Theaetetus* and *Cratylus* as a physical world can be. Behind the world revealed to us by our sense organs is an organization and stability which is due to intelligent production of the good.

Again the craftsman, while he cannot turn what is by nature created into something eternal, can, and does, endow it with an orderly and simple change "revolving according to number, an imitation of eternity" (37d–e). Likewise, as we've seen, the mathematical structure it embodies allows it to be indestructible. It is self-sufficient (68e), and possessed of every sort of measure, order and harmonious proportion (30a, 68b–d). In this way each aspect of the cosmos possesses a *summetria* both relative to itself and to everything else (69d2–5).[45] Finally, while he cannot remove the world of becoming from the realm of change altogether, he does his best. "He caused it to turn uniformly in the same place and within its own limits and made it revolve round and round; he took from it all the other six motions and gave it no part in their wanderings" (34a2–6).

The constant stress, then, on the creation of as Parmenidean a universe as possible is not a mistake – or if it is, it derives from a mistaken theory of goodness. Given the conception of the good that is operative, and given the goodness of the Demiurge (which we dare not deny!), his activities as characterized in the *Timaeus* are as we should expect.

Where does this concept of goodness as a mathematical ordering and unifying of a diverse plurality come from? We have seen it as a natural feature of the craft model. But the notion of order and unity is given a very precise meaning in the *Timaeus*. An ordering and unifying of elements is here achieved by creating relations of proportionality and commensurability among them and their changes.[46] And this is carried through in the production of mortals by the created, imitating gods[47] and in the transformations undergone by the solids which constitute earth, air, fire, and water. Plato did not have to invent the idea of a mathematical account of any of these domains. But for him that such accounts were possible itself required explanation. He accounted for the underlying measurability of the *cosmos* by identifying that measurability with the good aimed for by a divine craftsman.

It is this mathematical version of counterentropic goodness which the Demiurge seeks to achieve by his actions and is perhaps most explicitly articulated in the following comment of Socrates near the close of the *Philebus:* "[Surely no one is ignorant of this] that every compound which does not in any way partake of measure and the nature of proportion necessarily destroys both the mixture and first of all itself" (*Philebus* 64d9–11). Rather, you end up with, in the inspired translation of Hackforth "a miserable mass of unmixed messiness" (64e2–3). This is the power of the good (64e5) found in the nature of the beautiful, in that beauty and excellence turn out to be a matter of measure and proportion.[48]

Likewise, in a quiet reference to the *demiurgos* of the heavens in the *Republic* (530a3–b4), we are told that it is the astronomer who focuses on the nature of the commenserability which the heavenly movements exemplify as well as physical bodies can, who may hit on the nature of the beautiful and the good (531c5).

And indeed, apprehension of *this* good is the teleological explanation why our eyes interact with the physical world as they do.

> But for our part, let us speak of eyesight as the cause of this benefit, for the sake of these things:[49] the god invented and gave us vision in order that we might observe the circuits of intelligence in the heavens and apply them to the circuits of our own thought, which are akin to them, the orderly to the disorderly; thus by learning from them and taking part in correct calculations in accordance with nature, and imitating the completely stable circuits of the divine, we might stabilize the wanderings in ourselves. (47b5–c4)

Notice that sight is the cause of the good which results, *and* that we have eyesight *because* of the good which results. Vision is the mechanism by means of which we may discover the good. But this is not an end in itself. We are provided with vision in order that we might get our souls in shape. The Demiurge aims at this, of course, because we are a part of the visible world he wishes to be good.

NOTES

1. *Sophist* 265c–266c; *Laws* X, 889a–890a; *Philebus* 28d–3, *Timaeus* 46c–47c; 68e–69d.
2. Gallop 1975, 175; Vlastos 1975, 97. Both these authors overstate the extent to which the 'mechanical philosophers' of the seventeenth century reject Plato's vision. That the universe as a whole was a rationally designed artifact was seldom in doubt before the nineteenth century.
3. Boyle vol. 5, 409.
4. Cf. Brumbaugh 1976, 40–52.
5. Burnet 1911, 108, claimed his ideas on the *deuteros plous* were in agreement with Goodrich 1903, 381–385; 1904, 5–11. However, Goodrich convincingly links Socrates' disparaging remarks concerning his own method of explanation to the hoped for teleology of 97b8–99b2, and so unlike Burnet saw no irony in this remark.

6. There are four pieces of evidence that indicate that the sense of intrusion of the Anaxagorean discussion is intentional. (i) The use of the present at 97b6 generates anticipation that Socrates' random method will be discussed immediately. (ii) The discussion of the *deuteros plous* is *re*-introduced at the end of our passage. To quote Goodrich, "Ἔδοξε τοινυν μοι κ.τ.λ. . . . [Now it seemed to me] (99d4) links back immediately where the narrative had previously broken off, at 97b8. . . ." (Goodrich 1903, 382). (iii) The problems that had led to Socrates' dissatisfaction with natural science, discussed just prior to the Anaxagorean excursus, are shown to be resolved by the *deuteros plous* (100e5–103c4), but are not mentioned from 97b–99e. (iv) As I will discuss in detail shortly, the sorts of *explanantia* focused on in this passage are in striking contrast to those on either side of it.

7. Frede 1980, 217–249, esp. 222–223.

8. *Phd.* 97c1–2. Compare Aristotle, *Metaph.* A 984b15–17.

9. For problems with translating *strongulē* in the *Phaedo,* cf. Morrison 1959, 101–119.

10. *Phaedo* 98c5–99d4.

11. The pattern of this formulation, though not its content, was suggested by Larry Wright 1976.

12. Cf. *Metaph.* Z.7 1032a12–13, a25–32; *Ph.* II.5 196b23–26, II.8 199a3–8; *PA* I.1 639b15–21. The story is complex. Plato in *Laws* X discusses those he opposes as holding that the cosmos is due to nature and chance rather than to intelligence and craft. Ultimately, however, as Joan Kung reminded me, Plato wishes to insist that if nature refers to what is primary and an origin, then it is soul and intelligence that are nature and *their* products that are by nature (cf. 892c). Aristotle treats nature as *sui generis,* refusing to range either intelligence and craft on the one hand, or chance and spontaneity on the other, with it.

13. As is made clear at *Crito* 53b3ff; cf. *Phaedo* 99a1–2.

14. E.g., Burnet 1911, 106; Gallop 1975, 175.

15. Vlastos 1975, 30, wrongly claims that "he [Plato] reproaches them [the *physiologoi*] for deciding such a question as whether the earth is flat or round without first asking which of the two would be the 'better' (97e)." This reverses the order clearly recommended in the text.

16. Cf. note 6 above, for the evidence.

17. Indeed, the discussion has in a general way been about how to account for generation and destruction from 70d7. 71a10's "All things come to be in this way. The opposite *things* come to be from opposites . . ." is referred to at 103a4 when Socrates notes that it doesn't contradict the idea that opposites *themselves* don't come to be from their opposites.

18. *Phaedo* 97c6–7.

19. 102c3–103b5.

20. In the discussion of teleological explanation, purely mechanical accounts are held to state only the means for accomplishing various ends; such accounts fail to discriminate between various ends achievable by these means because they fail to inquire why the ends achieved are good (98b7–99c5). In the discussion of explanation *via* participation in hypothesized forms various commonsense explanations are criticized on three grounds, summarized by David Gallop (1975, 186), as follows:

(i) No opposite, F, can count as the 'reason' for a thing's having a property, if its opposite, G, can also give rise to that property (97a7–b3).

(ii) Nothing can count as a 'reason' for a thing's having a property, if its opposite, G, can also give rise to that property (101a6–8).

(iii) A 'reason' for a thing's having a property F, cannot itself be characterized by the opposite of that property, G (101a8–b2).

21. On the meaning of *deuteros plous* in this context I am following Goodrich 1903, 1904, and Hackforth 1955, 127n5. I have not been convinced by Shipton 1979, 33–53, that the issue here is whether Socrates can acquire a 'divinely revealed' and therefore certain account or whether he must proceed 'hypothetically'. The reference of this sort of cause at 99c7 is clearly to the good achieved by intelligence. It is *this* sort of explanation Socrates failed either to discover himself or learn from another, and compared to which what he goes on to state is a *deuteros plous*. On the other hand, the other uses of this term in Plato (*Philebus* 19c1–2, *Politicus* 300c4) do not *merely* imply a more laborious means to the same good (*pace* Dorter 1982, 125), but a considerably more modest approach to a subject. The explanation of each things' coming to be, being or ceasing to be F by means of its coming to be, being or ceasing to be related to what truly is F is a *deuteros plous* with respect to the epistemic desires of 97b–99c in just this way.

22. It is thus distressing that virtually every discussion of the passage focuses either on 96a–97c/100a–105, or on 97c–99d, as a glance at the various discussions referred to in these notes shows.

23. Cf. Vlastos 1969, 291–325; Burge 1971, 1–13; Burnet 1911, 103, 108.

24. Cf. Damascius 1, para. 417–418, in Westerink 1977; Bluck 1957, 21–31.

25. Those who take this approach typically read the ideas on explanation of *Republic* VI–VII or the *Timaeus* into the passage. Gallop 1975, 191, is properly cautious, as is Annas 311–326, esp. 318. Gallop himself interprets Socrates' enigmatic reference to God and the form of life (220–221) to suggest a form-explanation as a replacement for Anaxagorean teleology. While this is ingenious, it throws his earlier caution to the winds.

26. Compare *Phaedo* 100d6–7, 102c2 with Aristotle *Metaph*. A 991b3, *Metaph*. Z 1034a2–5, *GC* II 335b7–24. An effective reply to the claim (in Vlastos 1969) that Aristotle has misunderstood Plato in these criticisms is to be found in Annas 1982.

27. For excellent discussions of which, cf. Santas 1983, 232–263; White 1979, 171–181.

28. With the reference at 530a6 the strangely playful wording of *Republic* X has more force. It remains true, however, that the latter discussion is ambiguous: 596b9–10 says, "For surely none of the craftsmen craft the idea itself; for how could he?" but then asks what such a craftsman would be called (596b12). In the same vein, 596e5–9 suggests this craftsman is the one referred to at 530a6, but then hints that only a person with a mirror could produce all natural things, and only in the sense of producing *images* of them. But then the form of the craft product is reintroduced as perhaps the work of a god at 597b5, again with some hesitancy. Finally, at 597c1–d1 Socrates seems straightforwardly to assume such a deity.

29. Typically the craftsman 'looks toward' these paradigms; e.g., *Gorgias* 503d8,

504d5; *Cratylus* 389a5, c5, c7, d6, 390e; *Republic* V 472a4–7, *Republic* X 596b7, *Timaeus* 28ab.

30. These passages, which all suggest the good is a unity achieved through mathematical bonds, are a clue to the relevance of the increasingly mathematical course of study recommended in *Republic* VII to grasping the nature of the good. In a variety of Platonic texts it is suggested that goodness for a plurality is to be found in the principles which bind it into a unity, principles of proportion and commensurability or measure. The same language appears in the *Statesman*'s characterization of statecraft (308c–311a) and the *Timaeus*' characterization of the Demiurge's work (32b, 69c). Such passages make it less surprising that, as Aristoxenus relates, Aristotle reported that Plato's lecture "On the Good" turned out to be a discussion of arithmetic, geometry, and astronomy, culminating in the claim that "the limit that is good is unity." (Aristoxenus, *Elementa harmonica* 2: 30–31). Cf. Gaiser 1980, 5–37; Cooper 1977, 151–157, esp. 155. The development of this section of this chapter owes much to suggestions by David Furley and Joan Kung.

31. A related but somewhat different distinction is drawn in the *Statesman* between the overseer of a craft such as weaving or governing and those who supply the materials necessary for weaving. Cf. *Statesman* 281a–e, 287d; and compare Aristotle, *EN* I.2 1094a26–1094b11.

32. Aristotle's claim that Plato's desire to separate forms from particulars grew out of the influence of Heraclitus and Cratylus and their doctrine of radical flux has been used to shed light on the development of Plato's thought by Irwin 1977, 1–13; and Jordan 1983. *Theaetetus* 179d3–183b7 and *Cratylus* 439d–440 indicate Plato's concern with this doctrine, and it is common to suggest that Plato may have held some such view of the physical world. I believe that, at least from the *Timaeus* onward, Plato's view could be stated counterfactually as follows: If the physical world were not the product of a good and efficacious craftsman, it would be as the friends of flux describe it. The initial description of the Receptacle prior to divine craftsmanship is remarkably like the account of the flux doctrine in the *Theaetetus*, but it is important to recall that that passage does *not* describe the physical world as it actually is.

33. These expressions with final clauses are used at least 36 times in the *Timaeus*. Notice the virtual absence of these expressions, however, from *Timaeus* 49–69, where the effects of necessity are discussed without reference to intelligence.

34. Cornford 1937, 173–175; Vlastos 1975, 28–30.

35. *Statesman* 281a–3; 287d.

36. Cf. Allen 1965b, 55–58; Cornford 1937, 181; Sayre 1983, 249. Indeed all three writers talk as if 52c discusses the Receptacle as a mirror. It doesn't; *Republic* X 596c, uses the notion of mirror images, but with reference to the relationship between images and their imitations.

37. It is startling that the standard English commentaries on the *Timaeus* don't really face the issue of the nature of the basic elements which the craftsman encounters in the Receptacle. The triangles themselves are never explicitly said to be constructed – earth, air, fire, and water are constructed from them by god (53b), or traces or *pathē* of this chance to occur (52d, 53b, 69b–c). On the other hand, forms of earth, air, fire and water – accounts of their stereometric configurations, perhaps – are mentioned, but *not* forms of the two basic triangles used in the god's

stereometry. Plato leaves us with the material for two inferences, and I can't see any obvious means of deciding between them. The first, suggested by Mary Louise Gill at the *Princeton Ancient Philosophy Colloquium* on Plato's Natural Philosophy in "Matter and Flux in Plato's *Timaeus*" is that the triangles are the basic physical constituents of the Receptacle. Gill does not discuss the possibility of there being permanent images of Forms of the Right Angle Scalene and Isosceles Triangles. There are difficulties with either view. On the one hand, no such forms are mentioned. On the other hand, the radically indeterminate and unmeasured nature of the pre-crafted contents of the Receptacle is difficult to reconcile with the view that it is replete with geometrical objects.

In either case it remains true that all those *gignomena* (generated objects) for which forms are mentioned are *constructed out of basic elements*. This view of the physical world, furthermore, is detachable from the mythic imagery of the *Timaeus*, for it is mentioned in virtually *all* later dialogues.

38. Thus Plato himself seems to have answered Aristotle's critique of form-participation explanations in the *Phaedo* (discussed in note 26 above). He accepts the view that without the activity of a goal-directed agent, participation will not provide an account of coming-to-be. The Demiurge is a response to such complaints. Cf. Annas 1982, 313, 315.

39. As Cornford 1937 notes (27), the background is likely the distinction between true producers and mere imitators in *Republic* X 597–598. Another use of this distinction in the *Timaeus* is Plato's reference to the created gods, who base their mortal constructions on the Demiurge's created model, as *imitators* of his work.

40. Keyt 1971, 230–235; for criticisms in a similar vein, cf. Santas 1983.

41. Cf. Parmenides, fragment VIII.

42. This response was suggested to me by a comment from Richard Perry during the Princeton Colloquium; cf. his 1979, 1–10. I agree with Perry that the Demiurge's primary concern is to craft an orderly and harmonious perceptible world. But I believe the argument set forth here allows *that* without requiring us to abandon the standard account of the Living Thing Itself embracing its four *genē* as something like a relation between kind and subkinds.

43. Cf. 32c5–33b1.

44. For an interesting conjecture on the mathematical background to the passage, cf. Cornford 1937, 45–52.

45. Cf. the interesting account of *summetria* in *Republic* VII in Mourelatos 1980, 33–73. In particular, the important discussions of the parallel between *Republic* 530a–b and *Timaeus* 69b3–5, pp. 39–40 and 56–58.

46. Cf. *Timaeus* 73c1, 74c5, 85c5, 86c5, 87c–d, 90a2.

47. Cf. *Timaeus* 59a1, 62a3, 64d9, 66b1, 66d3, 67c7.

48. Compare *Republic* VII: 529e3–530b1: Aristotle, *Metaph.* M.3 1078a36–1078b6.

49. Following Cornford 1937 on 47b5–6; cf. 158n2.

Works Cited

Achinstein, P. 1977. 'Function statements', *Philosophy of Science* XLIV, 341–67.
Ackrill, J. L. 1963. *Aristotle's Categories and De Interpretatione*. Oxford.
1981a. *Aristotle the Philosopher.* Oxford.
1981b. 'Aristotle's theory of definition: some questions on *Posterior Analytics* II 8–10', in Berti 1981, pp. 359–84.
Albritton, R. 1957. 'Forms of particular substances in Aristotle's *Metaphysics'*, *Journal of Philosophy* LIV, 699–708.
Allen, R. E., ed. 1965a. *Studies in Plato's Metaphysics.* London.
1965b. 'Participation and predication in Plato's middle dialogues', in Allen 1965a, pp. 43–60.
Annas, J. 1982. 'Aristotle on inefficient causes', *Philosophical Quarterly* XXXII, 311–26.
Anton, J. P., ed. 1980. *Science and the Sciences in Plato.* New York.
Anton, J. P. and Preus, A., eds. 1983. *Essays in Ancient Greek Philosophy.* Albany.
Ayala, F. J. 1970. 'Teleological explanations in evolutionary biology', *Philosophy of Science* XXXVII, 1–15.
1978. 'The mechanics of evolution', in *Evolution: A Scientific American Book,* New York, pp. 14–29.
Balme, D. M. 1962a. 'Development of biology in Aristotle and Theophrastus: theory of spontaneous generation', *Phronesis* VII, 91–104.
1962b. 'ΓΕΝΟΣ and ΕΙΔΟΣ in Aristotle's biology', *Classical Quarterly* XII, 81–98.
1972. *Aristotle's De Partibus Animalium I and De Generatione Animalium I (with Pages from II.1–3).* Oxford.
1980. 'Aristotle's biology was not essentialist', *Archiv für Geschichte der Philosophie* LXII, 1–12.
1987a. 'The place of biology in Aristotle's philosophy', in Gotthelf and Lennox 1987, pp. 9–20.
1987b. 'Aristotle's use of division and differentiae', in Gotthelf and Lennox 1987, pp. 69–89.
1987c. 'Teleology and necessity', in Gotthelf and Lennox 1987, pp. 275–85.
1987d. 'Aristotle's biology was not essentialist', in Gotthelf and Lennox 1987, pp. 291–312.

1991. 'Aristotle's *Historia Animalium:* date and authorship', in *Aristotle: History of Animals* VII-IX, Cambridge, MA and London, pp. 1–13.

1992. *Aristotle. De Partibus Animalium I and De Generatione Animalium I (with Passages from II.1–3) with a Report on Recent Work and an Additional Bibliography by Allan Gotthelf.* Oxford.

Barnes, J. 1975a (2nd edition, 1993). *Aristotle's Posterior Analytics.* Oxford.

1975b. 'Aristotle's theory of demonstration', in Barnes, Schofield and Sorabji 1975, pp. 65–87.

1981. 'Proof and the syllogism', in Berti 1981, pp. 17–59.

Barnes, J., Schofield, M. and Sorabji, R., eds. 1975. *Articles on Aristotle I: Science.* London.

Beatty, J. 1980. 'Optimal design models and the strategy of model building in evolutionary biology', *Philosophy of Science* XLVII, 532–61.

Bekker, I., ed. 1831. *Aristotelis Opera.* Berlin.

Berti, E., ed. 1981. *Aristotle on Science: The Posterior Analytics.* Proceedings of the 8th Symposium Aristotelicum. Padova.

Bluck, R. S. 1957. 'ὑπόθεσις in the *Phaedo* and Platonic dialectic', *Phronesis* II, 21–31.

Bodson, L. 1990. *Aristote: De Partibus Animalium, Index Verborum, Listes de Fréquence.* Liège.

Bolton, R. 1976. 'Essentialism and semantic theory in Aristotle: *Posterior Analytics* II.7–10', *Philosophical Review* LXXXV, 514–44.

1987. 'Definition and scientific method in Aristotle's *Posterior Analytics* and *Generation of Animals*', in Gotthelf and Lennox 1987, pp. 120–66.

1993. 'Division, définition et essence dans la science aristotélicienne', *Revue Philosophique* II, 197–222.

Bonitz, H. 1862. *Aristoteles Studien.* Vienna.

1870. *Aristotelis Opera,* Vol. V. Berlin.

Boorse, C. 1976. 'Wright on functions', *Philosophical Review* LXXXV, 70–86.

Bowen, A. C. 1983. 'Menaechmus vs. the Platonists: two theories of science in the early Academy', *Ancient Philosophy* III, 12–29.

ed. 1991. *Science and Philosophy in Classical Greece.* New York.

Boyle, R. 1688. *A Disquisition on the Final Causes of Natural Things,* in Thomas Birch, ed., *The Works of the Honorable Robert Boyle,* Vol. 5. London.

Brumbaugh, R. S. 1968. *Ancient Greek Gadgets and Machines.* Westport, Ct.

1976. 'Plato's relation to the arts and crafts', in Werkmeister 1976, pp. 40–52.

Burnet, J. 1911. *Plato's Phaedo.* Oxford.

Burnyeat, M. F. 1981. 'Aristotle on understanding knowledge', in Berti 1981, pp. 97–139.

1992. 'Is an Aristotelian philosophy of mind still credible?', in Nussbaum and Rorty 1992, pp. 15–26.

Burnyeat, M. F. et al. 1981. *Notes on Book Zeta of Aristotle's Metaphysics.* Oxford.

Byl, S. 1980. *Recherches sur les grands traités biologiques d'Aristote: sources écrites et préjugés.* Brussels.

Bylebyl, J. J. 1979. 'The school of Padua: humanistic medicine in the sixteenth century', in Webster 1979, pp. 334–70.

Works Cited

Charles, D. 1990. 'Meaning, natural kinds and natural history', in Devereux and Pellegrin 1990, pp. 145–67.

1991. 'Teleological causation in the *Physics*', in Judson 1991, pp. 101–128.

1997. 'Aristotle and the unity and essence of biological kinds', in Kullmann and Föllinger eds. 1997, pp. 27–42.

Charlton, W. 1970. *Aristotle's Physics I and II*. Oxford.

Chroust, A.-H. 1962. 'The miraculous disappearance and recovery of the Corpus Aristotelicum', *Classica et Mediaevalia* XXIII, 50–67.

Cooper, J. 1977. 'The psychology of justice in Plato', *American Philosophical Quarterly* XIV, 151–57.

1982. 'Aristotle on natural teleology', in Nussbaum and Schofield 1982, pp. 197–222.

1987. 'Hypothetical necessity and natural teleology', in Gotthelf and Lennox 1987, pp. 243–74.

Cornford, F. M. 1937. *Plato's Cosmology*. London.

Cunningham, A. 1985. 'Fabricius and the "Aristotle project" in anatomical teaching and research at Padua', in Wear, French and Lonie 1985, pp. 195–222.

Detel, W. 1997. 'Why all animals have a stomach. Demonstration and axiomatization in Aristotle's *Parts of Animals*', in Kullmann and Föllinger 1997, pp. 63–84.

Devereux, D. and Pellegrin P., eds. 1990. *Biologie, logique et métaphysique chez Aristote*. Paris.

Dobzhansky, T. 1970. *Genetics of the Evolutionary Process*. New York.

Düring, I. 1950. 'Notes on the history of the transmission of Aristotle's writings', *Göteborgs Högskolas Årsskrift* LVI, 35–70.

1956. 'Ariston or Hermippus?', *Classica et Mediaevalia* XVII, 11–21.

1957. *Aristotle in the Ancient Bibliographical Tradition*. Göteborg.

Ferejohn, M. 1981. 'Aristotle on necessary truth and logical priority', *American Philosophical Quarterly* XVIII, 285–93.

1982/3. 'Definition and the two stages of Aristotelian demonstration', *Review of Metaphysics* XXXVI, 375–95.

1990. *The Origins of Aristotelian Science*. New Haven.

Fortenbaugh, W. W. and Sharples, R. W., eds. 1988. *Theophrastean Studies: On Natural Science, Physics and Metaphysics, Ethics, Religion and Rhetoric*. Rutgers Studies in Classical Humanities III. New Brunswick.

Frede, M. 1980. 'The original notion of cause', in Schofield, Burnyeat and Barnes 1980, pp. 217–49.

Freeland, C. A. 1987. 'Aristotle on bodies, matter, and potentiality', in Gotthelf and Lennox 1987, pp. 392–407.

1990. 'Scientific explanation and empirical data in Aristotle's *Meteorology*', in Devreux and Pellegrin 1990, pp. 287–320.

Friedlein, G., ed. 1873. *Procli Diadochi in primum Euclidis Elementorum librum commentaria*. Leipzig.

Furley, D. J. 1983. 'The mechanics of *Meteorologica* IV: a prolegomenon to biology', in Moraux and Wiesner 1983, pp. 73–93.

1984. 'The rainfall example in *Physics* 2.8', in Gotthelf 1985a, pp. 177–82.

Furth, M. 1988. *Substance, Form and Psyche: An Aristotelean Metaphysics*. Cambridge.

Works Cited

Gallop, D. 1975. *Plato's Phaedo*. Oxford.

Geiser, K. 'Plato's enigmatic lecture "On the Good"', *Phronesis* XXV, 5–37.

Ghiselin, M. T. 1969. *The Triumph of the Darwinian Method*. Berkeley.

Goodrich, W. J. 1903/4. 'On *Phaedo* 96a-102a and on the δεύτερος πλοῦς 99d', *Classical Review* XVII, 381–85/XVIII, 5–11.

Gotthelf, A. 1975. *Aristotle's Conception of Final Causality*. Columbia University Dissertation. Ann Arbor.

— 1976. 'Aristotle's conception of final causality', *Review of Metaphysics* XXX, 226–54.

— ed. 1985a. *Aristotle on Nature and Living Things: Philosophical and Historical Studies Presented to David M. Balme on his Seventieth Birthday*. Pittsburgh and Bristol.

— 1985b. 'Notes towards a study of substance and essence in Aristotle's *Parts of Animals* II-IV', in Gotthelf. 1985a, pp. 27–54.

— 1987a. 'First principles in Aristotle's *Parts of Animals*', in Gotthelf and Lennox 1987, pp. 167–98.

— 1987b. 'Aristotle's conception of final causality', in Gotthelf and Lennox 1987, pp. 204–42.

— 1988. '*Historiae* I: *Plantarum* et *Animalium*', in Fortenbaugh and Sharples 1988, pp. 100–35.

— 1989. 'Teleology and spontaneous generation in Aristotle: a discussion', in Kraut and Penner 1989, pp. 181–93.

— 1997. 'The elephant's nose: further reflections on the axiomatic structure of biological explanation in Aristotle', in Kullmann and Föllinger 1997, 85–96.

— 1997b. 'Understanding Aristotle's teleology', in Hassing 1997, pp. 71–82.

— 1999. 'Darwin on Aristotle', *Journal of the History of Biology* XXXII, 3–30.

Gotthelf, A. and Lennox, J. G., eds. 1987. *Philosophical Issues in Aristotle's Biology*. Cambridge.

Gottschalk, H. 1998. 'Theophrastus and the Peripatos', in van Ophuijsen and van Raalte 1998, pp. 281–92.

Gould, S. J. and Lewontin, R. C. 1979. 'The spandrels of San Marco and the panglossian paradigm: a critique of the adaptationist program', *Proceedings of the Royal Society of London: Biology* CCV, 581–98.

Grene, M. 1963. *A Portrait of Aristotle*. London.

— 1974. 'Is genus to species as matter to form?', *Synthèse* XXVIII, 51–69; repr. in M. Grene, *The Understanding of Nature*, Dordrecht and Boston, 1974, pp. 108–26.

Hackforth, R. 1955. *Plato's Phaedo*. Cambridge.

Hankinson, R. J. 1991. *Galen: On the Therapeutic Method, Books I and II*. Oxford.

Hassing, R. 1997. *Final Causality in Nature and Human Affairs. Studies in Philosophy and the History of Philosophy*, vol. XXX. Washington DC.

Heath, T. L. 1949. *Mathematics in Aristotle*. Oxford.

— 1956. *Euclid: The Thirteen Books of the Elements*, 3 vols. London.

Heiberg, J. L., ed. 1910–15. *Archimedis opera omnia cum commentariis Eutocii*, 2nd ed., 3 vols. Leipzig.

— 1919. *Euclidis Elementa*. Leipzig.

Herschel, J. 1830. *A Preliminary Discourse on the Study of Natural Philosophy*. London.

Works Cited

Hicks, R. D. 1907. *Aristotle's De Anima*. Cambridge.

Horn, J. J. et al., eds. 1979. *Analysis of Ecological Systems*. Columbus.

Hughes, J. D. 1988. 'Theophrastus as ecologist', in Fortenbaugh and Sharples 1988, pp. 67–75.

Hull, D. 1965–6. 'The effect of essentialism on taxonomy – two thousand years of stasis', *British Journal for the Philosophy of Science* XV, 314–66; XVI, 1–18.

1967. 'The metaphysics of evolution', *British Journal for the History of Science* III, 309–337.

1967–68. 'The conflict between spontaneous generation and Aristotle's *Metaphysics'*, *Proceedings of the Inter-American Congress of Philosophy* VII, 245–50.

Hultsch, F., ed. 1876–78. *Pappi Alexandrini collectionis quae supersunt*, 3 vols. Berlin.

Irwin, T. 1977. 'Plato's Heracliteanism', *Philosophical Quarterly* XXVII, 1–13.

Jordan, R. W. 1983. *Plato's Arguments for Forms*. Cambridge.

Judson, L., ed. 1991a. *Aristotle's Physics: A Collection of Essays*. Oxford.

1991b. 'Chance and always or for the most part in Aristotle', in Judson 1991, pp. 73–100.

Kahn, C. 1985. 'The place of the prime mover in Aristotle's teleology', in Gotthelf 1985a, pp. 183–205.

Keaney, J. J. 1963. 'Two notes on the tradition of Aristotle's writings', *American Journal of Philology* LXXXIV, 52–63.

Keyt, D. 1971. 'The mad craftsman of *Timaeus'*, *Philosophical Review* LXXX, 230–35.

Kirk, G.S., Raven, J.E., and Schofield, M., 1983. *The Presocratic Philosophers*, 2nd ed. Cambridge.

Kitcher, P. 1981. 'Explanatory unification', *Philosophy of Science* XLVIII, 507–31.

Knorr, W. 1986. *The Ancient Tradition of Geometrical Problems*. Boston and Basel.

Kosman, L. A. 1973. 'Understanding, explanation, and insight in the *Posterior Analytics'*, in Lee, Mourelatos and Rorty 1973, pp. 374–92.

1990: 'Necessity and explanation in Aristotle's *Analytics'*, in Devreux and Pellegrin 1990, pp. 349–64.

Kraut, R, and Penner, T., eds. 1989. *Nature, Knowledge and Virtue: Essays in Memory of Joan Kung*. (*Apeiron* XXII.4, special issue.)

Kullmann, W. 1974. *Wissenschaft und Methode: Interpretationen zur aristotelischen Theorie der Naturwissenschaft*. Berlin and New York.

1985. 'Different concepts of the final cause in Aristotle', in Gotthelf 1985a, pp. 169–176.

1990. 'Hintergründe und Motive der platonischen Schriftkritik', in Kullmann and Reichel eds. 1990, 317–34.

1997. 'Zoologische Sammelwerke in der Antike', in Kullmann, Althoff and Asper 1997.

Kullmann, W., Althoff, J. and Asper M., eds. 1997. *Gattungen wissenschaftlicher Literatur in der Antike*. Tübingen.

Kullmann, W. and Föllinger, S., eds. 1997. *Aristotelische Biologie. Intentionen, Methoden, Ergebnisse*. Stuttgart.

Kullmann, W. and Reichel M., eds. 1990. *Der Übergang von der Mündlichkeit zur Literatur bei den Griechen*. Tübingen.

Laks, A., Most, G. W. and Rudolph, E. 1988. 'Four notes on Theophrastus' *Metaphysics'*, in Fortenbaugh and Sharples 1988, pp. 224–33.

307

Works Cited

Lambros, S. P., ed. 1885. *Commentaria in Aristotelis Graeca: Supplementum Aristotelicum* 1.1. Berlin.

Lear, J. 1980. *Aristotle and Logical Theory.* Cambridge.

LeBlond, J. M. 1945. *Aristote, philosophe de la vie: Le livre premier du traité sur les Parties des animaux.* Paris.

Lee, E. N., Mourelatos, A. D. P. and Rorty, R. M., eds. 1973. *Exegesis and Argument: Studies Presented to Gregory Vlastos.* (*Phronesis*, suppl. vol. I.) Assen.

Lennox, J. G. 1980. 'Aristotle on genera, species, and "the more and the less"', *Journal of the History of Biology* XIII, 321–46.

1984. 'Recent studies in Aristotle's biology', *Ancient Philosophy* IV, 73–82.

1985. 'Aristotle, Galileo and the mixed sciences', in Wallace 1985, pp. 29–53.

1990. 'Notes on David Charles on *HA*', in Devereux and Pellegrin 1990, pp. 169–183.

1992. 'Teleology', in Evelyn Fox Keller and Elisabeth A. Lloyd, eds., *Keywords in Evolutionary Biology.* Cambridge, Mass., pp. 324–333.

1993. 'Darwin *was* a teleologist', *Biology and Philosophy* VIII, 409–422.

Lewontin, R. C. 1978. 'Adaptation', in *Evolution: A Scientific American Book,* New York, pp. 114–25.

1979. 'Fitness, survival, and optimality', in Horn et al. 1979, pp. 387–405.

Lloyd, G. E. R. 1966. *Polarity and Analogy: Two Types of Argumentation in Early Greek Thought.* Cambridge.

1968. *Aristotle: the Growth and Structure of his Thought.* Cambridge.

1990. 'Aristotle's zoology and his metaphysics: the status questionis', in Devereux and Pellegrin 1990.

1991. *Methods and Problems in Greek Science.* Cambridge.

1996. *Aristotelian Explorations.* Cambridge.

Lonie, I. M. 1964. 'Erasistratus, Erasistrateans, and Aristotle', *Bulletin of the History of Medicine* XXXVIII, 426–43.

Louis, P. 1961. *Aristote, De la génération des animaux.* Paris.

Loux, M. J. 1979. 'Form, species, and predication in *Metaphysics* Z, H, Θ', *Mind* LXXVIII, 1–23.

Mansion, A. 1946. *Introduction à la physique aristotélicienne,* 2nd ed. Louvain and Paris.

Marcos A. 1996. *Aristóteles y otros animales. Una lectura filosófica de la Biología aristotélica.* Barcelona.

Mattock, J. N. 1987. *Maqāla tashtamil 'alā Fusul min Kitāb al-Hayawān li-Aristu (Tract comprising Excerpts from Aristotle's Book of Animals) Attributed to Musā b. 'Ubaid Allāh al-Qurtubi al-Isrā'ili.* Arabic Technical and and Scientific Texts 2. Cambridge.

May, M. T. 1968. *Galen: On the Usefulness of the Parts of the Body,* 2 vols. Ithaca.

Mayr, E. 1963. *Populations, Species and Evolution.* Cambridge, MA.

1982, *The Growth of Biological Thought.* Cambridge, MA.

McKirahan, R. D., Jr. 1992. *Principles and Proofs: Aristotle's Theory of Demonstrative Science.* Princeton.

Meyer, J. B. 1855. *Aristoteles Tierkunde: ein Beitrag zur Geschichte der Zoologie, Physiologie und alten Philosophie.* Berlin.

Works Cited

Meyer, S. S. 1992. 'Aristotle, teleology, and reduction', *Philosophical Review* CI, 791–825.

Modrak, D. 1979. 'Forms, types, and tokens in Aristotle's *Metaphysics*', *Journal of the History of Philosophy* XVII, 371–381.

Moraux, P. 1951. *Les listes anciennes des ouvrages d'Aristote.* Louvain.

1985. 'Galen and Aristotle's *De partibus animalium*', in Gotthelf 1985a, pp. 327–44.

Moraux, P. and Wiesner J., eds. 1983. *Zweifelhaftes im Corpus Aristotelicum: Studien zu Einigen Dubia: Akten des 9. Symposium Aristotelicum.* Berlin and New York.

Morrison, J. S. 1959. 'The shape of the earth in Plato's *Phaedo*', *Phronesis* IV, 101–19.

Mourelatos, A. D. P. 1980. 'Plato's "real astronomy": *Republic* 527d-531d', in Anton 1980, pp. 33–73.

Nehamas, A. 1979. 'Self-predication and Plato's theory of Forms', *American Philosophical Quarterly* XVI, 93–103.

Nussbaum, M. C. 1976. 'The text of Aristotle's *De Motu Animalium*', *Harvard Studies in Classical Philology* LXXX, 111–59.

1978. *Aristotle's De Motu Animalium: Text with Translation, Commentary, and Interpretive Essays.* Princeton.

Nussbaum, M. C. and Rorty, A. O., eds. 1992. *Essays on Aristotle's De Anima.* Oxford.

Nussbaum, M. C. and Schofield, M., eds. 1982. *Language and Logos.* Ithaca.

Ophuijsen, J. M. van and Raalte, M. van, eds. 1998. *Theophrastus: Reappraising the Sources.* New Brunswick.

Owen, G. E. L. 1978/9. 'Particular and general', *Proceedings of the Aristotelian Society* XCIX, 1–21.

Pellegrin, P. 1982. *La Classification des animaux chez Aristote: statut de la biologie et unité de l'aristotélisme.* Paris.

1985. 'Aristotle: a zoology without species', in Gotthelf 1985a, pp. 95–115.

1987. 'Logical and biological difference: the unity of Aristotle's thought', in Gotthelf and Lennox 1987, pp. 313–38.

Perry, R. 1979. 'The unique world of the *Timaeus*', *Journal of the History of Philosophy* XVII, 1–10.

Pliny. 1967–71. *Natural History,* trans. H. Rackham, W.H.S. Jones and D.E. Eicholz. Cambridge, MA.

Popper, K. 1952. *The Open Society and Its Enemies,* 2nd ed., 2 vols. London.

Preus, A. 1975. *Science and Philosophy in Aristotle's Biological Works.* Hildesheim and New York.

1981. *Aristotle and Michael of Ephesus on the Movement and Progression of Animals.* New York.

Raalte, M. van. 1993. *Theophrastus: Metaphysics.* Leiden.

Randall, J. H., Jr. 1960. *Aristotle.* New York.

Repici, L. 1990. 'Limits of teleology in Theophrastus' *Metaphysics?*', *Archiv für Geschischte der Philosophie* LXXII, 182–213.

Rorty, R. M. 1973. 'Genus as matter: a reading of *Metaphysics* Z-H', in Lee, Mourelatos and Rorty 1973, pp. 393–420.

1974. 'Matter as goo: comments on Grene's paper', *Synthèse* XXVIII, 71–7.

Rose, V., ed. 1886. *Aristotelis qui ferebantur librorum fragmenta.* Stuttgart.

Ross, W. D. 1924. *Aristotle, Metaphysics: A Revised Text with Introduction and Commentary,* 2 vols. Oxford.

1949. *Aristotle, Prior and Posterior Analytics: A Revised Text with Introduction and Commentary.* Oxford.

1952. *The Works of Aristotle Translated into English, Vol. XII. Selected Fragments.* Oxford.

1955a. *Aristotle, Parva Naturalia.* Oxford.

1955b. *Aristotle's Physics,* revised ed. Oxford.

Ross, W. D. and Fobes, F. H. 1929. *Theophrastus' Metaphysics.* Oxford.

Santas, G. 1983. 'The Form of the Good in Plato's *Republic'*, in Anton and Preus 1983, pp. 232–63.

Sayre, K. M. 1983. *Plato's Late Ontology: A Riddle Resolved.* Princeton.

Schofield, M., Burnyeat, M. and Barnes, J., eds. 1980. *Doubt and Dogmatism: Studies in Hellenistic Epistemology.* Oxford.

Sedley, D. 1991. 'Is Aristotle's teleology anthropocentric?', *Phronesis* XXXVI, 179–96.

Sharples, R. W. 'Theophrastus as philosopher and Aristotelian', in van Ophuijsen and van Raalte 1998, pp. 267–80.

Shaw, J. R. 1972. 'Models of cardiac structure and function in Aristotle', *Journal of the History of Biology* V, 335–88.

Shipton, K. M. W. 1979. 'A second best: *Phaedo* 99bff.', *Phronesis* XXIV, 33–53.

Simpson, G. G. 1961. *Principles of Animal Taxonomy.* New York.

Sloan, P. R., ed. 1992. *Richard Owen: The Hunterian Lectures in Comparative Anatomy, May-June 1837.* Chicago.

Smith, J. A. and Ross, W. D., eds. 1912. *The Works of Aristotle Translated into English,* 12 vols. Oxford.

Smith, J. M. 1966. *The Theory of Evolution.* Baltimore.

1978. 'Optimization theory in evolution', *Annual Review of Ecology and Systematics* IX, 31–56.

Smith, R. 1989. *Aristotle: Prior Analytics.* Indianapolis.

Sorabji, R. 1980. *Necessity, Cause, and Blame: Perspective on Aristotle's Theory.* London.

Stent, G. 1971. *Molecular Genetics: An Introductory Narrative.* San Francisco.

Stubbe, H. 1972. *History of Genetics: From Prehistoric Times to the Rediscovery of Mendel's Laws,* trans. T. R. W. Waters. Cambridge, MA.

Thompson, D. W. 1971. *On Growth and Form,* abridged edn., ed. J. T. Bonner. Cambridge. (Original edn. Cambridge 1917; rev. edn. 1942.)

Tiles, J. E. 1983. 'Why the triangle has two right angles *kath 'hauto'*, *Phronesis* XXVIII, 1–16.

Tkacz, M. W. 1993. *The Use of the Aristotelian Methodology of Division and Demonstration in the De Animalibus of Albert the Great.* Catholic University of America. Washington, DC.

Torstrik, A. 1875. ''Περὶ τύχης καὶ τοῦ αὐτομάτου, Aristotle's *Physics* B, 4–6', *Hermes* IX, 425–70.

Vlastos, G. 1969. 'Reasons and causes in the *Phaedo'*, *Philosophical Review* LXXVIII, 291–325.

1975. *Plato's Universe.* Washington, DC.

Wallace, W., ed. 1985. *Reinterpreting Galileo.* Washington, DC.

Wallies, M., ed. 1909. *Philoponi in Aristotelis Analyticorum Posteriorum commentarium paraphrasis.* Berlin.

Wear, A., French, R. K. and Lonie, I. M., eds. 1985. *The Medical Renaissance of the Sixteenth Century.* Cambridge.

Webster, C., ed. 1979. *Health, Medicine and Morality in the Sixteenth Century.* Cambridge.

Wehrli, F. 1967. *Die Schule des Aristoteles.* Basel.

Werkmeister, W. H., ed. 1976. *Facets of Plato's Philosophy.* (*Phronesis,* suppl. vol. 3.) Assen.

White, N. P. 1979. *A Companion to Plato's Republic.* Indianapolis.

Whiting, J. 1992. 'Living bodies', in Nussbaum and Rorty 1992, pp. 76–91.

Wieland, W. 1962. *Die aristotelische Physik.* Göttingen.

Williams, G. C. 1966. *Adaptation and Natural Selection.* New York.

Wright, L. 1976. *Teleological Explanations.* Berkeley.

Zeller, E. 1897. *Aristotle and the Earlier Peripatetics,* tr. B. F. C. Costelloe and J. H. Muirhead, 2 vols. London.

Index

accidents
  essence versus, 44, 175–176
  material, 175–176
account, see *logos*
Achinstein, Peter (1977), 258 n15
Ackrill, John L. (1981b), 34 n12, 35 n15, 67
  n5
actuality, 234
Aelian, 116
affection(s)
  of bodies, 165, 166
  quantitative and qualitative, 175, 176
aitia, aition and, 282–285, 288, 290–291,
  294, *see also* cause
Albertus Magnus, 110, 123
Albritton, Rogers (1957), 158 n18
Alexander, 67 n15, 77–80, 94 n17
analogy, 161, 170
anankē, see necessity
Anaxagoras, 72, 282, 284, 285, 287, 300; see
  also *nous*
Anaximander, 157 n4, 181 n41
Andronicus, 66 n2, 116, 118
Annas, Julia (1982), 300 n25
anti-reductionism, Aristotle's, xxi
apes, 11, 18, 23, 24
appearances, *see* phenomena
Apollonius, 116
archē
  first principle, 142, 184, 188–189, 206–
    208, 211, 215, 219, 222 n17
  (hypothetical) starting point, 102, 206–208,
    209–211
  origin, 137, 299 n12
argument
  deductive, *see* syllogism (syllogistic)
  dialectical versus demonstrative, 74
Aristophanes of Byzantium, 116, 117, 118
Aristotelianism, Paduan, 220
Aristotle
  ancient lists of Aristotle's works, 116

biological corpus: disappearance of, 110–
  125 passim; relation to *APo.*, 32, 90
Categories (Cat.), 73, 161, 163–167, 180
  nn6, 18
De Anima (de An.), 119, 131, 137, 155,
  157 nn3, 8–10, 158 nn13, 22, 182, 198,
  202 n14, 245 n13, 258 n9, 276 n6
De Caelo (Cael.), xxi, 157 n7, 247 n27
De Motu Animalium (MA), 111, 232, 245
  n12
Dissections, 35 n16; *Selections from the
  Dissections*, 35 n16, 115, 118
Eudemian Ethics (EE), 257, 258 n3
Generation and Corruption (GC), xxi, 119,
  131, 134, 137, 138, 157 nn3, 6, 158 n17
Generation of Animals (GA), xxii, 35 n16
  46, 66 n2, 111, 119, 131, 133, 137, 141,
  145, 149, 153–155, 157 n3, 158 n23,
  159 n30, 176, 181 n39, 200, 203 n15,
  208, 214–216, 221 n6, 222 nn14–16,
  225, 229–234 passim, 242, 244 n5, 245
  n15, 246 n18, 248 n34, 258 nn1–2, 273,
  274, 276 n6
Historia Animalium (HA), 35 n16, 98, 101,
  111, 114–122 passim, 161, 162, 172,
  179 nn5, 12, 180 n14, 181 n40, 229,
  233, 236, 242, 270, 273, 276 n13, 277
  nn18–19, 23–24, 26, 28, 278 nn33, 38;
  aims of, 2, 15–25, 40, 53, 54; content of
  Books II–IV 18–19; method in, 39–71;
  as pre-demonstrative science, 53–64
  passim; relation to *APo.*, 15–25 passim,
  39–71; relation to other treatises, 39; relation to *PA*, 7, 24–29 passim, 31; title
  of, 39, 66 n2
Locomotion of Animals (IA), 41–42, 101,
  111, 137, 188, 195, 203 n16, 206, 208,
  214–218 passim, 222 n15, 223 n19, 247
  n24, 267, 268, 278 nn33–34, 43
Metaphysics (Metaph.), 34 n7, 96 n27,
  111, 131, 135–136, 140–153 passim,

313

Index

Index

in Aristotle's predecessors, xxi
and art, xxi, 235, 245 nn8–9, 246 n17,
280–302 passim
contrasted with matter, 185
does nothing in vain; does best, 205–223,
259, 276
extensional scope of, 189
formal and material, xxi, 130, 182–204,
207, 214, 220
as goal and as mover, 202 n14
as inherent source of motion (and rest),
209–210
necessary and definitional, 185, 201–202
n8
number of occurrences of in *PA*, xxiii, 182,
206
(un)qualified, 94 n17
teleological conception of, 208, 212
necessity
absolute, 141
and *aitia*, 282
hypothetical (conditional), 37 n38, 112,
138–140, 141, 186, 192, 201 n3, 220
n2; along with unqualified, 100; and biological generation, 138; and demonstration, 102; and elemental materials, 201
n3; and form, 112–113; in the organic
world, 139–140; pre-hypothetical (conditional), 187, 195, 196 relation of material causation or necessity, 26–27,
112,195
material (Democritean), 202 n12; and the
'good', 36; relation to hypothetical necessity, 26–27, 194–195
number of occurrences of in *PA*, xxiii
types of, 186
universal, 237
neck, 59
Nehamas, Alexander (1979), 34 n6
Newton, Isaac, 96 n31, 98
Nile, increased flow of, 86–87
noēsis, see conception
nomenclature, 50
lack of a zoological, 30
nous, Anaxagorean, 282, 285, 287
Nussbaum, Martha (1978), 245 n7, 246 n18,
276 n1, 278 n41

Ogle, William, xxiii n1, 104, 123
opposites, 280
orexis, 271
organ(s), *see* parts
organism
dualizing, 154, 155
produced by formal replication, 242, 244
origin, see *archē*
ousia, see substance
ovipara, 22, 59, 197, 198, 213

Owen, Gwilym, E. L., 158 n24; (1966) 157
n5; (1978–79) 147

Pappus, 91, 93, 96 n37
paradigm, eternal, 288, 294, 295, 300 n29
Parmenides, 72, 132, 302 n41
participation
in the eternal and divine, 134, 137
in forms, 294
parts, (non)uniform animal, 18–19, 50, 54, 113,
121, 166, 191, 194, 195, 197, 198–199
and function, 220 n3, 222 n17
pathēma, pathos, see affection(s)
Peck, Arthur L., 158 n12; (1965–70) 246 n16
Pellegrin, Pierre, 1, 3, 25, 127; (1982) 16, 69
n24, 160n, 168, 178, 180 n26; (1985) 69
n24, 180–181 n27; (1987) 69 n24, 168,
180 n26
perceptibles, 142
Perry, Richard (1979), 302 n42
Phainias of Eresos, 124 n4
phainomena, see phenomena (appearances)
phenomena (appearances), 42–43, 69 n24
phenotype, 216
Philoponus, 12, 69 n22, 81, 83, 87, 94 n14, 95
n18, 248 n32, 258 n10
physis, see nature
Plato (Platonism), 73, 119, 208–302
dialogues: *Cratylus*, 281, 287–290, 292, 297,
301 nn29, 32; *Crito*, 299 n13; *Gorgias*,
281, 287–289, 300 n29; *Laws*, 280–281,
286, 298 n1, 299 n12; *Parmenides*, 141;
Phaedo, 34 n6, 66 n4, 151–153, 157 n5,
180 n16, 247 n26, 281, 282 (relation to the
Timaeus, 36 n38, 290); *Phaedrus* 66 n4;
Philebus, 94 n6, 161–163, 181 n30, 281,
286, 297, 298, 298 n1, 300 n21; *Republic*
91, 157 n5, 281, 286, 287, 289, 290, 298,
301 nn29–30, 36, 302 nn39, 45, 48; *Sophist*, 75, 287; *Statesman*, 75, 281, 287,
293, 300 n21, 301 nn30–31, 35; *Symposium*, 157 n5, 158 n14; *Theaetetus*, 32,
37 n44, 292, 297, 301 n32; *Timaeus*, xxi, 36
n38, 157 n5, 228, 268, 269, 277 n30; 280–
302 passim
division, kind, form and differentia in, 102
on forms, 132, 136, 141, 147, 280–302
interest in scientific understanding, 32–33,
37 n44
on 'the more and the less', 161, 162, 180 n16
on teleology (and causation generally), 36–
37 n38, 280–302
Pliny, 115
pneuma, 230, 233, 234, 242, 267
Popper, Karl, 178–179 n1
Porphyry, 51–54, 239, 247 n29, 251–254 passim, 257
posit (*thesis*), 209

319

organisms as, 47, 53, 132, 153, 182
as a 'such' (being), 136
substrate (*hupokeimenon*)
  kind as, 170
  matter as (potential being), 165, 167
sullogismos, see syllogism (syllogistic)
summetria, 297, 302 n45
sunaition, see cause, cooperative
supposition, 209–211
  constraint on the formation of, 219
  and definitions, as starting points of expla-
    nation, 33
  demonstrative, 215
  distinct from definition, 209, 222 n11
  presupposition, 211–215, 221 n9
  teleological, 219
syllogism (syllogistic)
  meaning of, 94 n5
  in science, 14, 33, 34 nn11, 13, 36 n37, 69
    n94

taxonomy, absent in Aristotle's biology, 20,
  24–25, 40, 64–65, 115–116, 174, 178
teeth, 193, 196, 208, 211–215
  differentiation of, 214
  lack of upper, 49
teleology, 225, 228, 251, 255, 258; *see also*
    cause, final
  absence of, in Empedocles and
    Democritus, 243
  Anaxagorean, 300 n25
  anthropocentric, 201 n5
  chance and, 226, 229–249 passim; chance,
    spontaneous generation and, 205–223
    passim
  cosmic, 201 n5, 275
  and definition, 175–176
  etiological concept of, 215
  limits of, 259–279 passim
  and necessity, 101, 112, 280, 294
  in Plato: *Phaedo,* 281–287; *Timaeus,* 228,
    280–302 passim
testacea, 19, 61, 229, 232, 248 n40
Thales, 157 n4, 245 n13
that for the sake of which, *see* cause, final;
  teleology

Themistius, 248 n32
Theophrastus, 3, 67 n7, 110, 114, 124 n1,
  225–227, 259–279 passim
theory, versus scientific practice, 98–99
Third Man, regress, 154, 181 n35
Thompson, D'Arcy W. (1971), 160, 179 nn2,
  3
Tiles, J. E. (1983), 34 n7
Torstrik, Adolfus (1875), 251, 258 n4
Tredennick, Hugh, 83
triangles, 8–10, 47

understanding
  demonstrative, 53
  scientific, 32, 111, 129
  as translation of *epistēmē,* 7, 8, 33, 34 n4
  unqualified versus incidental (sophistic), 7,
    13, 29, 31, 33, 34 n4
vessels, blood, 19, 22
vivipara, 21–22, 25, 59, 187–188, 195, 213;
    see also quadrupeds
Vlastos, Gregory (1969), 159 n31; (1975)
  291, 294, 299 n15
voyage, second best (in the *Phaedo*), 284–
  286, 298 nn5–6, 300 n21

water dwellers, 31, 55, 56
Waterlow, Sarah (1982), 140
Weber, Bruce, 127
Wieland, Wolfgang, 251; (1962) 258 n5
Williams, George C. (1966), 258 n16
Wilson, Malcolm, 221 n9
windpipe, 21, 58–60, 70 n32, 196–198, 262,
  266–70, 277 n32
Wright, Larry (1976), 248 n38, 257, 299 n11

Xenophanes, 181 n41

Zabarella, Jacopo, 94 n15
zoology
  absence of Hellenistic, 110–125
  as an axiomatic system, 33
  as research program, 111–112, 114, 122
  as a special science (Aristotle's invention),
    110